Late Roman Fortifications

Late Roman Fortifications

STEPHEN JOHNSON

B T Batsford Ltd London

© Stephen Johnson 1983
First published 1983

All rights reserved. No part of this
publication may be reproduced in any
form by any means, without permission
from the Publisher.

Photoset by Keyspools Ltd, Bridge Street, Golborne.
and printed in Great Britain by
Ebenezer Baylis and Son Ltd.,
The Trinity Press, Worcester and London
for the publishers
Batsford Academic and Educational
a Division of B. T. Batsford Ltd
4 Fitzhardinge Street
London W1H 0AH

British Library Cataloguing in Publication Data

Johnson, Stephen
   Rome against the Barbarians.
   1. Fortifications, Roman
   I. Title
   623 .12 0937   UG401

ISBN 0-7134-3476-7

Contents

List of Plates 6
Maps, Plans and Reconstructions 7

1 Roman Imperial Fortifications 9
2 The New Design in Late Roman Defences 31
3 Contemporary Records 55
4 The Pressures on Rome: Barbarian Invasions and Tactics 67
5 Town and City Fortifications 82
6 The Frontiers I: The Rhine 137
7 The Frontiers II: The Danube 169
8 The Frontiers III: The North Sea 196
9 The Defence of Italy and Spain 215
10 Local and Rural Protection: Hill-top Defences 226
11 Late Roman Frontier Policy 245

Appendix 1 Selected Groups of Gallic City Walls 262
Appendix 2 Late Roman Watchtowers (*burgi*) in the Rhine and Danube Areas 270
Appendix 3 Hill-top Fortifications of the Late Roman Period 280

Bibliography 291
References 297
Index 311

The Plates

(between pages 112–113)

1. Le Mans, Tour de la Magdeleine
2. Le Mans, detail of patterning on the Tour du Vivier
3. Le Mans, La Grande Pôterne
4. Beauvais, interior face of Roman wall
5. Sens, exterior face of walls
6. Périgueux, foundation courses of wall
7. Périgueux, La Porte Normande
8. Die, La Porte St Marcel
9. Carcassonne, La Pôterne du Moulin d'Avar
10. Carcassonne, Tour de la Marquière
11. Barcelona, projecting rectangular tower
12. Gerona, walls and projecting tower
13. Zaragossa, late Roman wall and towers
14. Sópron, late Roman wall and tower
15. Susa, Porta Savoia
16. Burgh Castle, projecting tower

Figures in the text

Page numbers shown in italics

1 Map to show distribution of Augustan walled cities   *12*
2 Outline plans of Augustan city walls   *14*
3 Autun, Porte St André   *16*
4 Plans of Augustan city gates   *17*
5 Plans of gates of 'Fréjus' type   *18*
6 Outline plans of city walls built after Augustus   *19*
7 Distribution map of walled cities in second and third centuries   *21*
8 Plans of gates with rectangular projecting towers   *22*
9 Plans of gates with semicircular projecting towers   *23*
10 Plans of second- and third-century fort gateways   *25*
11 Fort plans, late second and early third centuries   *26*
12 Plans of forts on the eastern frontiers   *28*
13 Reconstruction drawing of a Gallic town wall   *34*
14 Le Mans, drawing of demolished Tour Huyeau   *39*
15 Le Mans, detail of Tour Magdeleine   *41*
16 City illustrations in late Roman documents   *42*
17 City illustrations on late Roman coins   *43*
18 Rome, plans of gates on the Aurelianic Wall   *45*
19 Plans of late Roman gates in Spain   *46*
20 Plans of late Roman gates with U-shaped projecting towers   *47*
21 Plans of late Roman towers of square or rectangular shape (Andernach gateways)   *48*
22 Plans of late Roman gates with polygonal projecting towers   *49*
23 Barbarian invasions AD 250–60   *71*
24 Barbarian invasions AD 268–80   *75*
25 Map of late Roman walled towns in Gaul   *83*
26 Late Roman walled sites: Lugdunensis I   *85*
27 Late Roman walled sites: Lugdunensis II   *87*
28 Late Roman walled sites: Lugdunensis III   *89*
29 Tours, postern gate   *90*
30 Late Roman walled sites: Lugdunensis III   *92*
31 Late Roman walled sites: Belgica II   *96*
32 Late Roman walled sites: Belgica II   *98*
33 Late Roman walled sites: Lugdunensis IV (Senonia)   *100*
34 Late Roman Paris   *102*
35 Late Roman walled sites: Lugdunensis IV and Aquitanica I   *103*
36 Late Roman walled sites: Viennensis   *105*
37 Late Roman walled sites: Aquitanica II   *107*
38 Late Roman Périgueux   *108*
39 Périgueux, the Porte de Mars   *109*
40 Late Roman walled sites: Novempopulana   *110*
41 Dax, opening in the walls   *111*
42 Constructional grouping of late Roman town walls in Gaul   *115*
43 The late Roman Walls of Rome   *118*
44 Map of late Roman fortifications in Italy and the East   *120*
45 Late Roman walled sites: Hungary   *122*

46 Map of late Roman fortifications in Spain *124*
47 Late Roman walled sites: Spain *126*
48 Late Roman walled sites: Spain *127*
49 Late Roman walled sites: Britain *133*
50 Map of official establishments in the western empire *134*
51 Map of the lower Rhine frontier in late Roman times *137*
52 Late Roman walled sites: Germania I *139*
53 Late Roman *burgi* in the Rhineland areas *140*
54 Late Roman *burgi* and landing stations in the Rhineland *141*
55 Late Roman walled sites: Belgica I *144*
56 Late Roman walled sites: the lower Rhine *147*
57 Late Roman walled sites: the lower Rhine *149*
58 Late Roman walled sites: the middle Rhine *151*
59 Section through the fort walls at Alzey *154*
60 Late Roman road-posts in the Rhineland *156*
61 Late Roman road-posts in the Rhineland *157*
62 Map of the Upper Rhine and Danube frontier areas in the late Roman period *159*
63 Late Roman walled sites: Maxima Sequanorum *160*
64 Late Roman walled sites: The Upper Rhine frontier *162*
65 Late Roman *burgi* in the Upper Rhine area *164*
66 Late Roman walled sites: The Upper Rhine area *165*
67 Late Roman walled sites: Bavaria and Austria *170*
68 Map of late Roman forts in Noricum *176*
69 Late Roman walled sites: Noricum and its region *177*
70 Map of the late Roman Pannonian frontier *181*
71 Late Roman walled sites: Pannonia *183*
72 Map of the late Roman frontier at the Danube bend *184*
73 Late Roman walled sites: Valeria *186*
74 Late Roman *burgi* on the Danube frontier *187*
75 Map showing distribution of fan-shaped towers and corner *burgi* on the late Roman Danube frontier *190*
76 Map of the coastal areas of Britain and Gaul, early third century *197*
77 Site plans of coastal defences in the early third century *198*
78 Map of late Roman coastal sites in Britain and Gaul *200*
79 Late Roman walled sites: The British coast *203*
80 Late Roman walled sites: The British coast *205*
81 Late Roman *burgi* in British coastal areas *213*
82 Map of the area of the Julian Alps in the late Roman period *217*
83 Late Roman walled sites: The Julian Alps *219*
84 Map of late Roman Morocco *222*
85 Late Roman hill-top sites: Belgica and the Rhineland *228*
86 Late Roman hill-top sites: Belgica *229*
87 Map of hill-top sites in the Rhineland and Belgium defended in the late Roman period *232*
88 Map of late Roman defended hilltop sites in the Upper Rhine and Danube areas *237*
89 Late Roman hill-top sites: Switzerland and Bavaria *238*
90 Late Roman hill-top sites: Austria *241*
91 Late Roman walled sites (villas?): Dalmatia *243*
92 The development of frontier defences: the Gallic Empire *248*
93 The development of frontier defences: Aurelian and Probus *250*
94 The development of Frontier defences: the Tetrarchy *252*
95 The development of Frontier defences: Constantine and his successors *254*
96 The development of Frontier defences: Julian *256*
97 The development of Frontier defences: Valentinian *258*

I

Roman Imperial Fortifications

Roman frontiers in both portions of the Roman empire came under heavy pressure from barbarian invaders in the third century AD. After Rome's rapid expansion in the first century, which led to the control of much of western Europe, her frontier lines were soon established. The siting of most of these had been the choice of emperors of the early second century, Trajan and Hadrian, who recognised the military problems caused by the control of so much land. The frontiers, roads, palisades, or walls, protected by permanent forts and garrisoned by full-time troops, represented the fringe of Roman control over provincial territories. Although Rome's influence extended beyond these limits, control was indirect and was maintained through intelligence services, a watchful military presence and in places through economic or market forces.

Pressure on these established frontiers from outside forces was to all appearances slight for much of the second century. There were major campaigns against invaders from outside, or further attempts to improve frontier control. These did not, however, result in any substantial change of any of the frontier zones. The Roman policy of containment appears to have remained the same, and only the troubles in the Danube area with the German tribe of the Marcomanni in the 170s and 180s heralded the sorts of problems which were to recur with such force.

Barbarian pressure on the Roman frontiers, at times developing into full invasions with large-scale effects on the homes and livelihood of the provincials, began to increase in the third century. Time and again, foreign tribesmen took advantage of political, military or economic instabilities of the Roman world to invade. Waves of tribesmen swept past the frontier defences to reach deep into the civilian parts of the provinces. The pickings here were easy. Except in the frontier zones themselves, Roman cities, towns and villages were normally open, unwalled settlements. Their lack of protective walls was a sign of the peaceful climate of Roman provincial life in the first two centuries of our era, but it left them defenceless now.

At times of crisis, the government is forced to act, but, due largely to internal instabilities, the Roman government of the day was only able first to suffer these raids, then to respond, at first with military force, and finally by strengthening the defences of the empire. Not only was it necessary now to refurbish the frontiers themselves with new or rebuilt forts, but many of the major towns and administrative centres within the provinces also required protection. Major programmes of fortification of cities and towns were begun at this period in Gaul, Spain, the Danubian provinces, and probably in Italy too—the ultimate target for much of the barbarian raiding.

The scale of the task of fortification which now faced the Roman authorities was enormous. Only in Britain previously had anything approaching a comprehensive series of city and town defences before been constructed. In terms of restoration of cities ravaged by barbarian invasion, the defences themselves, to modern eyes possibly the most tangible evidence for Roman concern, were only a partial answer: other buildings too will have needed repair and replacement quite apart from questions of social redress for losses suffered in the invasions. There were probably incentives offered to cities for schemes of wall-building, if the labour for such projects was not actually offered from the Roman army itself. It is not certain how far the initiative for construction of defences was left to local cities and their councils, or whether it was directly controlled by central government. When one considers, however, that 70 cities or so in Gaul alone, possibly 40 in Spain and maybe another 70 in Italy all required this form of protection, quite apart from those sites on the frontiers where new forts were also required, the task must have seemed Herculean.

This form of major response to barbarian pressure came in the last 40 years of the third century. City and town walls built now thus belonged to what is commonly known as the Later Roman period. They began a trend in defensive design which was to form a major advance and produce some spectacular examples of skill and ingenuity. In many ways, too, the Roman model was the direct forerunner of medieval castle design, only finally overtaken by the new demands of artillery fortification in the sixteenth and seventeenth centuries.

The style of building of such city, town and fort walls is thus of considerable interest. The walls were of massive height and thickness, built of solid concrete masonry. They were normally provided with projecting towers, higher then the walls, and the gates were also well protected and sometimes surprisingly elaborate structures. The walls encapsulated the latest forms of defensive technique available to Roman military engineers of the day. They were a translation into Roman shape from Hellenistic ideas and principles which had evolved on the eastern fringes of the ancient Mediterranean world through contact and conflict with the vast Persian empire. Thus although in one sense these defences were a culmination of effort, in Roman terms, they did have recognisable antecedents. In order to assess both what they owed to earlier architecture in the same styles and their own unique place at least in the western Roman world, it is necessary to trace the development of Roman fortifications in the imperial period.

Many early Roman ideas about fortification had been taken over from the Greeks: Etruscan towns were fortified originally with walls constructed from great blocks of irregular polygonal stone in a style which is a familiar prehistoric style in other parts of the world. As fortifications in Greece became more and more sophisticated, reaching their peak in Hellenistic times, so Italian styles also changed, and Hellenistic techniques, notably the construction of projecting towers and the arrangements for arrow and ballista shooting, began to find their way into the Italian repertoire. Italian fortifications, however, never matched Hellenistic work for subtlety, and it was mostly as a consequence of later contact with the Greek East and tribes and peoples of Asia and elsewhere, who were used to Hellenistic types of warfare and fortifications, that Rome had to take notice of this sophistication of technique and use.[1]

Up to the time of Caesar and Augustus, the majority of colonisation by the Romans had been within Italy, though occasionally cities had been founded in other areas. In the earliest years of Rome's growth, the implantation of citizen colonies or of colonies with slightly less generous 'Latin rights' in various parts of Italy had been used a tactical move to widen Rome's power-base and immediate sphere of interest. Starting in about 100 BC, there was a greater tendency to establish colonies for retired legionary veterans on the captured land of some former enemy, thus rewarding faithful troops with a new home, and with some land to till. The dual purpose of military ascendancy and implantation of Roman lifestyle was in this way maintained.

The implementation of such an overt policy of domination within Italy had understandably led to considerable political difficulties. By the time of Julius Caesar, most new colonies were established abroad, many of them distinguishable by the name *Colonia Julia*. Augustus continued the policies of his adopted father, and was able to claim in the *Res Gestae* that he had founded colonies in most of the existing provinces. These were normally for retired veterans, and their implantation was in areas where some degree of Roman control or watchfulness was desirable. Most of these areas were well away from the theatres of conflict with barbarians: they might include the area of Asia Minor, but they would not as yet be placed in any portion of Gaul or Germany outside the Mediterranean strip of Provence.

Such colonies were intended by Augustus as the showpieces for the organisation, superior administration, and the quality of Roman life. This entailed the provision of substantial and spacious administrative and official buildings, among which the defences were often of paramount importance (fig. 1). In the peaceable areas where these colonies for the most part lay, there can have been little call for such walls purely as a defensive measure. The possession of walls was a sign that a colony had a certain standing: it was the mark of a regularly founded city, and gave it a particular religious and civil status.[2] A city surrounded by solid walls could scarcely be mistaken for a chance settlement, and a Roman colony was intended to be as prestigious as possible.

The established Roman city had its *pomerium*, the sacred area within which its walls were built. Originally this was the limit of a colony traditionally marked out in virgin soil by the founder's plough. Later the term was transferred to all city boundaries, though by the later Roman period the situation had become

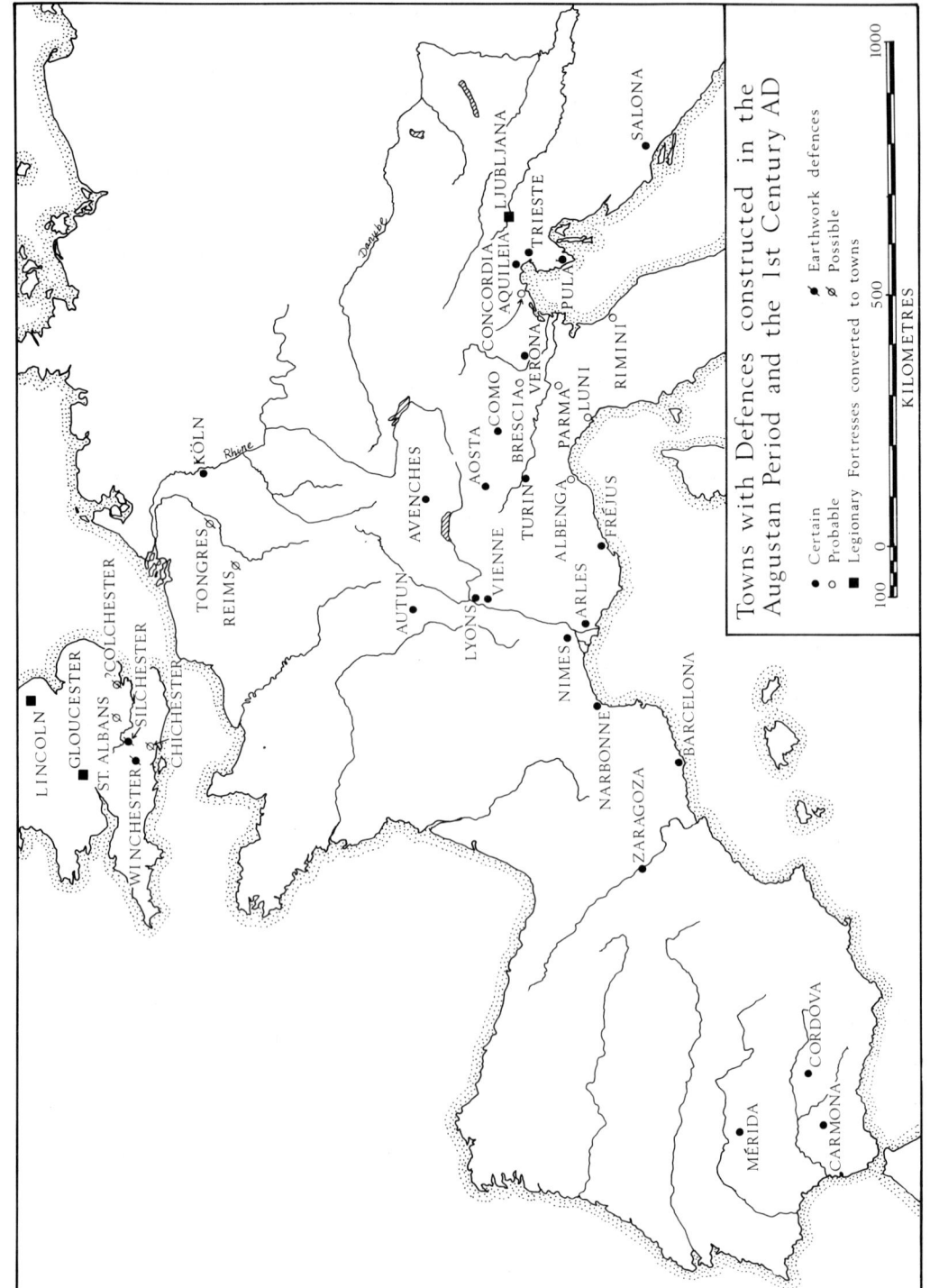

1 Map to show distribution of walled cities of the Augustan period

exceedingly complex, and such tradition may have ceased to be as binding as before. In any case, city walls and gates were always *res sacrae*, under public ownership and the tutelage of the gods. They were thus, particularly in the earlier periods, as much civic monuments as functional passageways and might be expected therefore to have been highly ornamented.

Augustan city walls were built on an expansive scale (fig. 2). Some of the colonies, like those at Aosta, Ljubljana (Emona) or Barcelona had a basically rectangular plan, which reflects the military nature of the veterans' retirement settlements. Although it is not certain where the Roman military predilection for a regularly planned camp of rectangular shape came from, this was clearly the normal and most convenient way of organising permanent bases for the army. Aosta's plan was strictly regimented in 16 blocks, each 180 by 140m, within enclosing walls 720 by 560m. Much of the wall-circuit, built of small blockwork, survives, with four gates and rectangular towers astride the walls at intervals. The gates have a single wide carriageway, flanked by rectangular towers: only the western gate (fig. 4), leading to Turin and Rome, was a more monumental affair, which, with a large central carriageway flanked by pedestrian passages, finds ready parallels with several other examples of monumental Augustan gateways.[3]

Not all the Augustan colonies were so regularly planned. Fréjus (Forum Julii), although probably founded by Caesar, was given colonial status by Augustus. The walls enclosed a most irregular area which is in total only some 40ha. Their style of construction was in very precise small blockwork throughout, with occasional patches of opus reticulatum and mosaic work. On the northern side there were round towers astride the walls, with two storeys of three windows in each, facing the exterior. The lower register of windows was of the slit type, but the upper row, of keyhole style, was larger.

In basic design, Augustan walls were freestanding, normally with a carefully finished facing of small blockwork. While the layout might reflect the current rectangular planned area of a legionary camp, mostly it did not, and the Augustan and other early colonies of Gaul are marked by both their expansiveness and their irregularity of plan. The largest were Nîmes, Autun and Vienne, all of which had wall-circuits enclosing over 200ha (approximately 500 acres): all three incorporated low-lying flat portions of land, together with ground which formed a kind of acropolis. The Augustan towers were either rectangular or round, and stood astride the walls. Essentially these walls were plain and unfussy, functional without being ostentatious.

It was at the gateways that the monumental aspects of the wall construction were most obviously stressed (fig. 4). Though by no means standard, Augustan gates were normally built in a spacious, expansive style, with double or single passageways for vehicles, usually flanked by one or more passageways for pedestrians. There were projecting towers, in most cases semicircular or 'U'-shaped, on both sides of such a gateway. Above the portals, to judge from examples where there is enough of the superstructure remaining, there was an arcaded gallery at rampart-walk level. Most reconstructions of the gates carry the towers themselves up one storey higher. There were windows in the towers at

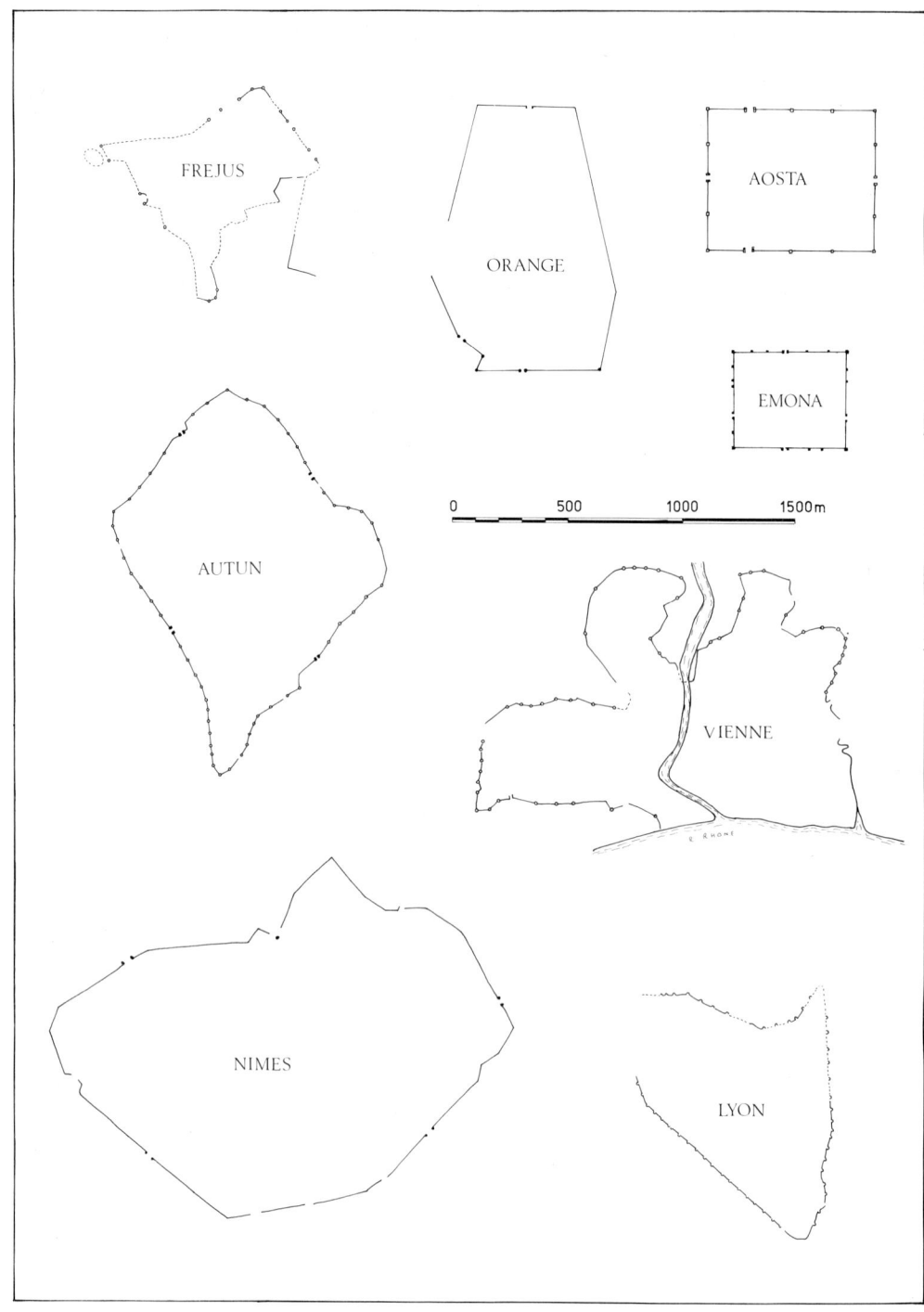

2 Outline plans of Augustan city walls (*scale 1:30,000*)

least at the height of the ramparts. Most of our knowledge of the upper portions of the gates is gained from few surviving examples—at Autun there are arcaded galleries in both the Porte St André (fig. 3) and the Porte St Honoré, although there is some evidence that at least parts of these arcades were rebuilt in the later Roman period. Remains of Augustan gates similarly provided with arcades are also known from Turin, Verona, Aosta and Fano.

The gate façades at Autun and Nîmes were built of large blocks of stone decorated with several forms of engaged ornament—cornices, mouldings, architraves and pilasters. By contrast, their flanking towers were in plain and regular small blockwork. There are examples of similar differential treatment in Italy, too. At Turin and Spello, where gates and towers remain in a good state of preservation, the tower facing had less decoration than the arcade above the gate.

Outside Gaul, Augustan gates display a remarkable diversity of plan, in particular. The main distinctions can be drawn in comparing tower shapes and the presence or absence of an inner court behind the gate façade. The majority of flanking towers were semicircular or 'U'-shaped, though polygonal or rectangular examples are also known. Aosta and Emona have square or rectangular gate towers: Barcelona, Salona, Como, Zara and Turin have polygonal towers, the last on a massive scale. Some sites have round towers: examples are known at Ventimiglia, Valence and Ravenna.[4]

Perhaps the most interesting form of Augustan gate is the so-called 'Fréjus' type, in which the gate is set at the rear of a courtyard formed in a re-entry of the city walls (fig. 5). It is sometimes supposed that a monumental arch occupied a central position in the courtyard thus formed outside the defences, but as yet there is little evidence that this was so in every case. Gates of this type are known at Fréjus, Arles, Aix les Bains, Tipasa in Africa and even in a simplified form at Neuss, enabling a parallel to be drawn between developments on the Rhine frontier and Augustan town defences in the southern part of Gaul.

After Augustus' death, there was less expansion of Roman influence by founding new colonies, although this continued to be a device used to great effect in the early stages of Roman annexation of new territory. Augustus had given up any thoughts he may have had about the conquest of the whole of Germany after the disaster of AD 9, when three legions under Quintilius Varus were annihilated in the Teutoburg Wood, near Mainz. At this date, and for some time afterwards, military affairs in Germany and on other battlefronts were fluid. Army encampments were of turf-and-timber, and only later translated into stone. What is known of campaign forts of the time of Augustus and his immediate successors suggests that they were extremely irregular in plan. The rectangular plan of fort or camp with rounded corners, the hallmark of Roman forts and camps on Trajanic and Hadrianic frontiers, seems to have been universally adopted only in the later part of the first century.

The first major permanent fortress to be built after the death of Augustus was the camp of the Praetorian Guard at Rome, built under the emperor Tiberius. This was rectangular in plan, with projecting small rectangular towers at widely spaced intervals and flanking the four gates. The facing was of tile, rather than

3 Autun, Porte St André from the exterior

stone, its design undoubtedly owing much to the rectangular form of fortress-colony like Aosta or Turin, and it translated into masonry the form of military architecture which might have been seen on the expanding frontiers of Rome's influence.[5]

The colony of Köln was founded in AD 50 by the emperor Claudius (fig. 6). It was provided with a broad, but rather irregular, circuit of freestanding walls and circular towers on the Augustan model. The gates were of similar plan to Augustan city gates at Aosta and other Italian towns. One of the best known of the Roman monuments of Köln, the 'Römerturm', is one of the round towers of the colonia wall. It bears on its exterior face a series of bizarre decorations in mosaic work, rather similar in effect to the patches of mosaic on the Augustan walls at Fréjus.[6] Probably slightly later in date are the walls of Avenches. These have semicircular towers which project inside the city and there is also a gateway which echoes the Augustinian plan, with a double-carriageway entrance with pedestrian side-passages flanked by a pair of polygonal towers. The carriageways lead into a circular space which forms a central courtyard, a feature paralleled in a gate at Aquileia.[7]

Avenches, elevated to the status of colony under the Flavians, was among the

ROMAN IMPERIAL FORTIFICATIONS

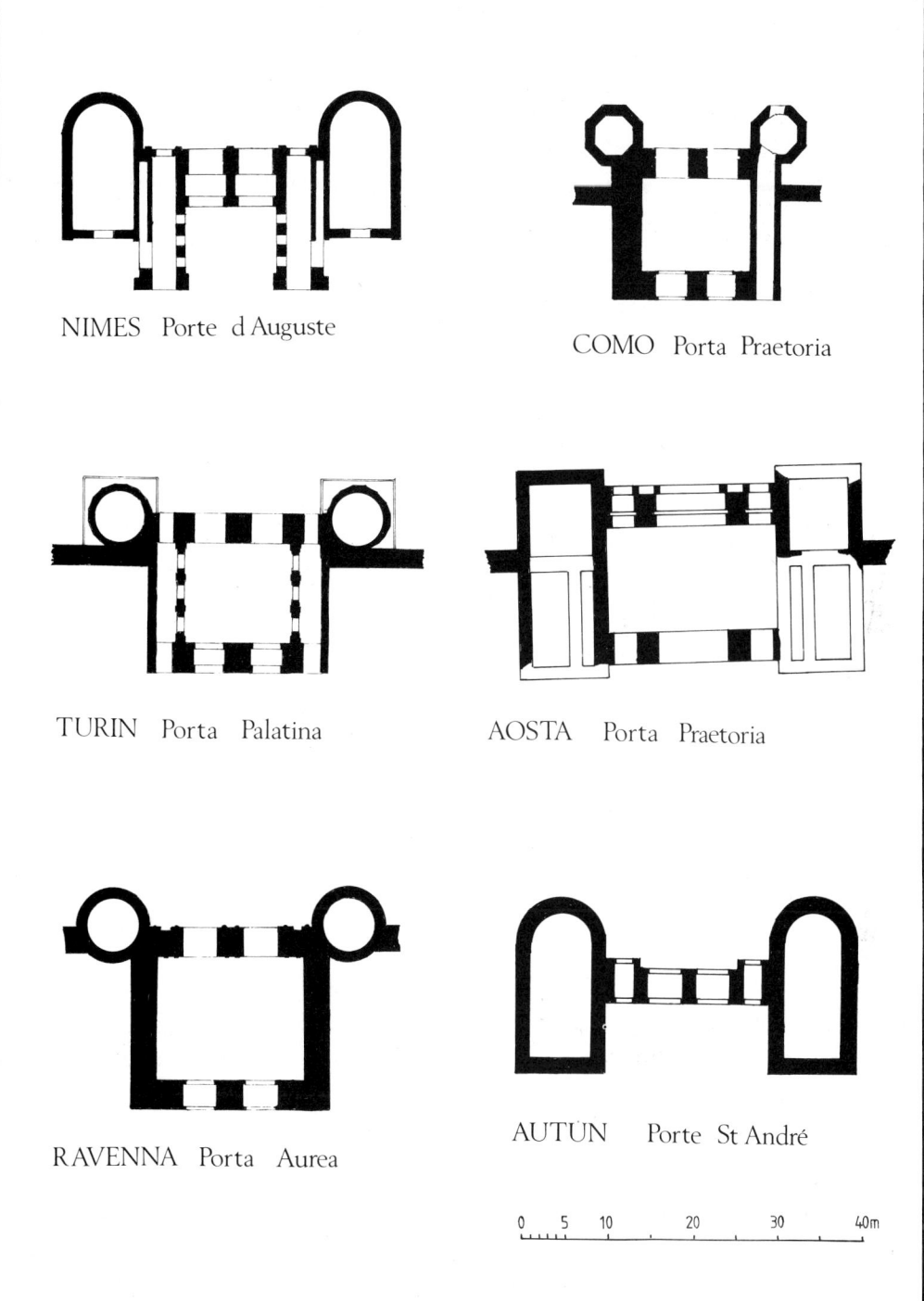

4 Plans of Augustan city gates (*scale 1 : 800*)

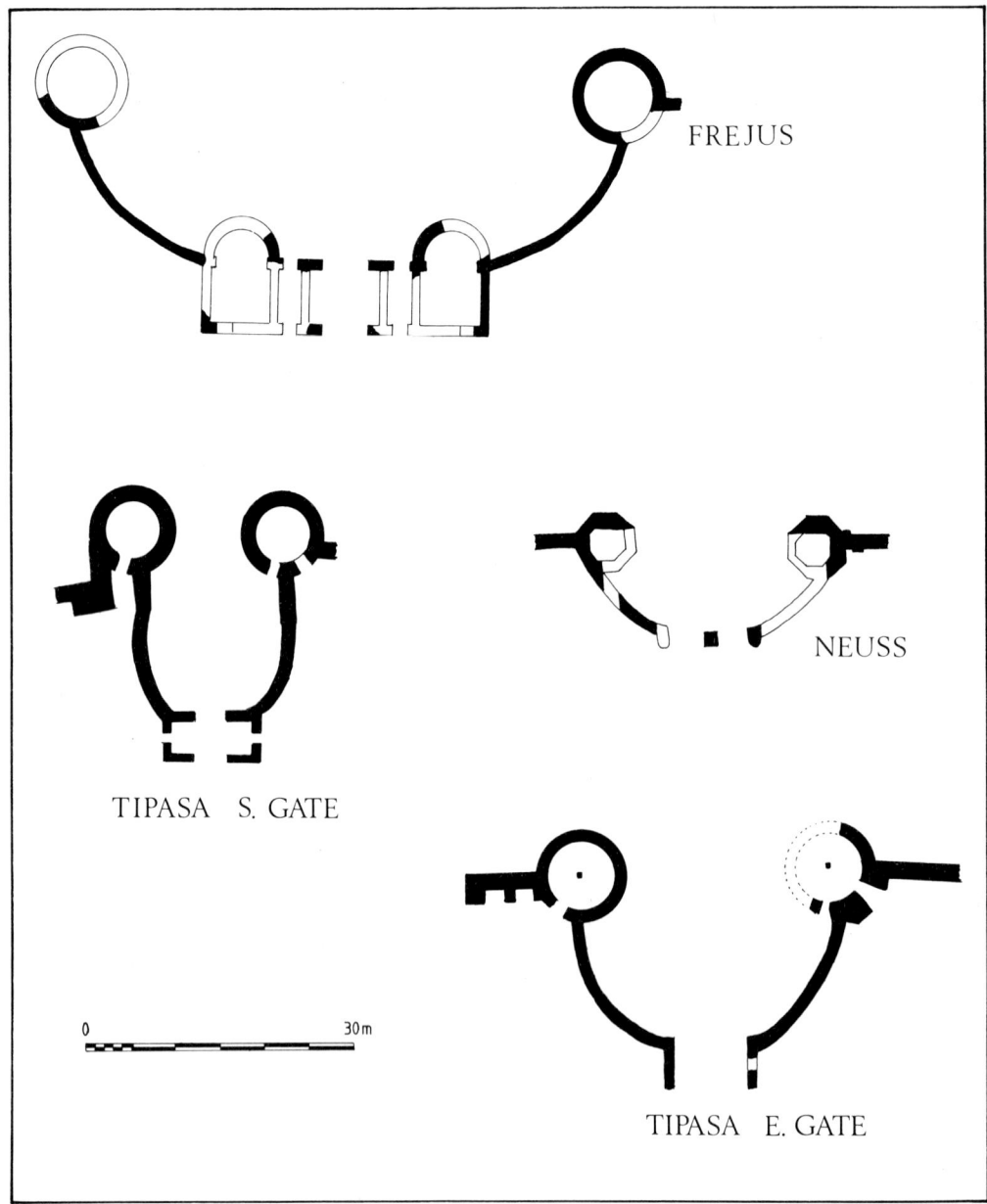

5 Plans of gates of 'Fréjus' type (*scale 1:800*)

last of western cities to receive a large and expansive set of city defences on the Augustan model as a matter of course. By the end of the first century, Roman troops in Germany and Britain had pushed into barbarian territory as far as they could be expected to go: the limits of the empire were set, at first loosely by military zones linked with the communications system, then eventually by solid and visible artificial frontier lines. Forts and fortresses, formerly of earth-and-timber, were rebuilt in stone. Their design was the familiar playing-card shape,

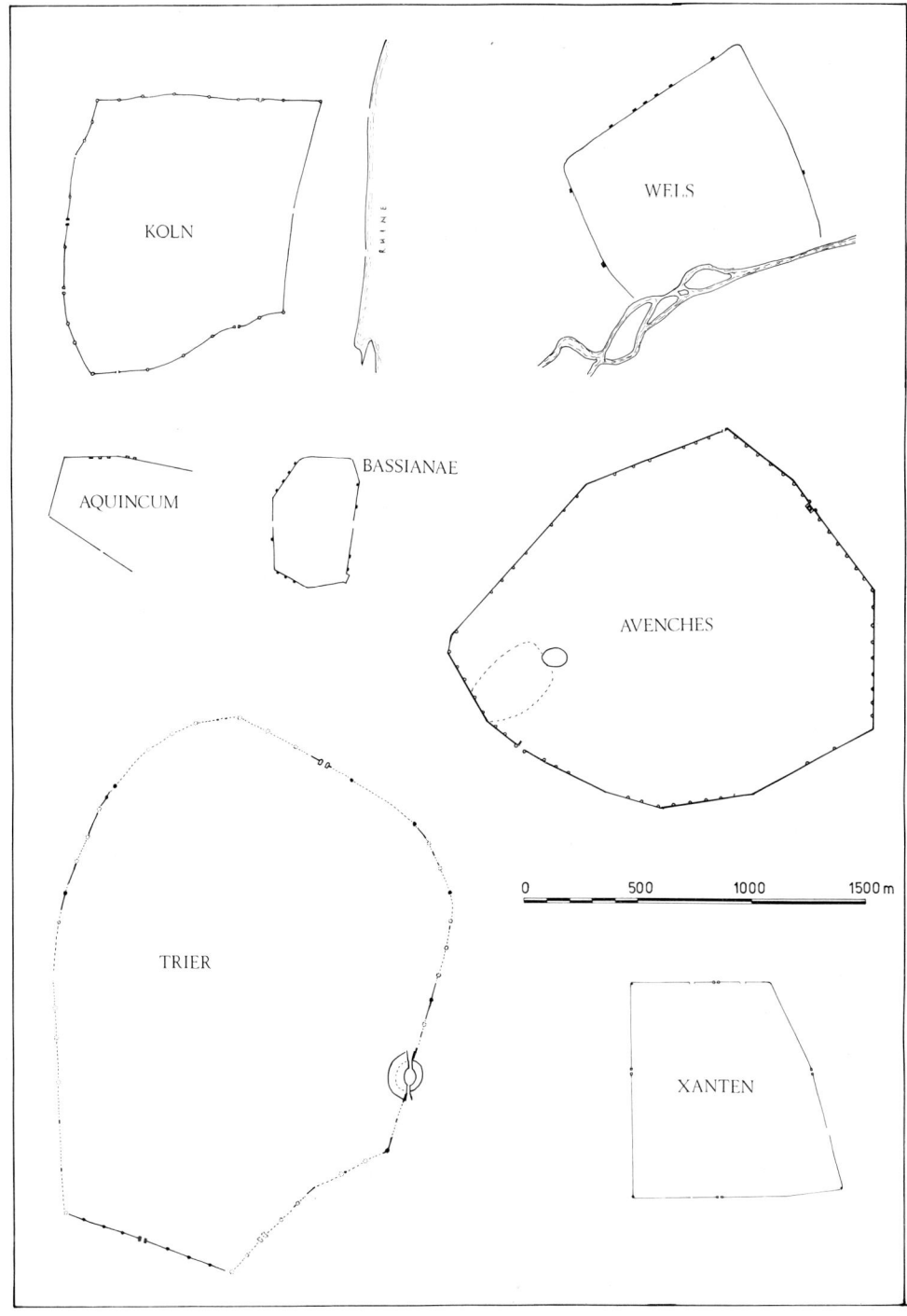

6 Outline plans of city walls built after the reign of Augustus (*scale 1 : 30,000*)

with double-portalled gateways and rectangular gate, angle, and interval-turrets.

Perhaps as a consequence of so much concentration on the frontiers in the reigns of Trajan and Hadrian, there was a lull in the construction of city walls. The majority of Gallic cities developed as open settlements and were left undefended: with frontier defence assured, there was little point in going to the considerable expense of providing a city wall. Accordingly, cities and towns provided with walls in the early second century were relatively few (fig. 7). In Britain, the colonies of Colchester, Lincoln and Gloucester all had walls: the latter two were legionary fortresses whose defences were taken over by the civilian colonial settlers (presumably retired veterans) when the legions moved on after conquest was assured. Colchester was the only colony specifically implanted: this city too had walls which are not precisely dated, but could have been constructed (somewhat after the design of a military camp) as early as the late first century.[8] Elsewhere, the military architecture of fort and fortress was making its mark. At Aquincum civil town (fig. 5), Xanten (fig. 5) and Tongres (fig. 55, p. 144) the walls are assigned to an early second-century date: apart from Tongres, which had a large irregular oval circuit with round towers astride the walls, the others are of quadrilateral plan with simple gateways, and defences very much after the style of a military camp.[9] This type of defensive architecture became most prevalent in the frontier zones: it was later used at Ladenburg and Heddernheim, two Roman settlements east of the Rhine which grew up out of the former *vici* of Roman forts and which were walled in the early third century.[10] There is even a group of cities in Spain, including Italica and Baelo, which have defences and gates in the same style (fig. 8), and which, on the grounds that Italica was a Hadrianic colony, are dated to the same period.[11]

Among the stranger examples of city wall construction are two cities where the defences were left uncompleted. At Augst, two portions of city wall cut off the south eastern and south western approaches.[12] Like the walls of Avenches, its nearest sizeable neighbour town, the two short stretches have semicircular interval towers which project inwards. The only dating evidence from the wall foundations suggests that they were built after the reign of Hadrian, but the view that their construction was set in train or even interrupted by the great Alemannic invasions of 233 is too simple a view to take. The walls flanked the two main routes from the south to the plateau on which the city stands, and they could easily be seen as 'token' defences on either side of main gateways to give the city's approaches a more impressive appearance. A similar unfinished wall, running in a straight line for more than 400m and incorporating a gate, has been found at Lienz (Aguntum).[13] This is tentatively dated to the reign of Hadrian, and an additional peculiarity is that it appears to leave most of the main known town buildings outside its protection.

The only signs of a consistent approach to the construction of town defences came in Britain. Here most major cities were enclosed with earthen ramparts within the last few decades of the second century, followed soon thereafter in most cases by the addition of stone walls to front these defences.[14] None of these ramparts or walls is dated by any means other than archaeological, and, in the

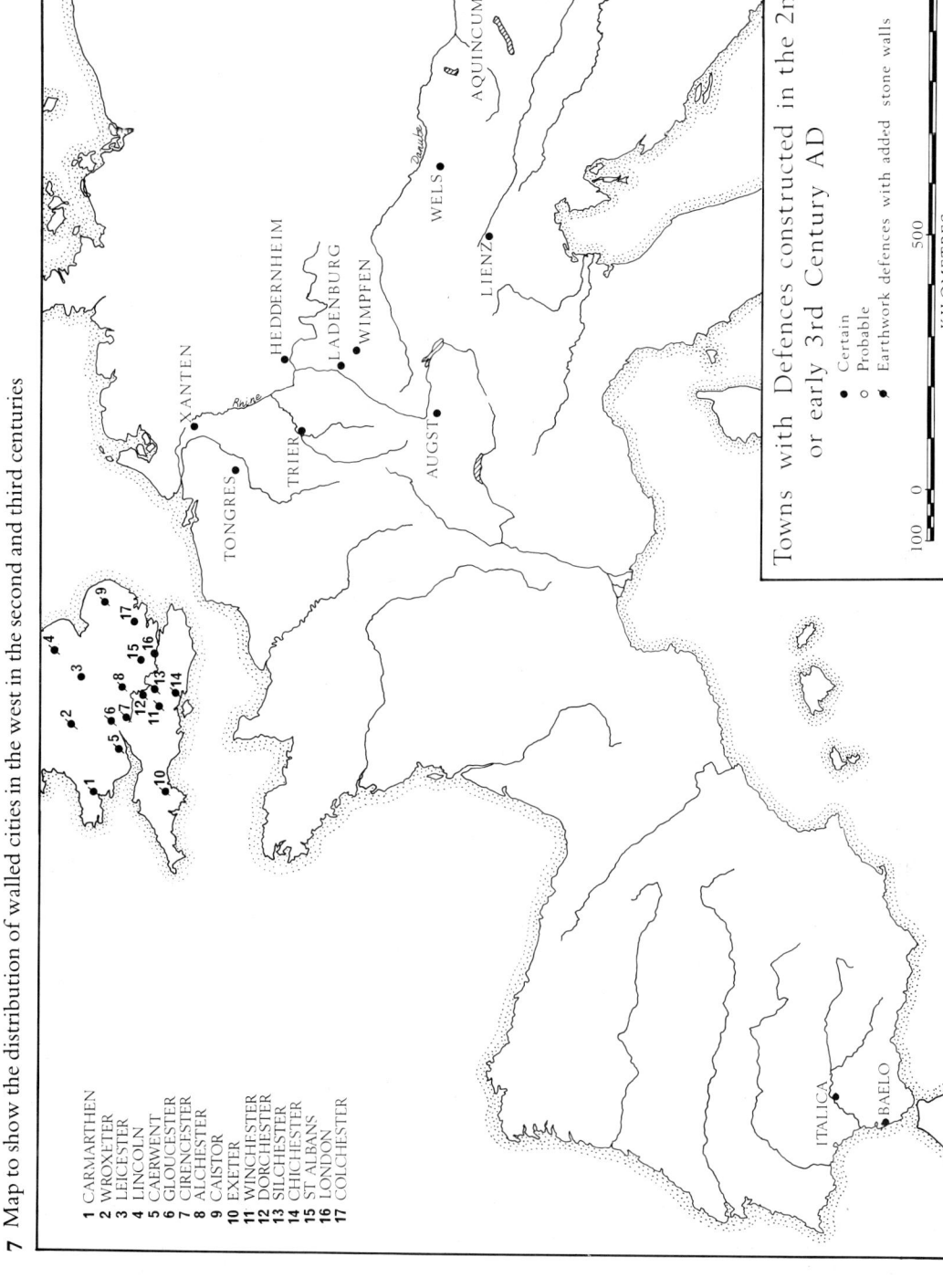

7 Map to show the distribution of walled cities in the west in the second and third centuries

1 CARMARTHEN
2 WROXETER
3 LEICESTER
4 LINCOLN
5 CAERWENT
6 GLOUCESTER
7 CIRENCESTER
8 ALCHESTER
9 CAISTOR
10 EXETER
11 WINCHESTER
12 DORCHESTER
13 SILCHESTER
14 CHICHESTER
15 ST ALBANS
16 LONDON
17 COLCHESTER

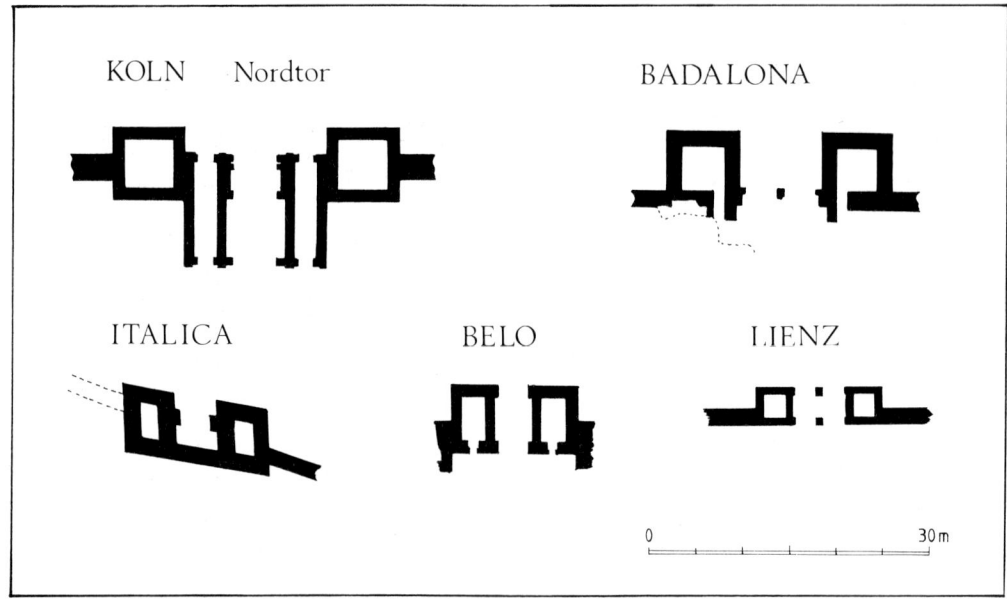

8 Plans of gates with rectangular projecting towers (*scale 1:800*)

absence of even one dated inscription, it is dangerous to posit a single comprehensive building campaign as a reaction to the scantily known historical events of the late second century. There is evidence from Cirencester and Verulamium that large masonry dual-carriageway gates, with pedestrian entrances and flanked by projecting semicircular or segmental towers, may have been built before the main wall circuits, but probably contemporaneously with the earthen rampart (fig. 9).

There is little in the western part of the Roman empire to parallel this British phenomenon. Only in the eastern European provinces is there anything comparable. Here, after a series of invasions by tribes from the north who crossed the Danube, aiming for Athens, a number of Thracian cities were provided with walls, among them Philippopolis, Pataulia, Serdica and Beroe. Although there appears to be little standardisation in these defences, there are problems in distinguishing late second-century work from later additions to the walls. Coins struck by the Antonine emperors depict some of these new fortifications. Several of the wall circuits were rectangular, with towers astride the walls: the walls of Serdica, for example, enclosed about 16ha with walls 2.15m thick faced in brick on foundation courses of large ashlar. At the angles and at intervals between them there were round towers astride the wall. At Philippopolis, an inscription confirms the construction of the defences in the reign of Marcus Aurelius. It has an irregular layout, with a circuit probably about 3.5km long. The wall is of similar thickness to that of Serdica, but as yet no towers have been found.[15]

These fortifications, both the British and the Bulgarian examples, do not diverge significantly from the style of defences in current use on the European frontiers at the time. Gradual changes, however, were taking place in the design

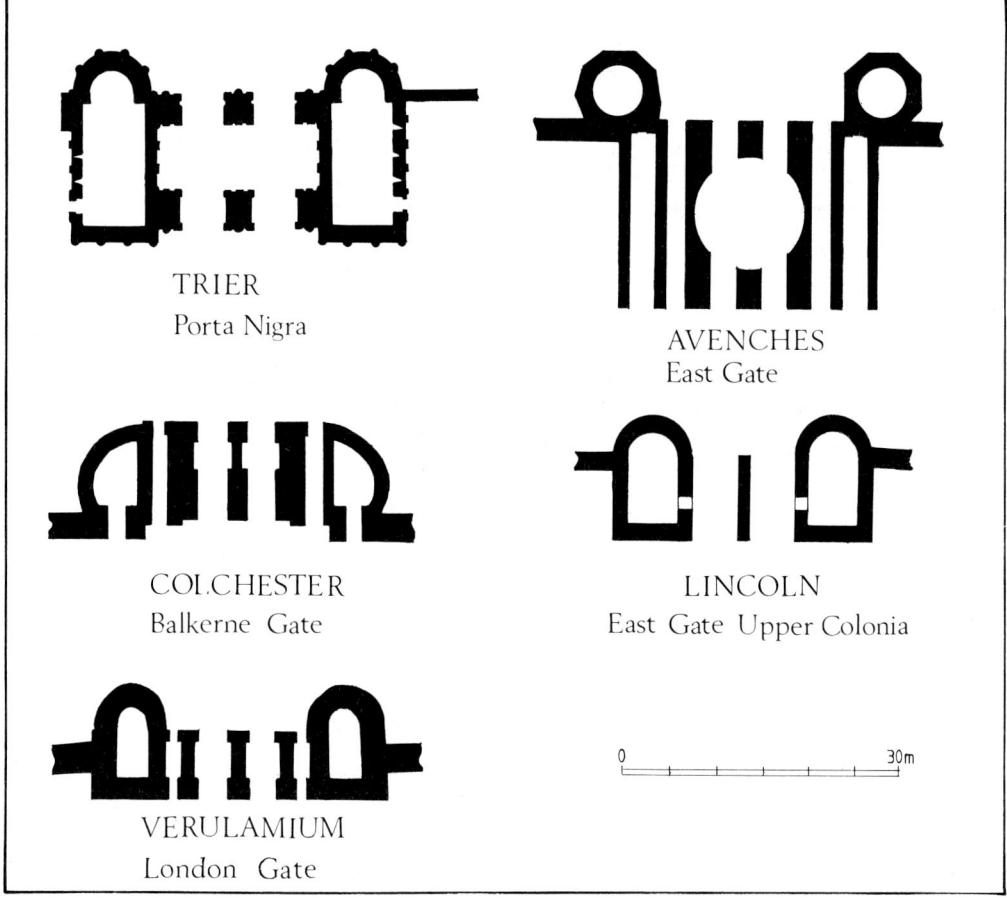

9 Plans of gates with semicircular projecting towers (*scale 1:800*)

of fort gateways. At Regensburg, a legionary fortress established in 179 for the *Legio I Italica*, the Roman Porta Praetoria (fig. 10), has now been identified as part of the original design, although it was once claimed to be late Roman in date.[16] The gate was double-portalled, and flanked by a pair of projecting semicircular towers. Although much of the facing stone is now lost, the surviving portion shows that it was constructed of fine ashlars, with moulded string courses. At rampart walk level, in the surviving eastern tower, there were three large windows of inverted U-shape.

At about the same time, the large and important city of Trier was also receiving a circuit of defences. Only the Porta Nigra, the largest of its gates, survives above foundation level.[17] It, too, was double portalled, and flanked by a pair of large U-shaped towers (fig. 9). Behind the gates themselves was a large enclosed courtyard similar in design to those in Augustan gates at Ravenna, Aosta, and Turin. Also in the Augustan mould was the arcading of large U-shaped arches above the gate passage, and carried round the towers at rampart walk height: there were in fact at least two storeys of such arcading. Like the Porta Praetoria at Regensburg, the

Porta Nigra was decorated with a rich array of moulded detail. Perhaps significantly, another shared feature of both gates is that they were never completed.

Traditional fort design on the western frontiers in the late second or early third century was also beginning to see some slight modifications. In the normal Hadrianic fort, none of the turrets had projected forward from the line of the wall. Towards the end of the second century, the towers flanking the gates would often stand proud of the wall line, and eventually some forts were designed with pairs of round fronted towers flanking the four main entrances (fig. 10).[18] Examples of auxiliary forts which have gate towers of this design have been found in several areas of the Roman world. In the area near Regensburg, they date from the late second century onwards, perhaps adopting the pattern of the gates of the newly built legionary fortress. In Britain and Dacia (Rumania), the design appears to belong to the early third century. This style of gate, dated to 198, is also found at the otherwise rather irregular fort of Castellum Dimmidi in Africa.[19]

In general, therefore, by the early or mid-third century, fortifications in the western empire were still very much within the traditional mould. Unless there were overriding reasons to the contrary, forts were designed in the normal rectangular plan (fig. 11), and even, as at Castellum Dimmidi, where the fort plateau was such as not to allow such a layout, the interior buildings were still as regularly planned as possible. Fort angles were still rounded, and only in rare instances did the interval or angle turrets project at all outside the line of the wall. More frequently, the gates, still double portalled, were flanked by a pair of projecting towers, sometimes semicircular, sometimes rectangular and occasionally polygonal, as in the gates of sites like Bu Ngem in Africa and Theilenhofen in Germany.[20]

The design of city walls, too, had scarcely progressed beyond the Augustan model. Though generalisations are scarcely possible, city walls were of medium thickness (1–3m), and any towers, square or round, would normally sit astride the walls. There was a tendency for city walls to be quadrangular in plan, and, in Britain in particular, to be backed by an earthen rampart. The gateways might be more massive, with projecting towers giving flanking cover to the double portalled entrances. None of this, however, was in any way out of character with or beyond the traditions of Augustan fortifications. The hallmark of Roman defences up to the mid-third century was essentially that of an inherent conservatism. This was probably adopted, not through positive choice, but because in the *pax Romana* of the second century there was no overriding need for strict defensive measures to protect major cities. The pressure to produce new defensive designs was not therefore being applied.

Any assessment of Roman defensive techniques in the first three centuries AD in the eastern provinces of the empire—Arabia, Syria and Mesopotamia—is hampered at present by lack of secure dating evidence. Roman contact with her eastern neighbours, in particular the state of Parthia, had been largely one of intermittent conflict and uneasy peace. The boundary was recognised as the river Euphrates, and natural topography demanded that the approach to this area was

10 Plans of second- and third-century fort gateways (*scale 1 : 800*)

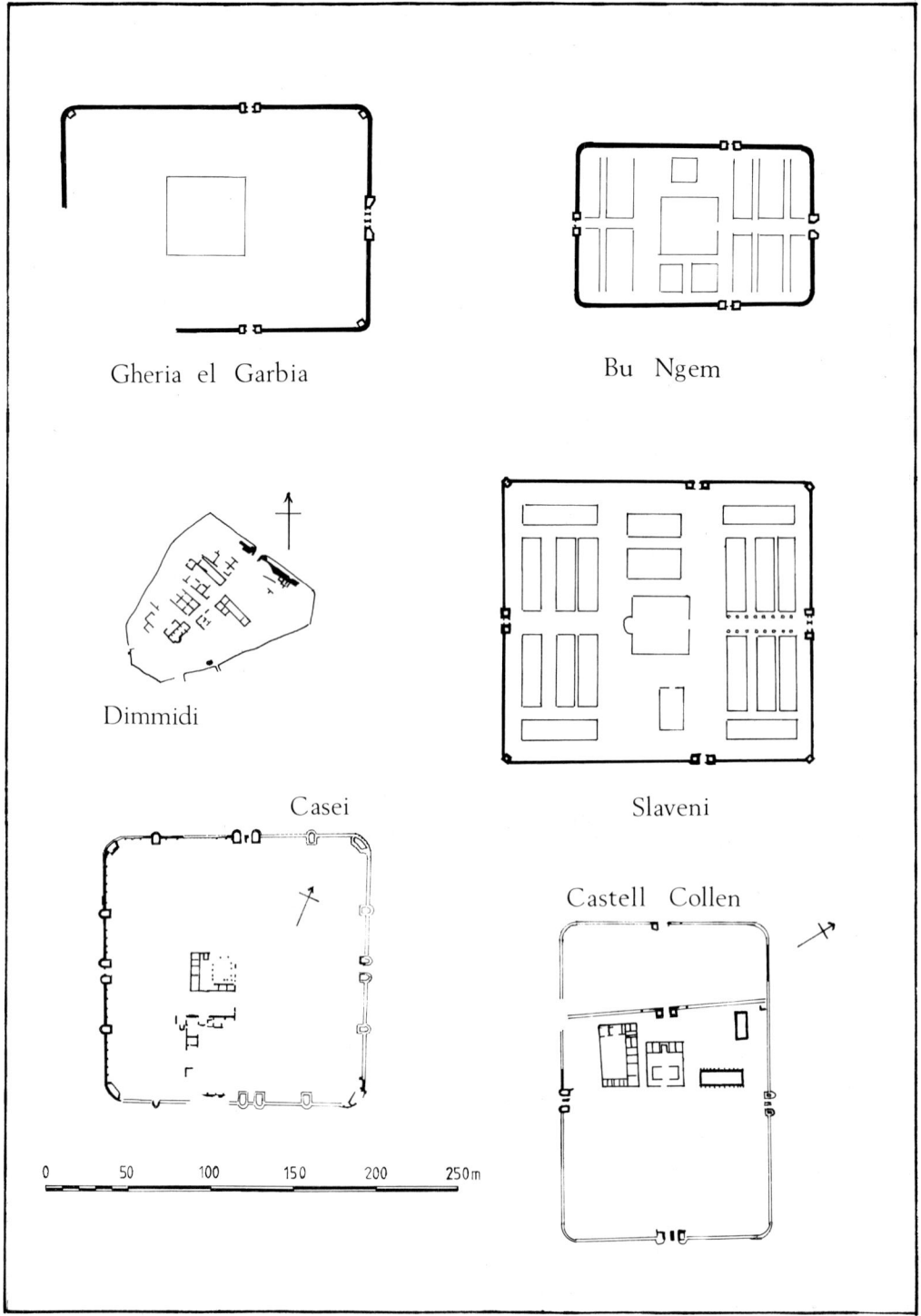

11 Plans of late second- and early third-century forts (scale 1:4000)

always via the fertile crescent of lands at the northern end of the Euphrates and Tigris. From Flavian times onwards, however, it appears that the Roman army had a permanent presence in a loosely defined frontier zone bordering on the desert lands to the south and east of Palestine and Syria.

In this area, there was a need to establish Roman forts not only in strategically important places, but also where a water supply was readily available. Forts, fortlets and watchtowers clustered along main frontier roads and their subsidiaries. Only a few have produced any dating evidence, and it is clear from milestones found along the frontier road and inscriptions discovered at some of the forts themselves that a continued Roman presence in this area led to rebuilding of the forts at one or more different periods.

Several basic types of fort have been identified and classified according to plan by earlier investigators and on this basis assigned dates.[21] Despite the dangers of this approach, this classification may help initially to understand the types of fort in use on the eastern frontier. The simplest is the type which consists of a four-square walled courtyard enclosed by plain walls with casemate buildings all round. This type of post, usually relatively small, seems to have been in use at least as early as the third century BC in the Negev area and in Nabatean Judaea.[22] It has been suggested that the Roman forces found and adopted this type of small fortification when they began to assume military control in the first century AD. Its use on the Roman frontiers of Arabia, however, is not closely dated even though pottery evidence has suggested that some of the fortlets there may also be of pre-Roman origin.

Two other fort types which have also been identified and were originally assigned a date in the second or early third century AD are of basically rectangular plan, with projecting towers (fig. 12). In the Syrian desert area there are several small posts with thick walls and projecting corner towers in a rather bulbous fan-shape. Two sites further south, at El-Leggun and at Odruh, also exhibit such corner towers. These are much larger in size, and are far closer to a size normal for auxiliary units of the early imperial forces. Both forts are quadrangular (Odruh is slightly trapezium-shaped), and both have projecting U-shaped interval towers. There are gates approximately midway along all sides, and the corner towers form a distinctive bulge: from surveys carried out at the turn of the century it can be clearly seen that the corner towers incorporate a staircase round a central pillar. El-Leggun has a marked number of interior buildings capable of definition, mainly barrack blocks with ranges of small mess rooms, all grouped about a central building which appears to be the *principia*. At Odruh, on the other hand, little beyond the walls was still definable on the ground. One solidly built structure near the north entrance at El-Leggun was designated the *armamentarium*—the arsenal.[23]

One further type of fort originally assigned to the second century in this area is that represented by the remains at El-Kastal. This is basically a fort built in the local 'four-square' tradition, with a cluster of rooms within an almost square enclosing wall and giving onto a central courtyard. This fort, however, only some 60m square, bears round corner- and interval-towers.[24] The single portalled gate is

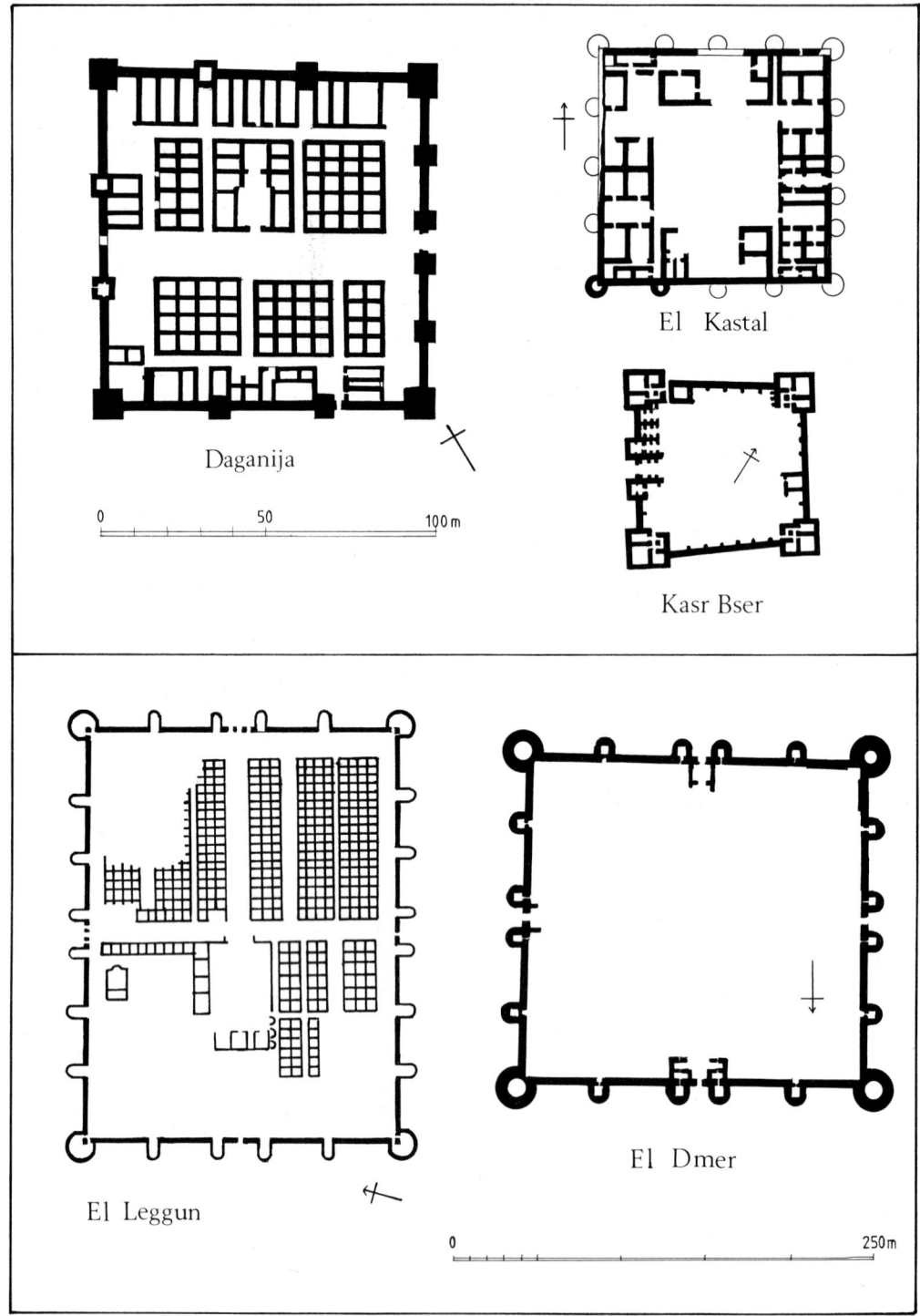

12 Plans of forts on the eastern frontiers (*scale 1:2,000 and 1:4,000*)

protected by a pair of similar towers, but such was the state of survival of this fort that only two of the towers were identified and drawn on the original survey. The dating here is crucial: in among the stones lying near the building identified as the arsenal was found a stone block with an inscription recording building under Marcus Aurelius, in 162. Another stone was found built into the arsenal itself, recording construction under the Casesar Valerian (253–9). If the construction of the projecting towers is to be assigned a date on the evidence of one of these inscriptions, it has to be established with which of the two phases of building they belonged. The natural assumption has been to assign the date of the fort, its towers and the majority of the standing buildings, to the earlier date (162), but this is by no means certain. Nor is it certain that either known inscription necessarily dates the defences.

The dating evidence for the surviving remains at El-Leggun and at Odruh is even more shaky. There are no inscriptions to date the buildings, and it is assumed that both these sites were originally intended as forts to hold a wing of cavalry. El-Leggun, however, as its name implies, in the late third century served as the headquarters of the *Legio IV Martia*, a sign of the increasing pressure on the cultivated strip of land along the eastern sea coasts of the Mediterranean from nomadic tribes in the desert. This pressure had clearly begun under the Antonine emperors, and Marcus Aurelius in particular had been the first to respond with fort construction.

If surviving remains at Odruh, El-Leggun and El-Kastal represent the second-century arrangements of the forts, they are remarkably more advanced than sites of the same date on other comparable frontiers. In Tripolitania, for example, the fort at Bu Ngem was built within the normal tradition with rounded corners and no towers, except at the gateways as late as the early third century.[25] Later in the same century, however, a small square type of fort of 'courtyard' plan with projecting corner towers (very similar, by its description, to El-Kastal) made its appearance on this African frontier, though a fort of similar plan (known as the quadriburgium shape) at Eski Hisar in Syria is now dated to the early third century by an inscription.[26]

The design of this trio of Eastern forts is so strikingly different from anything otherwise clearly assignable to the Roman military at such an early date, that it is difficult to believe that this type of fort sprang fully fledged into the military repertoire in the early second century. Projecting towers of U-shape or fan shape are not known elsewhere in the Roman world for another century or so, and it seems preferable to believe, as excavation and analysis is beginning to show, that these forts were founded in the Antonine period, but had undergone substantial modification of their defences under later emperors. The corner towers, in particular, with their distinctive quadrilateral staircase around a central pillar, and giving off into separate rooms, some of them on mezzanine floors within the towers, are exactly paralleled by the corner towers of Kasr Bser (fig. 12), a smaller fort in the same area, which a surviving inscription in situ over the gate proclaims to be of Diocletianic (286–305) date.[27]

It is clear that contact and conflict with her neighbours often acted as the spur for Roman progress and advancement of architectural forms. The Parthian empire had by the second century AD a large repertoire of defensive works. At the most obvious points of contact with Rome, the cities of Dura Europus and Hatra had fortifications built to a design very similar to the Roman norm, and based also on the Hellenistic model, in the wake of Alexander's conquest of the area. Dura Europus itself, originally a Seleucid city founded *c* 300 BC, was an important trading post between Romans and Parthians in the early centuries AD. Rome seized military control of Dura under Trajan, but it was not fully embraced by Rome until AD 165. Although at Dura, the city plan is based on the grid system with surrounding walls and projecting rectangular towers, there are other Parthian cities and towns where an overall design for the fortifications took on a less taut and strictly 'Hellenistic' appearance. At Hatra, the wide sweep of walls forms an irregular ring, also protected by projecting rectangular towers and gates of complex design. These date from the turn of our era. Possibly of similar date, however, are other wall circuits such as that at Koi-Krylgan-Kala, with thick defences and projecting semicircular towers.[28]

The general progress of Roman fortification therefore up to the mid-third century AD can be summed up as follows. Augustan civic pride aimed at the provision of a prestigious wall circuit for most of the new city and colonial foundations. The peaceful times after Augustus' death to a large extent rendered the provision of city walls superfluous. Defence instead was concentrated on the frontiers, and here traditional forms of fort construction began to find permanent expression, first in earth-and-timber, and a little later in stone. At first, fort ramparts incorporated few defensive features of any sophistication, but gradually grander elements became current in the military architecture of the frontier regions.

Rome's gift of adapting and encompassing for her own use the ideas and traditions of other peoples may have led in localised areas to the takeover of Nabataean, Parthian or other ideas from fortifications derived from the east. The basic pattern of Roman fortifications throughout the empire by the mid-third century AD remained roughly that of the time of Augustus. City and town wall designs were based on Hellenistic or Augustan antecedents, and military forts had changed little in design since the early second century, when the 'playing-card' design was converted on the mainly static imperial frontiers to stone. The new and insistent pressures of the early third century and after were to produce a radical Roman response and a complete reappraisal of defensive techniques.

2

The New Design in Late Roman Defences

Contemporary sources serve us badly when recording the strategic or architectural thinking which led to the siting of town or fort walls at any period within the Roman era. In his instructions for the emplacement of an ideal town, Vitruvius, writing in the time of Augustus, stressed the importance of choosing terrain suitable for the purpose. All aspects—defence, health, availability of water—must be fully considered.[1] This is possibly a writer's commonplace since Vegetius, writing some 300 years later, stipulates almost exactly the same conditions when he says 'cities must be defended either by nature or by skill' and elucidates this with a list of conditions which are to be given full consideration. He concludes with the statement that the best situation for a town is to be so well defended naturally that it has no need of walls.[2] If walls are necessary, however, both authors were unanimous that the area enclosed should not be rectangular, and Vegetius volunteers the information that it should have 'not straight, but angular walls'. An oval, round, or polygonal shape is to be preferred, especially when provided with adequate towers.

On the actual method of constructing a town wall, only Vitruvius has anything explicit to say. Foundations must be dug down to the natural subsoil, and should be broader than the superstructure of the walls. When built, the wall should be thick enough for two fully armed men to pass each other on the rampart walk: its facings should be bound together by frequent tie-beams of charred olive wood. Towers, of round or polygonal shape, should project from the walls and be about a bowshot apart. To make the defences doubly safe, the addition of an interior earth rampart is to be recommended, since it makes the walls virtually impregnable against battering ram and undermining.

Both Vitruvius and Vegetius, the latter to some extent at least following the former, treat their descriptions of town-wall construction as idealised schemes to be undertaken on virgin soil. They are, therefore, more appropriate as instructions to colonists going to previously uninhabited lands than as a general architects' guide: their relevance to the situation in the late third century, when the task of maximum priority was to enclose existing sites and settlements with walls, is therefore limited. Indeed, Vitruvius, writing at about the turn of our era, seems to disregard the majority of Augustan colonies, such as Emona (Ljubljana) Como and Barcelona, all of which were laid out in a rectangular pattern with a regular street grid after the fashion prescribed by the *agrimensores*.[3] Vegetius' *De re militari*, on the other hand, is very much a mixture of details about military affairs of all periods, and his description of town and fort wall building cannot be said to date precisely from one period rather than another: both descriptions of town siting and building should perhaps be regarded in the light of the literary tradition of the 'Ideal State', begun in the works of Plato and Aristotle rather than as a builder's (or even archaeologist's) manual of factual detail and dimensions. Only Vegetius made any mention of fort building, and his account is disappointingly scant.[4] The fort he described is merely a temporary marching camp used by troops for overnight stays rather than a permanent walled fort. By the late third and fourth centuries, there are indications that the temporary camp—the marching camp used by troops on campaign—was by no means as widespread as in earlier, rather more expansionist days. Although we know from the work of ancient historians of places where punitive or expeditionary campaigns were mounted in later centuries, it has not been possible to identify positively a single late Roman marching camp. Admittedly, these might look precisely like earlier Roman marching camps, and thus impossible to date with any degree of certainty. Moreover, the character of a marching camp does not guarantee significant results from even large-scale excavation, nor can one expect the deposition of groups of material (other than possibly the odd chance breakage) which will facilitate dating.

Only one temporary camp, at Ermelo in Holland, has been claimed as a fourth-century example of a marching camp[5]: the evidence is scant, and appears to rely mainly on the site's position in the area where Julian was known to have campaigned in the mid fourth century.[6] In southern Scotland, however, where Constantius undertook campaigns at the very beginning of the fourth century, no sites have been claimed as fourth century in date. There is evidence, however, to show that, on campaign, troops might more readily re-fortify former sites of forts abandoned as Rome's frontiers contracted, and use these sites as the base of their operations.[7]

For the walls built in the late Roman period the construction method was as follows: the foundation trench, usually straight-sided, was first dug through the site, and filled with a stiff, durable material. Often this was clay and cobbles, though from area to area it might vary from puddled chalk to river boulders set in mortar. In some places the foundations were virtually non-existent, but elsewhere they may have been more than a metre deep. Occasionally, extra care was taken to empty all pre-existing pits and features underneath the projected line of the wall and these were packed with foundation material in order to prevent the wall sinking in that part of the circuit.

Usually the foundations ran through an area which had been comprehensively built over before: this meant that there were sometimes walls of earlier buildings which could be used as a basis for the foundation of the defensive circuit; one case of this is provided in Rome itself, where portions of Aurelian's walls have whole facades of houses of an earlier period incorporated in their facing. At Angers, there are remnants of a villa or possibly a town-house and arcading, possibly belonging to an earlier building, incorporated within the defensive walls: at Bavai and elsewhere in Belgica Secunda walls leant in part against the remains of the forum and cryptoporticus.

At many sites, particularly those where the conditions were marshy, vertical piles were used to underpin the foundations. This feature has been found to be far more common than at first suspected, and has been variously interpreted. Originally, when first found at Altrip, these wooden piles were interpreted as a device to foil tunnelling, thus rendering the walls impregnable. This is scarcely likely and in view of what Vitruvius says about the use of timber in marshy conditions, it is probably that such timber stakes were in fact structural.

The use of timber in the lowest levels of the walls was not confined to vertical stakes. Above the foundations and before the beginning of the footings and the wall itself, there is at some sites a layer of horizontal beams, visible as a series of grooves in the foundation course. This feature is particularly noticeable at Pevensey and at other Saxon Shore forts, but it is also found in widespread use on the continent. At Dax, where the subsoil conditions are marshy, there were layers of wooden beams under the walls, and at Strasbourg, the external towers of the later wall were tied in with the rest of the wall by such beams. Since the only surviving trace of these beams is the empty holes which show their former position, it is impossible to tell what sort of wood was normally in use for such construction. Vitruvius recommends alder for the piles underneath buildings and charred olive as tie-beams, but olive at least is hardly likely to have been in use all over the western empire.

Late Roman town walls are particularly remarkable for the amount of reused material in their bottommost courses, and it is from this rich source of large inscribed and sculpted stone that many of the most notable Gallic examples of monumental art have come (pl. 6). Immediately above the foundations was spread a layer of mortar, and then came the footing courses of the walls. Throughout almost the whole of Gaul, and probably elsewhere too, large stone blocks were laid dry over the foundation. Sometimes this was plain stone quarried from nearby, but it was more likely to be parts of large monuments—architectural fragments, column drums, bases and capitals, funerary monuments and inscriptions, culled usually from demolished buildings within the earlier town on the site and from its surrounding cemeteries. At all events, wholly or partially destroyed monuments formed a very useful quarry for secondary building stone.

The superstructure of the wall was built normally one vertical course at a time. An inner and outer face were set in position to form a shallow trough. The space between the wall facings was then packed with builders' rubble, and capped with a liberal amount of liquid mortar. This formed a flat platform ready for the construction of the next course. There were two areas of potential weakness in such a method. The first was the vertical adhesion between the superimposed

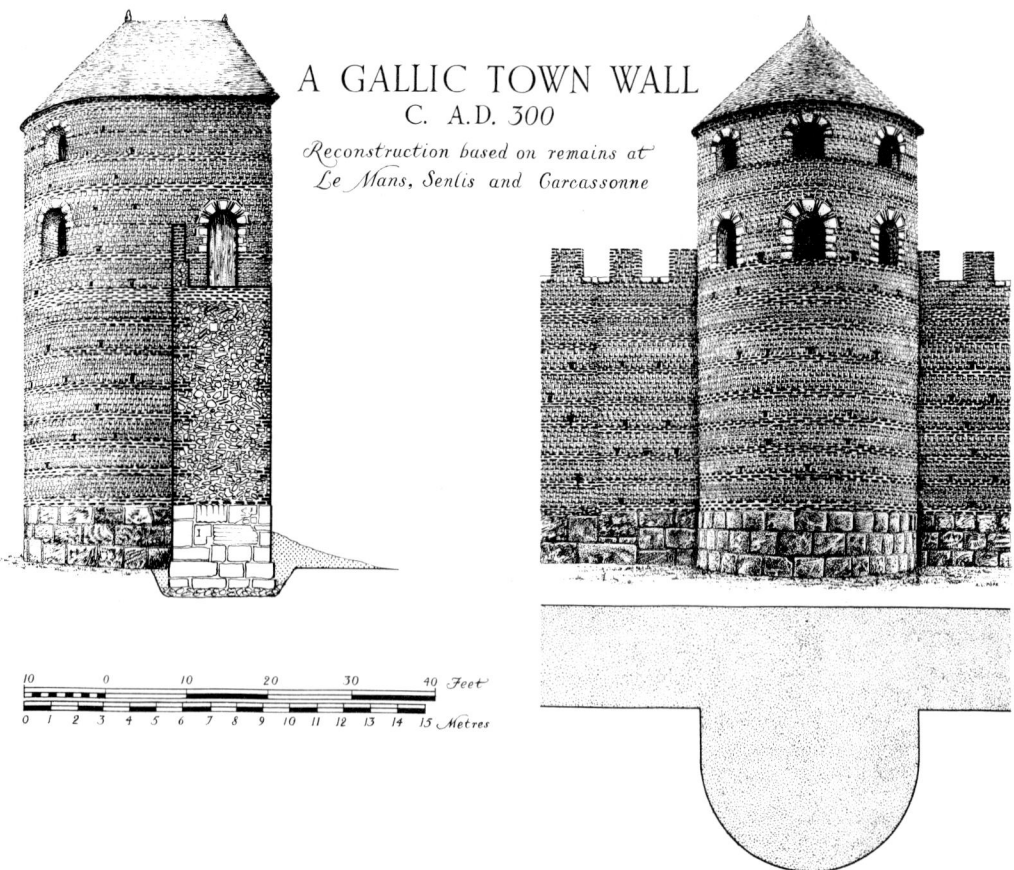

13 Reconstruction drawing of a Gallic town wall, based on surviving examples at Senlis, Le Mans and Carcassonne

layers of wall core. This does not seem to have been a problem, though occasionally where a fort or town wall has now toppled, it has on landing split, as it were, into slices. The second weakness occurred at the junction of the facing stones and wall-core. Although the builders often used material for their facing which had a long tail which could be well held by the wall mortar, as an extra security, bonding courses were also used at intervals. The material used in these courses was normally deep flat stones or tiles. These reached further back into the wall-core than the facing stones themselves, and helped to key in the facing more securely. They also may have served, especially when (as at one or two sites) they seem to have spanned the whole thickness of the wall, as levelling courses.

Construction of a wall by this method may have been facilitated by the use of shuttering to ensure that the wall face could not bulge when the liquid mortar or grout was applied to the interior and particularly so if rapid building were in progress, with the laying of a new course on top of one which was not yet properly dry. After construction, the facing still needed pointing. This was often done in a hard mortar with a much greater content of crushed tile that the core-mortar. The reasons for this are unclear, but the addition of the tile may have made the mortar set quicker, or ensured that it made the mix more waterproof.

The facing used could be virtually any kind of stone or building material. In Spain, where good stone is readily available, most walls were faced with square *opus quadratum*[11], large dressed rectangular stones laid as headers and stretchers (pl. 13), but there also occurs a much more fragmentary thin slate-like stone, which may equally well have been used in Roman times as in Medieval. Large stones were also used elsewhere, usually, as at Regensburg, on sites of military origin, but also at some town sites lying back from the frontier, such as Sopron (pl. 14), and many of the Italian towns. Large stone of this kind has very often been a popular target for stone-robbers up to the present day, and many examples of walls of this kind have doubtless disappeared.

Throughout northern and western Gaul, the style in use consisted of small squared blocks of stone. In walls of the Augustan period, the facing (as at Arles, Fréjus, and Autun) was usually of this type without bonding courses. A similar style was used in the walls of Köln, Tongres and Trier, and many second-century walls of towns in Britain. Ocasionally, and increasingly in the later period, courses of tile were used as bonding (pls. 4,5). The earliest example of this style used in a defensive context seems to be in the walls of the legionary fortress at Strasbourg, which is probably of Trajanic date. The use of tiles as bonding courses seems to have developed out of a similar use in public buildings of all kinds, and there are examples of early *basilicae* and *fora*, for example at Arles and Bavai, which use tiles in their structure in very much the same way as they are also found in fortifications.

This mixed style of construction, using stone and tiles (known as *opus mixtum*) was a favourite style for areas where good building stone was not readily available, since it affords the chance to use stone and tile in an ordered fashion, in whatever proportions are available. The type of stone used is immaterial: in East Anglia for example, the Roman walls of such sites as Burgh Castle, Caistor St Edmund, Bradwell, and Colchester all employ tile and split flint or septaria nodules as their facing; here the tiles act as an ideal keying agent for the flint nodules, which, by their very character, can never form a perfect bond with the wall-core. A further advantage of these materials is that they provide an acceptable method of using second-hand building materials brought to the site or found there from demolished buildings. The facing stones can simply be re-used or re-squared as necessary. The tiles used in defensive construction are not always the flat *lateres*, or building tile. There are several examples of roofing tiles (*tegulae*) to be found in late Roman walls, and it is clear that this construction method formed a useful and convenient way of giving this material a second life.

Tiles themselves sometimes formed a facing material for a wall of normal rubble construction. At Rome, continuing a much earlier tradition of brick building, Aurelian's wall was built of flat *lateres* split almost diagonally so as to present a flat outside surface, but so that the core was bonded to the triangular projection of tile jutting into it. The only other known western example outside Italy of a wall faced in tiles, at least in its upper portion, is that at Toulouse, where the lack of stone made the task of completing the massive circuit in small blockwork too onerous. In Italy, tiles were a more common building material at all periods for buildings of all types and the walls of such Augustan colonies as Como and Turin were faced in this material.

Many of the late Roman defences show signs of having been built in sections by separate gangs of workers. Where this type of construction was at its most efficient it is difficult and probably impossible without actually dismantling the wall to spot the points where one gang finished work and another started. This is as one would expect, for if all went according to plan, the tile courses and the facing material contributed by different gangs should cohere into a homogeneous whole. Indications of gang-work divisions, therefore, are often slight: at Portchester they consist of a group of levelling stones at or near foundation level at regular intervals round the circuit, indicating that a projecting tower together with a stretch of walling 30m long was the standard ration.[8] At Pevensey and Richborough, however, the pattern was by no means standardised. At Richborough, the south and west walls show no sign of gang-work at all, but the north wall bears the marks of no fewer than four gangs involved in its construction. The breaks here between sections average about 20m, but are by no means standardised. What is more, they are remarkably obvious, including one in which the tile courses do not correspond, the styles of wall-facing are markedly different, and the wall changes direction by a couple of degrees.[9] At Pevensey, though one can begin to see a pattern of construction by gangs of a length of wall and a tower, this is not consistent throughout the circuit.

As the wall progressed, so the builders needed scaffolding from which to work. This was timber-framed and supported either on horizontal poles which ran right through the wall core (or deeply enough embedded to take builders' and materials' weight) or by a series of horizontal poles linked to the walls by tie-beams and thus providing a framework on which planks could be laid. The latter type is possibly the more usual: putlog holes by which the short poles could be tied in to the wall seem seldom to run much deeper than 1m and more usually only half this distance: they are occasionally in vertical lines, which might lead one to suppose that they were tied by horizontal linking bars to vertical posts. Paintings in the early fourth-century catacomb of Trebius Justus in Rome show scaffolding of this free-standing style in use for house construction.[10] As the wall grew, the scaffolding poles will have become embedded in the wall-core, and will have needed sawing off or withdrawing. The resulting hole could be plugged with a cube of stone. Where the timber was left in position and rotted or the hole was left available for later use, it left the distinctive holes which are now visible on some sites.

These putlog holes occasionally enable something of the constructional history of the wall to be known: in at least one instance at Richborough, there appear to be two separate sets of holes, corresponding with an upper and a lower set of scaffolding.[11] This would suggest two stages of building work on the wall itself. In addition, the positioning chosen for putlog holes may also be seen as the signature of a gang or gangs of workmen: they might choose to locate these holes in the levelling courses, or at a particular point within the blockwork facing.

At the top of the wall, one would normally expect to find a rampart walk and breastwork. In only a few cases does any late Roman wall survive to this height, but the top of the ramparts would normally be at the height of the first storey of windows in the projecting towers. In places where the curtain wall itself has not survived, doorways leading from towers to the rampart walk have established the original height of the wall. At Trier, for example, the doorways leading out of the Porta Nigra give an original wall height of 6.1m; at Rome they are between 8–9m, at Autun, 8m. Other figures, for example at Carcassonne, Le Mans, Senlis, Beauvais, and other places where towers still stand to a requisite height, ought to be more readily available.

At rampart-walk height, the exterior face of the wall was normally marked by a proliferation of levelling courses, though often if the wall has survived to this height, there is a greater chance that later additions and improvements have here been carried out. At Le Mans, where the walls are of small blockwork and tile courses, there is a concentration of tile bonding courses forming a base for the rampart walk and the parapet fronting it (pl. 1). When found in a wall of larger stone blocks, as at Chester, a plinth course occurs at this height. At two sites, Rome and Barcelona (pl. 10), traces of crenellations survive in the walling, but at several earlier fort-sites, the capping stones for crenellated breastworks have been found: these appear to have formed 'h'-shaped niches, intended to protect the defenders on the wall-top on as many sides as possible from a missile attack. Although few examples of actual crenellations exist, representations of city walls on many ancient sources—mosaic borders, coin and medallion pictures, and on artistic representations of the period—all show that crenellations were normally to be found and form part of the normal portrayal of the walled town or fort (see figs. 16 and 17, pp. 42–3).

The walls of the later Roman period were mostly built to a thickness of some three or four metres—on average roughly twice that of defences of earlier days. They were freestanding, without an earth rampart behind them. The combination of narrow stone wall and earthen bank had been a product of the military architecture of the Trajanic and Hadrianic periods. By the late Roman period economy of space within the walled enclosures dictated that there was no longer room for a spread earthen rampart which was too wide, took up too much space (normally at least a 6m width) and prevented the positioning of lean-to buildings against the inside of the curtain wall. The massive width of most of the walls made provision of an earth bank unnecessary, at least for defensive purposes.

Although it is the facing stones of the wall which normally are distinctively recognisable as Roman work today, it is possible that the whole of the exterior face of the wall was rendered in Roman times and a protective covering thus given to the wall face. This would explain, for example, why in some places building gangs were unconcerned about the uneven effects of their efforts to tie in their work to each other. It is highly unlikely, however, that all late Roman walls would have had such a rendering: if it had been common practice, evidence should be more widespread. Contemporary evidence against the wholesale use of this technique comes from Toulouse. Here the walls were built largely of tiles (as are those at Rome), and the city was described by Ausonius as having 'baked walls':[12] if it was not evident then that they were of tile, the force of this poetic epithet would have been lost.

Another curiosity of late Roman walling, and one which has a bearing on the question of finish given to the exterior facing, is the problem of the patterning. In early imperial times, at Fréjus and at Köln the Roman walls bore decorations varying from the decorative to the bizarre: those on the Römerturm at Köln, for example, form a curious mosaic of black-and-white shapes, so bizarre that their Roman origin has sometimes been questioned. Examples of a patterned finish to Roman walls are again found in the later Roman period. At its simplest, this technique consists in the use of dark and light stone ashlars to provide a chequerboard or diaper pattern. This is to be found, for example, at Richborough in the north wall. Much more elaborate patterning in the facing is found at Le Mans (fig. 14, pls 1-2), where each band of small blockwork between the tile courses bears a different pattern of white chalk blocks on the brown sandstone of the background. A variety of stripes, losenges, circles, V-shapes and other patterns is represented and some of these designs are also to be found in other late Roman walls in the neighbourhood, notably at Rennes. A number of other features in the wall facing of various sites may also be of decorative rather than functional use: in some of the tile-courses at Carcassonne, for example, there are small inverted U-shaped patterns.

A particularly distinctive feature of late Roman wall building is the external, or at least externally projecting towers, placed at intervals along the walls. Walls and ditches were made less easy to attack by the provision of projecting towers with large windows in which defenders could be mounted to provide flanking fire for the walls. At several British sites, the change in defensive technique has been ably demonstrated by the discovery that a single broad and rather irregular ditch, suitable for exposing attackers to the fire from these towers, replaced the usual narrower double ditches of earlier days close to the walls[13].

Late Roman projecting towers are of widely differing shapes: possibly the most usual is the U-shaped or semicircular, but many variations upon even this theme are found—circular, tangential, quadrants of a circle, pear- or fan-shaped (pls. 10, 12-14, 16). All have the same purpose: to gain distance from the walls so that the side windows could be used to attack any enemy who, in trying to storm the walls, exposed his flanks. The fan-shaped towers of the Pannonian and Norican frontiers, perhaps the most unusual shape of all, were produced in an effort to combat the difficulties of enfilading the walls of a fort of Trajanic 'playing-card' shape from the corners. For a fort of this type, a normal-sized angle tower would not do: it had to project far enough from the curve to dominate the straight walls on either side, and hence the rather incongruous shape and plan, which, though almost always seeming peculiar, is yet the only one which adequately performs the task for which it was intended (see fig. 69, p. 177, fig. 71, p. 183 and fig. 73, p. 186).

Polygonal towers also occur, but they are rarer, and their use at such places as the south-west front of the legionary fortress at York and the great gates at Split (fig. 22, p. 49) perhaps suggest that they are decorative as much as functional.

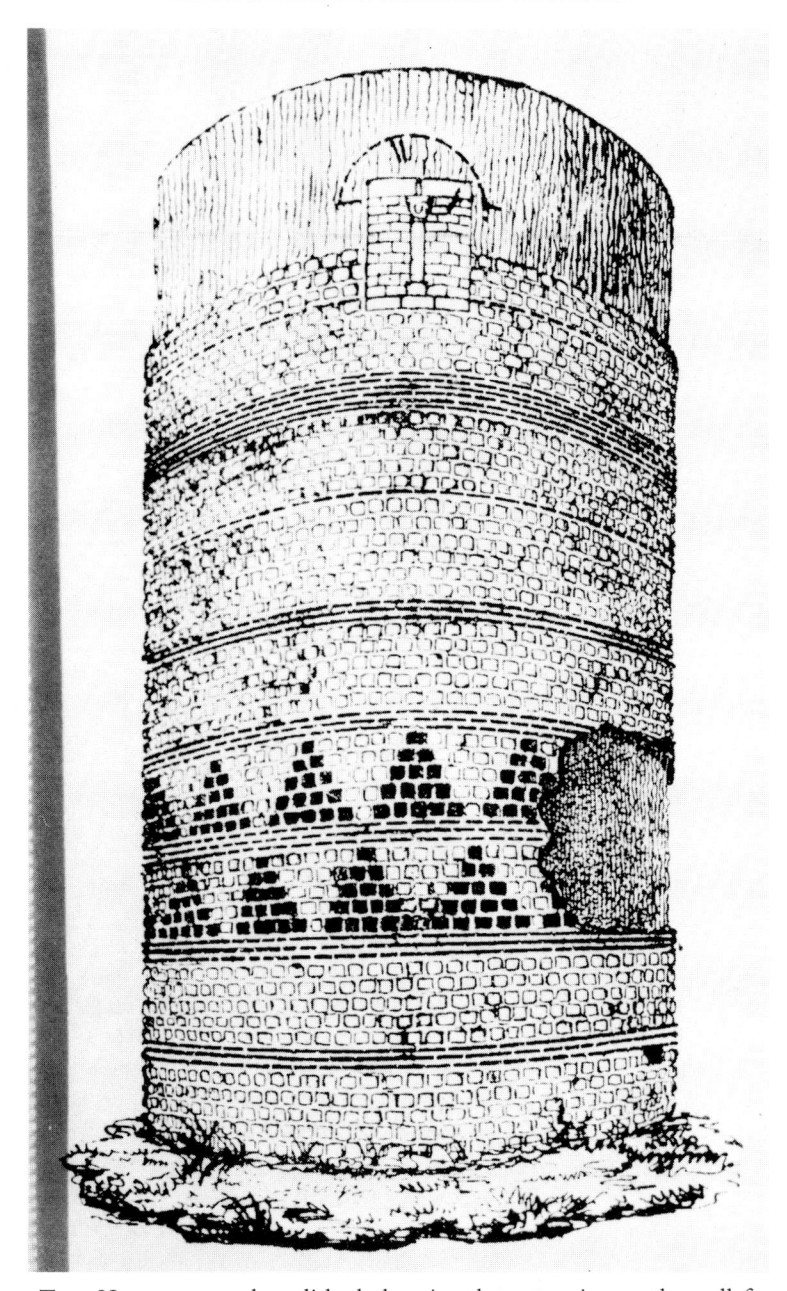

14 Le Mans, Tour Huyeau, now demolished, showing the patterning on the wall-face

They do occur, however, at forts and towns very widely scattered. It has been suggested that the construction of polygonal towers was solely the prerogative of a single highly mobile building gang, though this is scarcely likely since they do not all appear to be contemporaneous.[14]

Rectangular towers seem perhaps to be a throw-back to an earlier style of defence (pl. 11). They were used on very few sites and predominate in gates (the so-called 'Andernach gate' type) and posterns, in particular the 'clavicula' type. Vegetius had already suggested that the use of rectangular towers was not to be recommended, since, if under assault, towers with sharp edges can fall easy victim to the battering-ram.[15]

Above the height of the rampart-walk, towers have survived better than parapet walls. They seem usually to have been solid below this level, and with at least two floors above, each normally provided with large windows with semicircular arches (fig. 5, pls.1, 10, 12). This is in direct contrast with what is known of Augustan windows which (from the examples at Fréjus) have only narrow slit-type windows. The wider openings in later town walls are more consonant with the use of artillery, which to be effective needs a wide arc or fire. Usually there appear to have been three windows in a normal semicircular of 'U'-shaped tower: exceptionally there were four. At Rome, where there were rectangular towers, study of the fields of fire has shown that adequate cover could be assured;[16] where polygonal towers were involved, there might be a single window in each facet of the tower.

Fort towers were less standardised than those on city walls. On the Rhine frontier, the round tower astride the walls is the most common type taking as its model the towers built at the construction of the first century walls of Köln. These are sometimes hollow, sometimes solid, and some have postern gates through them, with entrances into the hollow central chamber both inside and outside the fort. Towers of larger diameter are sometimes found at the corners. More often forts were re-adapted to meet new needs at this period, with the result that towers of a more irregular shape were designed to attempt streamlining of the old-fashioned fortification types.

Many forts of the later period were of more irregular plan than for many years previously, when the playing card shape was usual, and tended to dominate even terrain which was relatively unsuited to its shape and design. Later Roman forts were almost always well adapted to the terrain on which they lay. This sometimes produced in them an extremely irregular ground-plan. Some were set on the crest of a suitable hill, their shape adapted to the demands of the contours, to heighten the defensive capabilities of the site. The terrain may thus have sometimes made the provision of towers impossible.

The type of roof such towers would have had could have been either crenellated and flat, or covered with a conical roof of tiles or slates.[17] The illustrations in the oldest surviving manuscript of the *Notitia Dignitatum* show small vignettes of forts with towers which have crenellated tops (fig. 16, C–E). These are also found in representations on coins and on pictorial depictions in various manuscript works which bear small pictures as, for example, the corpus of the Agrimensores (fig. 16, B), or in fifth and sixth century *vici* in the Vienna Genesis (fig. 16, F) or in the Utrecht Psalter. In these cases, as can be seen from the depiction of Avenches, in the *agrimensores*' corpus, the use of a foursquare, or (as used on coins or in the *Notitia*) a hexagonal turreted enclosure seems to bear little actual pictorial relationship to the site portrayed (see fig. 17, p. 43). The use of such an enclosure was therefore probably little more than a pictorial convention, the

15 Le Mans, windows at rampart-walk level and above in the Tour Magdeleine

equivalent of a historians' commonplace.

    There are representations, however, in similar sources which show towers bearing conical roofs. The manuscript of the *Notitia* normally regarded as being the closest copy of the original fourth-century text, that called Munich II, has the normal hexagonal style of fort with their corner towers crowned with a triangular shape suggesting a conical tiled roof. These are also found on coins and,

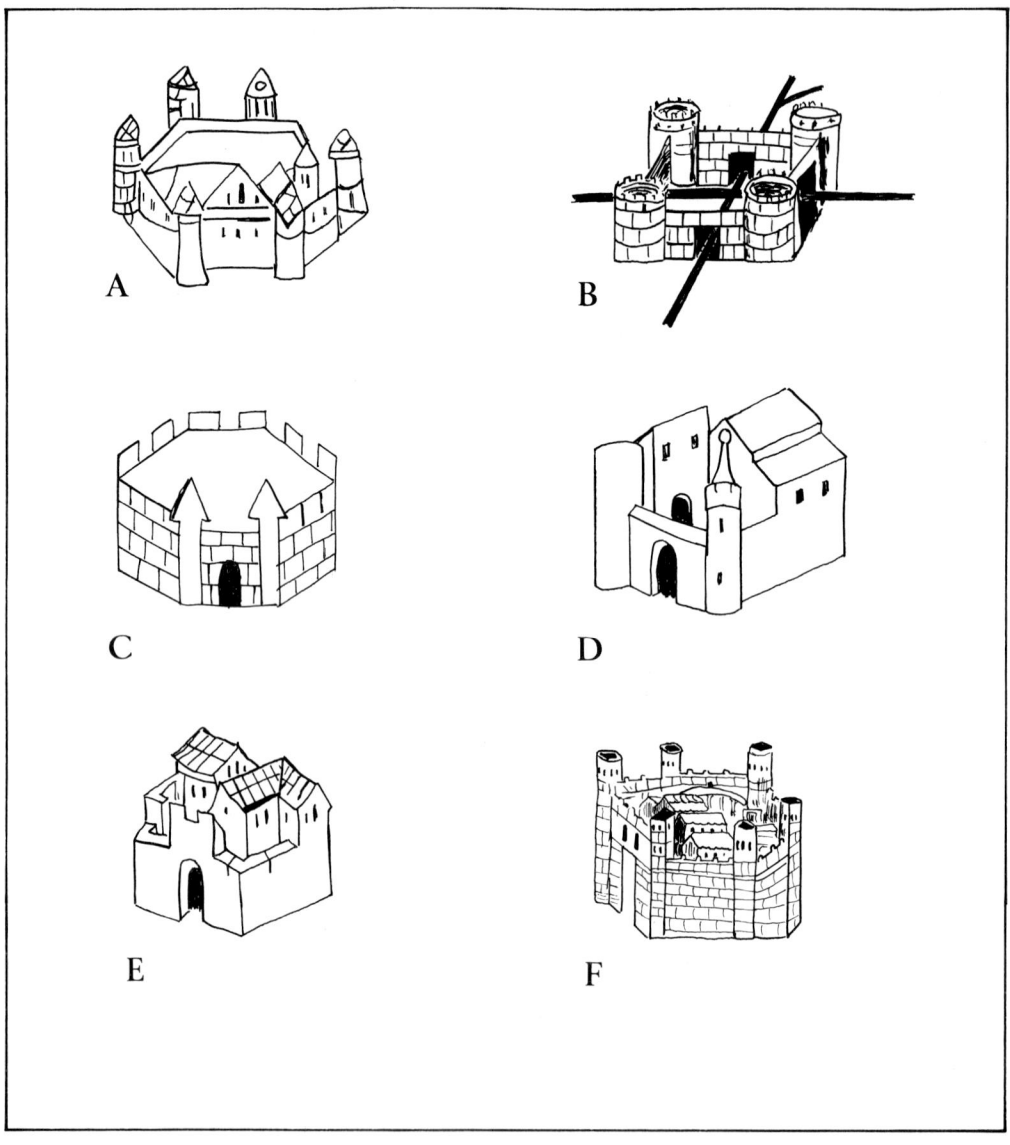

16 Sketches of city illustrations in late Roman documents:
A: Aquileia, from the *Peutinger Table*, segment IV 4
B: Avenches, from the Agrimensores' *Corpus*, MS P
C: Maxima Caesariensis from *ND* (Codex M2) Occ XXIII, 3
D: Anderidos (Pevensey) from *ND* (Codex P) Occ XXVII,
E: Anderidos (Pevensey) from *ND* (Codex C) Occ XXVII,
F: City of Naher, from the *Vienna Genesis*, picture 13

in particular, on the views of Mainz and London on the Lyons and Arras Medallions. There is also a series of coins which shows fort gateways flanked by a pair of round towers crowned by peculiar tripod arrangements, which may be the representations of conical roofs or some form of brazier.[18] An archaeological find—a clay moulding—from Strasbourg has been suggested to be the topmost

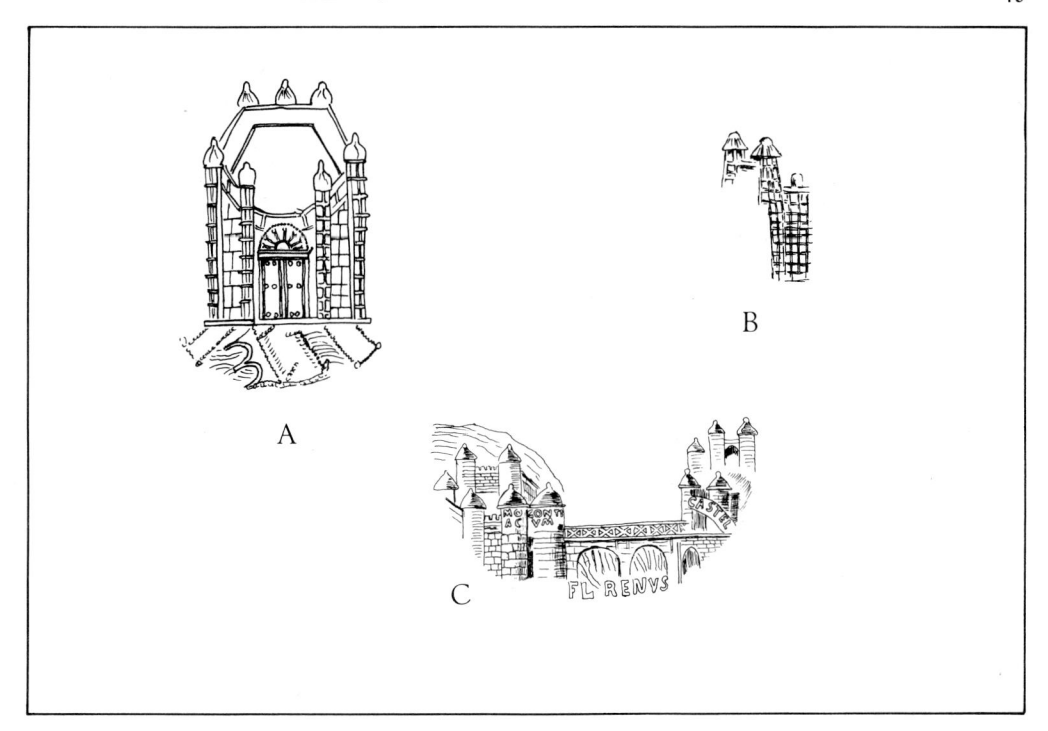

17 Sketches of city illustrations on late Roman coins
A: Unnamed city from the reverse of a coin of Constantine
B: London (?) from the Arras medallion
C: Mainz (l) and Kastel (rt) linked by the Rhine bridge, from a lead medallion, found at Lyons

pinnacle of one of these towers.[19] In those examples which still survive to their fullest height, most of the semicircular towers have conical roofs: Le Mans and Carcassonne are perhaps the best Gallic examples (see fig. 13, p. 34; pl. 1). At Carcassonne, the so-called Visigothic towers are no doubt largely rebuilt, especially in their upper regions, but there is no reason to suppose that their tops have been drastically altered. Rectangular towers, on the other hand, probably did have flat roofs: those on Aurelian's wall in Rome seem to have been of this type, and the large gate towers which still survive at Porta Appia, though U-shaped, have a crenellated top. Clearly, there is no hard and fast rule, and both types may have been current at various periods.

The use of timber for wall-walks and for breastworks on the tops of towers is another feature which merits consideration. Timber framing was used not only for floors in towers, but also for giving extra strength to the upper portions of the walls, as can be seen from Richborough, where there were beams in the walls at wall-walk level. Nowhere can one add further definite evidence to that from Boppard and Andernach for beam holes which supported timber fighting platforms.[20] Very few other forts and town walls stand to an adequate height for comparison: in Britain, on the Saxon Shore, there are fortifications in a comparable state of preservation, but their parapets and wall-walks seem to have been of stone. Only at Burgh Castle, where there are holes in the tops of the towers at their surviving height, is there any trace of timber beams or posts at a comparable level. These holes are usually interpreted as being the mounting poles for ballistae or swivelling onagri. Such machines would be very cumbersome on top of such a small tower. The strain on the posts would be immense, and the machine would lose much of its effectiveness by being fixed in a single arc. These central posts possibly supported roofs, and there may even have been some timber breastwork.

The development of late Roman gates and their typology is a complicated subject. Gates definitely assignable to the later Roman period are astonishingly few in number: in many Gallic towns the gateways have remained main thoroughfares since late Roman times, and Roman gateways and arches have been often rebuilt, first in the medieval period, and later swept away completely to cater for the greater volume of modern traffic. In Spain, however, several late Roman gates survive and the normal type is well illustrated: it is a narrow single passage, defended by a pair of large, usually semicircular, towers. This pattern was probably relatively standard throughout the empire. In Gaul, such examples of gates as exist are of this type: only a very few gates with double carriageways are known in later Gallic town walls, and some of these may have been earlier gates or triumphal arches incorporated into the wall circuits.

The prototype of late Roman defences in the west was provided by the walls constructed by Aurelian at Rome, and the style of Rome's gates is thus of great importance in viewing the origins and development of late Roman gateways as a whole (fig. 18). A substantial number of gates belonging to the Aurelianic Wall still survive. The main gates, such as the Porta Appia, the Porta Latina and the Porta Ostiensis East, were double portalled arches, flanked by projecting semicircular towers. At first-floor height in the towers there were large U-shaped windows; similar windows were also ranged along the upper storey of the wall above the gate passageways, after the same style as earlier Augustan examples. Less important access routes were single carriageways between a pair of projecting semicircular towers. Gates of this type are the Porta Nomentana, The Porta Pinciana and the Porta Tiburtina. The last two are additionally set at a slight angle through the walls with the result that any attacker would have had to expose his flanks in making an assault on the gates. Other gateways are simple passageways through the walls, protected only by the normally spaced rectangular towers and by an arcaded gallery in the wall provided with slit windows to watch the entrance. The Porta Metrobia and the Porta Asinaria were of this type.[21]

Most of the gateways of Rome were much modified and improved by later additions to the city defences and in no case can the original plan or aspect of Aurelian's wall and its original gates be clearly seen and appreciated. Additions and improvements by Maxentius, Honorius and Arcadius resulted in refacing in stone (instead of brick), the substitution of larger semicircular towers, the heightening of the walls and consequent raising of the towers by at least one storey. In some cases, as at the Porta Tiburtina and the Porta Appia, later builders

Porta Appia   Porta Asinaria

ROME

0   30m

Porta Ostiensis

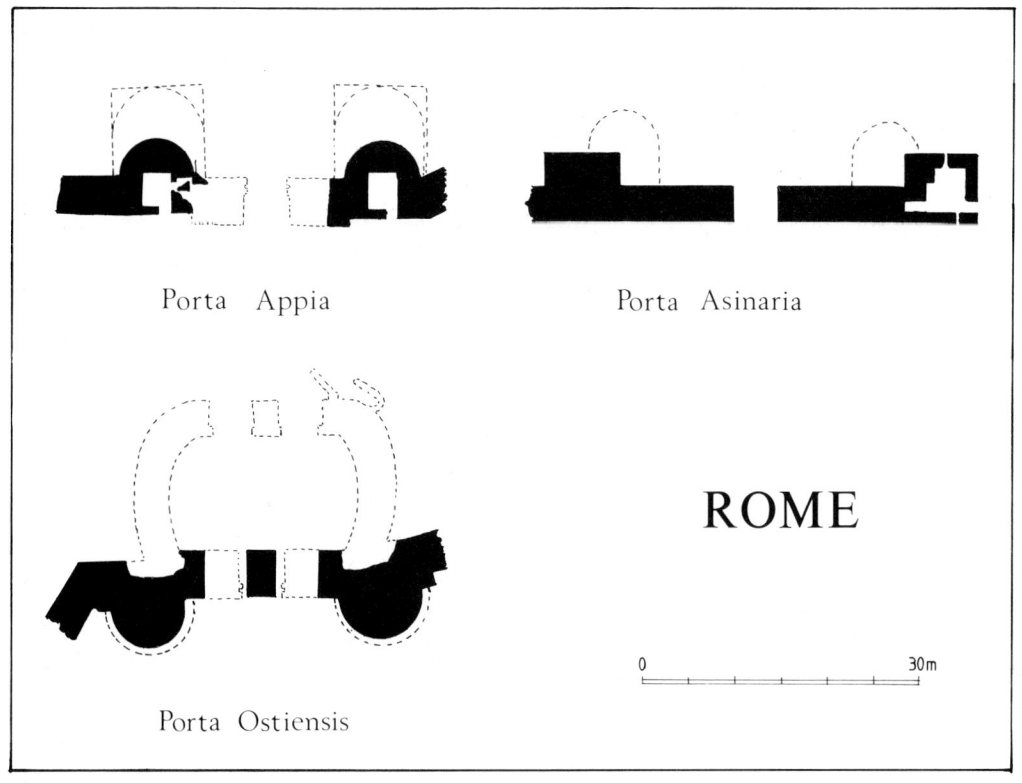

18 Plans of gates in the Aurelianic Wall at Rome (*scale 1:800*)

added a large courtyard behind the gate passage, forming a feature similar to the large courtyards of Augustan gates at Aosta and Nîmes.

The only Gallic gates which are of double portals are earlier triumphal arches or gateways. The poems of Ausonius describe the gate of Bordeaux as 'famous entrances'. Although it was common poetic licence to use plural or singular indiscriminately this may mean that one at least of the main gates into the late Roman city was double-portalled. At any event, there was a water gate to the city, for the poems of Paulinus of Pella record that the harbour lay actually inside the city.[22] At Le Mans, too, there is slight evidence that the main city gate was double-portalled.

The commonest form of late Roman gateway was a narrow single portal between U-shaped flanking towers (fig. 19). Few even of this sort of gate have survived in Gaul: only at Die (pl. 8), where the Porte St Marcel incorporates a monumental arch and may have been rebuilt anyway, Nantes, Périgueux (fig. 39, p. 109) and Bourges are there visible remains or records from excavation of this type of gateway. This single-portalled type of gate is commonest in Spain, where the late Roman walls of Lugo preserve two fine examples of very narrow gates protected by immense projecting towers. Similar gates also survive at Barcelona (Puerta de San Angel, complicated by the fact that an aqueduct also entered the Roman city at this point), and at Iruña. Though most of its facing is medieval, it is

CONDEIXHA-A-VELHA   INESTRILLAS   GERONA

BARCELONA   CORIA   GERONA

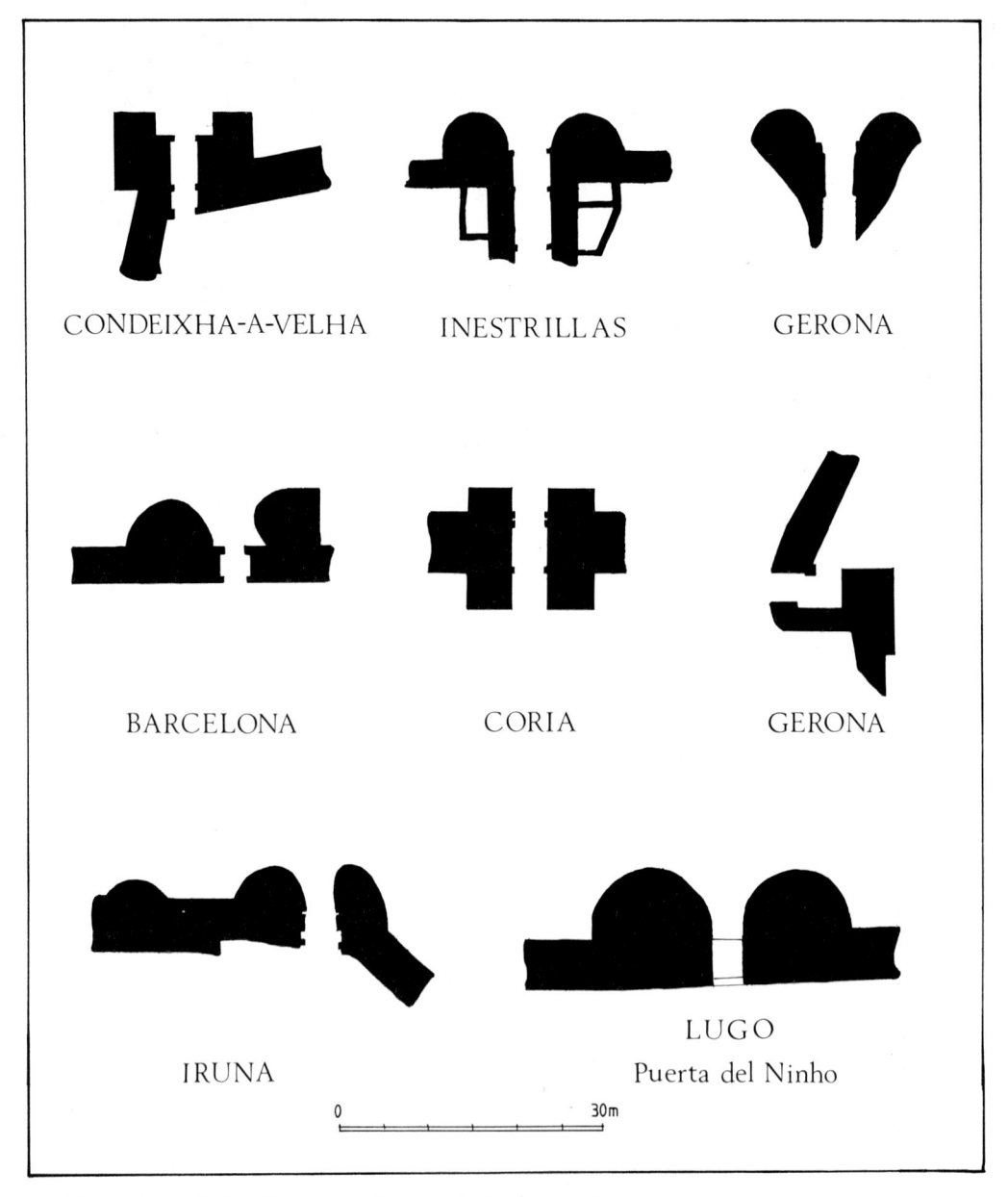

IRUNA

LUGO
Puerta del Ninho

0   30m

19 Plans of gates in late Roman walls in Spain (*scale 1 : 800*)

probable that the walls of Gerona also had at least one gate of similar type.

Gates of similar design (fig. 20) are also well attested from some of the Rhineland and Danubian fort sites—at Deutz, Yverdon and Isny, on the Rhine, and at Pilismarot, Tokod (fig. 71, p. 183), Mautern and Traismauer (fig. 69, p. 177) on the Danube. This type of gate was also normally used in hill-top fortifications, for example at the Moosberg near Murnau and at Cora, near Dijon. It is also found at the hill-top site of Inestrillas in Spain. The dating of the

PEVENSEY   PORTCHESTER   NANTES
            Landgate

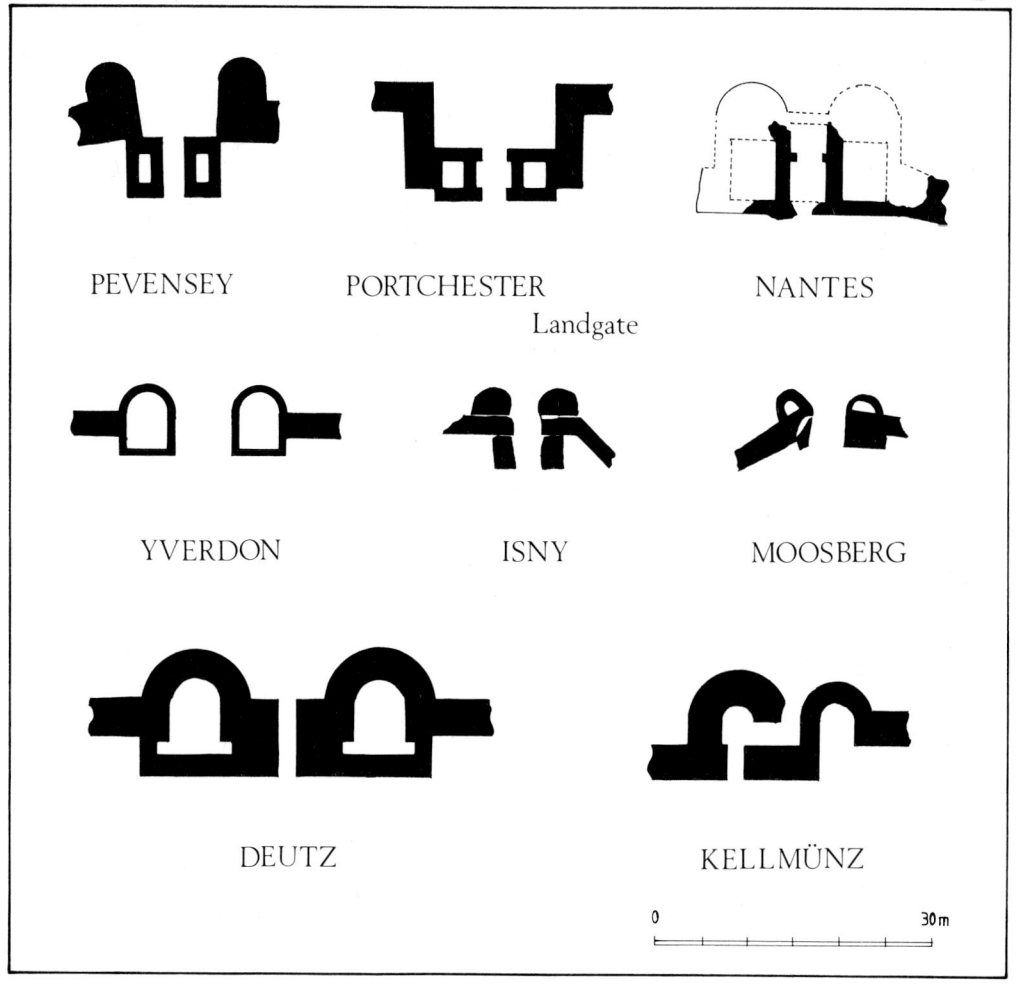

YVERDON   ISNY   MOOSBERG

DEUTZ   KELLMÜNZ

0   30m

20 Plans of late Roman gates protected by U-shaped projecting towers (*scale 1 : 800*)

majority of these examples is probably Diocletianic, but this is by no means standard, for Deutz is known to have been built under Constantine, and Tokod was a Valentinianic foundation.

A variant on this type of gate has a single entrance-passageway with round flanking towers: this is usually a local form based on the type of flanking tower in use for the projecting towers along the rest of the walls: there is a certain example at Windisch and one may expect it to have been used at Toul and in the group of cities in the area of Lugdunensis Prima (fig. 26, p. 85)—at Anse, Nevers, and at Autun, where the plans of the late Roman walls show a gate of this type connecting the large Augustan circuit of walls to the smaller late Roman acropolis. There are also reports of a gate of this kind at Orléans.

By no means an insignificant number of late Roman defended sites had gates flanked by polygonal towers (fig. 22). Two of these are fortunately closely datable: at Burg bei Stein-am-Rhein (fig. 64, p. 162) there is an inscription which

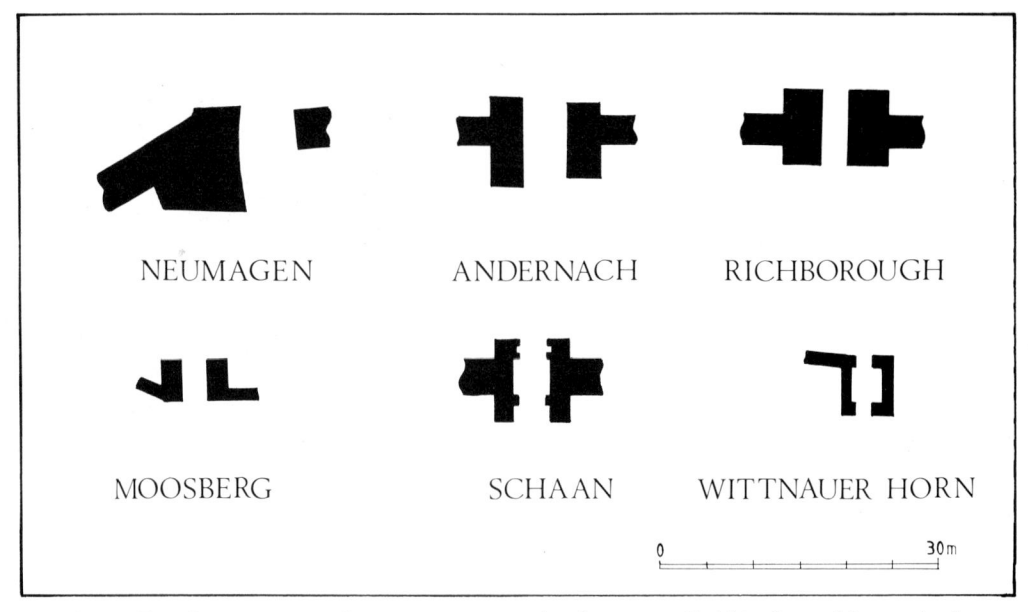

21 Plans of late Roman gates of square or rectangular shape so called 'Andernach' gates (*scale 1 : 800*)

shows that the fort was built in 294.[23] At Split, the palace was built for Diocletian's retirement, and thus must have been completed by 305. The fact that the only two datable examples of polygonal towers at the gateways of forts or 'fort-type' buildings can be closely tied to the Tetrarchic period should not automatically lead to the conclusion that all such gates and towers can be similarly dated. Polygonal towers flanking gateways are also known at Oudenburg (fig. 56, p. 147), Cardiff (fig. 79, p. 203) and Le Mans, and they may reasonably be assumed to have existed at York, where the entire river-frontage was embellished with polygonal towers early in the fourth century. Cardiff was built in the early years of the fourth century, and it seems that Diocletian's choice of polygonal towers at Split created a fashion which was followed in far-flung parts of the empire. The gateways at Split, as befits a palatial establishment, have much in them which harks back to the Augustan style of gateway, with a courtyard behind the main gate-passage and elaborate staircase arrangements and guard-chambers.

The architectural form—in elevation—of most late Roman gates, apart from those at Rome, is virtually unknown. At Rome, the Aurelianic gates had an arcade of windows above the gate passageways. At only one other example of a late Roman gate, at Susa (pl. 15), does enough stand to show the arrangement of windows not only in the towers flanking the gates, but also in the wall-facing above the gate-passage itself.

Late Roman gates with semicircular or polygonal towers belong to a type of ornamental architecture which one would normally associate with municipal or decorative gateways. It is thus no surprise that the usual form of gateway for forts (in the west at least) was a style based on a large rectangular tower (fig. 21). The most common type is the so-called Andernach gate, which occurs in the stone-

SALONA
Porta Caesarea

SPLIT

WINDISCH
West Gate

CARDIFF

OUDENBURG

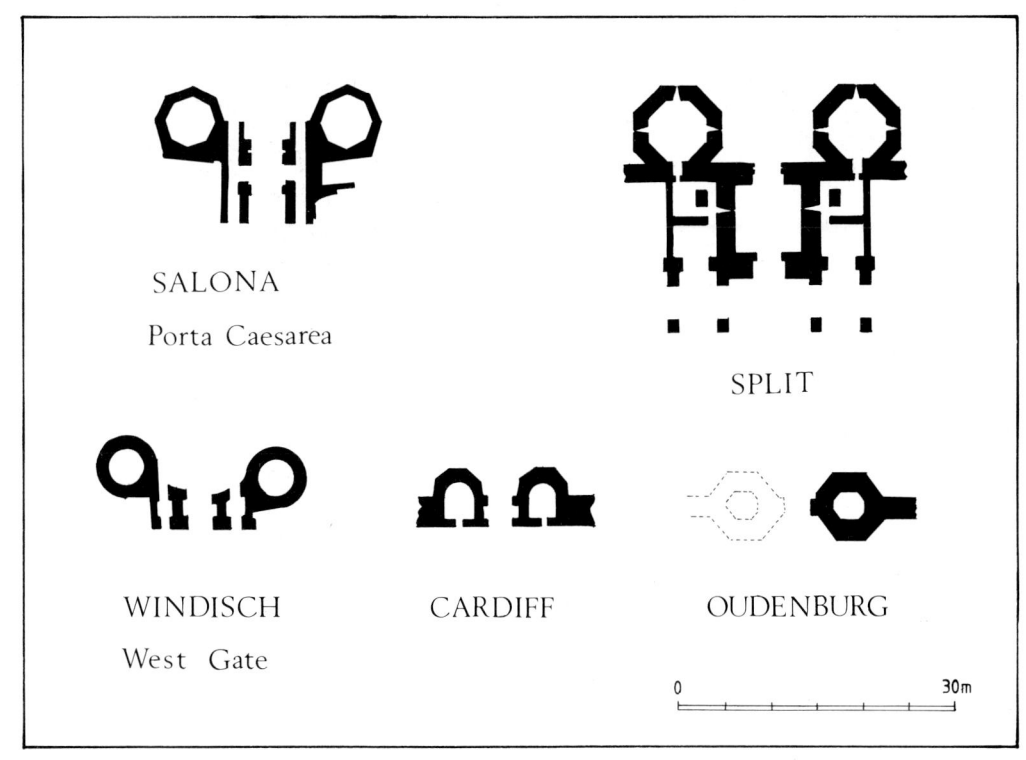

0　　　　　30m

22 Plans of late Roman gates protected by projecting towers of polygonal shape (*scale 1:800*)

built fort of Richborough c.275. This consists of a single entrance set deep between a pair of rectangular projecting towers. Gates of this sort are found at many sites throughout the European frontiers—at Andernach, Horbourg, Bad Kreutznach, Alzey, Breisach, Kaiseraugst and Wilten. A variant on this type of gate occurs at Portchester, where the gate is set at the back of a courtyard in an inturn of the fort wall (fig. 20). At Pevensey there is a developed form of the same sort of gateway, with double guard towers set back between two massive semicircular projecting towers. The Portchester gate, unique in the late Roman world, is almost a throwback to the Augustan idea of a monumental courtyard in front of the gate. It manages to enfilade the gate entrance without breaking the symmetry of the U-shaped flanking towers on the landward and seaward sides of the fort. The fact that these gates are known only on two sides of the fort, whereas on the other two sides there are simple posterns not afforded the defence of any projecting towers at all, prompts the question whether Portchester's gates were in essence defensive or decorative. Given the fort's foursquare layout and the symmetry within that pattern of the tower placings, the recessed gateways, which relieve the heaviness of the fort wall flanked with bastions, seem more of an architectural than a strictly defensive feature (see plan, fig. 79, p. 203).

No example of the Andernach type of gateway stands now above the bottom few courses: one cannot therefore determine whether there were twin rectangular towers flanking a recessed gate-passage. There are examples, at the

Wittnauer Horn, at Schaan, and at Zürich, of gates formed by what appears from the foundations to have been a single gate-tower with a passageway through it. Much discussion has centred on the gate at the Wittnauer Horn, in an attempt to prove that the type is medieval, but the presence of a similar gate at the Lindenhof in Zürich (fig. 66, p. 165) has established that the gate type was certainly known in the late Roman period.[24]

It is perhaps dangerous to label the type of gate with rectangular towers 'military' and its counterpart with semicircular or U-shaped towers 'civilian'. Both types are found on both types of site: many examples of semicircular or U-shaped towers can be found at forts, but there are fewer town gates with rectangular towers. This gate-type was current throughout the Tetrarchy, and the latest datable example is at Alzey (fig. 57, p. 149), either of the late Constantinian or of the Julianic period, and there is a possible later Valentinianic example at Breisach (fig. 63, p. 160). In Pannonia, at Hidegleöskerestz (fig. 71, p. 183), a gate with square towers is probably also datable (by a building inscription which may refer to the fort) to the Valentinianic period.

A simpler form of gate was the postern: several are of the type known as the *clavicula*—a passage which turns through two angles in the thickness of the wall. Posterns in general, however, were simple straight passages through the walls, usually only about 1m or 1.5m wide, and often flanked by a tower to left or right (pl. 9). Theoretically the best situation for such a tower was on the right of a potential attacker, since thus his exposed side would be open to attack from the walls.

It is not possible in general to ascertain the date of construction of a gate by the consideration of its ground plan. All forms of gate types were current in the late Roman period, and gates could be given rectangular, U-shaped, round or polygonal flanking towers. This choice would normally be dictated by local fashion rather than by the imposition of an overall style of building from some central source. Except in rare cases, no clear link can be established between the construction of towers of the same shape at widely differing parts of the empire: the use of polygonal towers in both the fort at Burg-bei-Stein and in Diocletian's own palace at Split may suggest that the emperor had some hand in making the choice for the towers on the Rhine fort. Without the evidence of a building inscription, however, no-one would ever suppose that the curiously shaped fort at Winterthur (fig. 66, p. 165) was its exact contemporary, and probably built as part of the same programme.

Important though late Roman fortifications are in displaying the defensive ingenuity and the changes in technique which had been forced on Roman engineers and military tacticians in the latter part of the third century, these changes were, in Roman minds, only means whereby towns and forts could be protected from surprise or considered attack. What was within the walls, therefore, is arguably of greater importance than the walls themselves. As far as towns and cities are concerned, the excavation and definition of buildings with specifically late Roman use within the defended areas is a task of some magnitude, which in the context of heavily built up medieval and modern urban centres cannot be expected to bear fruit, at least on a comprehensive scale. It is unlikely, however, that urban buildings of the fourth century will have been substantially different in form or function from those of earlier days, except in one or two specific instances. These might include churches, or such specialised buildings as the arms or equipment factories known to have existed in certain centres from the documentary sources of the *Notitia*. It has been suggested that a pair of long narrow double-aisled, and perhaps double-storied, buildings in Trier, known as the *horrea* (granaries) might have been such a factory.[25]

In the study of fortifications built for military purposes, one might expect to see alterations in the format and layout consequent upon changes in the garrisons in late Roman times. Unfortunately, however, traces of interior buildings of late Roman date are known at less than fifty of the forts of this period, and comprehensive excavation has taken place at very few sites. Broadly speaking, forts used in the late Roman period can be divided into two main groups, incorporating those which were in continual use normally from the beginning of the second century through until the fourth, and those built new in the late Roman period to protect specific areas where frontier defence was now necessary to supplement the existing cover.

Existing frontier forts where adaptations and alterations took place in the late Roman period are found on two main frontiers—that of Hadrian's wall in northern Britain, and the Danube frontier in Austria and Hungary. In Britain, campaigns of excavation at Housesteads and Wallsend in particular have elucidated the progression of barrack accommodation from the long strip-buildings of the fort—granaries, commandant's house, hospital and *principia* 'chalets' in the fourth, their plans and layouts markedly different.[26] This development was the most marked, and its significance in terms of the use being made of the fort interiors at the time is not understood. Apparently the main buildings of the fort—granaries, commandants' house, hospital and *principia* (with alterations)—were still used, and at Housesteads at least, there was no noticeable increase in the amount of space immediately within the walls taken up with buildings. Indeed, at Housesteads, there is evidence that the earth ramparts behind the defences were removed towards the end of the second century or in the third, only to be reinstated, at least in some areas, in the fourth. Alterations in the *principia* of Housesteads, Chesterholm and South Shields[27] seem to have been designed for the creation of more office space and a limited amount of living accommodation, but the administrative need for this alteration is not understood, and may depend as much upon local exigency as central government directive.

Elsewhere in the Empire where existing forts were used in the late Roman period, parallel developments might be expected, but only at Zwentendorf, Ulcisia Castra (Castra Constantia) and Intercisa have late Roman buildings been encountered over any appreciable area, and at none of them have 'chalet'-style barracks been recognised. At Zwentendorf, there were two buildings of this period encountered in the part of the fort which has been excavated—one was a courtyard building, and the other was a barrack of strip-type. At Castra Constantia (fig. 73, p. 186), the recorded buildings are of irregular shape and dimensions: the *principia* can be recognised, and so can a courtyard building, but neither of these is necessarily of late Roman date. Excavation at Intercisa (fig. 73) has been rather more thorough, and apart from the principia itself, a substantial late Roman basilical building, orientated east-west with an apse to the west, lay in the middle range of the fort, south of the principia. A number of halls and rooms lined the western end of this structure, which is of uncertain purpose. Against the fort wall, a series of small rectangular rooms was found to date from the Valentinianic period. Smaller-scale excavations at Carnuntum confirm the picture shown in fragments, that the barrack-blocks in the fortress retained the same or similar plans at least until the Valentinianic period. Thereafter, there was a complete change in the arrangement of buildings, some of which were now built over roads, and some failed to follow the regular fortress layout. These buildings, which use portions of earlier structures in their foundations, have been suggested to be the living quarters of civilian settlers filtering into the fortress. Despite the 'irregular' nature of such buildings, some of the houses were capable of some sophistication, with hypocaust heating systems, and bread ovens which would have done credit to a legionary fortress. It is possible therefore that these buildings at Carnuntum are the legionary equivalent of the Hadrians Wall 'chalets'.

By and large, elsewhere there was a considerable reduction in the area covered by newly built forts of the late Roman period as compared with earlier cohort and *ala*-forts of the frontiers. More evidence from these purpose-built forts gives some indication of the form and function of the interior buildings: for the most part, except where, as at Alzey and Lympne, the sites are re-used earlier *vici* or forts, any buildings found in the interior of the walled enclosures can reasonably be assumed to belong to the fort in question. There is, however, a remarkable lack of standardisation, and many of the interior buildings of forts of this date may well have been of timber. At Richborough and Oudenburg the excavators suggested that the main buildings were a series of wooden sheds, but no clear definition of the detail was possible at the former, while at the latter, too small an area was examined to enable firm conclusions to be drawn.

It is difficult to define any building within a late Roman fort which clearly occupies the position and performs the function of the *principia* of earlier days. One such building, on the site of the triumphal monument at Richborough (fig. 79, p. 203), has been identified, but nothing is known of the levels from which it was built, nor of the state of the monument at the time. The surviving remains of the walls of this building lie a good deal lower than the remains of the road crossing under the former quadrifrons, and the consequent problems of interpretation are immense. At Oudenburg also (fig. 56, p. 147), there is a stone building on the central axis of the fort which may have served the same function, but it does not lie centrally within the fort. At Lympne, too, (fig. 80, p. 205) a building very similar in plan to the *aedes* and administrative rooms of late Roman *principia* has been found in a central location within the fort. Because of the severe disruption to ground levels in the area of the fort at Lympne, and because it seems clear that there must have been an earlier fort under the late Roman remains, it cannot be certain that this building belongs to the late Roman fort.

A building type which is found in several late Roman forts is the aisled hall, variously interpreted as granaries, but more likely to be a store building for equipment of all types. These are normally rectangular, ranging from the dimensions of 62.5 by 17.5m externally for the three buildings at Wilten (fig. 67, p. 170), to the 26 by 14.5m of the building at Cuijk (fig. 56, p. 147). Buildings of this type are recorded also at Kaiseraugst, Tokod, Pilismarot (Ad Herculem) (fig. 71, p. 183) and at the Lorenzberg (fig. 89, p. 238). At none of these examples are there external buttresses such as would normally be expected in a granary, and at Ad Herculem and Kaiseraugst, the store-buildings were attached to subsidiary rooms, at the latter at least provided with heating systems and suggesting living accommodation. In fact at Kaiseraugst, the aisled building in the south-west portion of the fort is only one of a series of large buildings in that area of the fort.

Traces of buildings at other sites include courtyard-type arrangements at Schaan, Irgenhausen (fig. 66, p. 165), Bad Kreutznach and possibly at Alzey (fig. 57, p. 149), small rectangular buildings with porticoes (the so-called guild rooms) at Richborough, and more fragmentary buildings whose original shape and purpose cannot yet be determined at Alzey, Breisach (fig. 63, p. 160), and in parts of Kaiseraugst. Baths are known within the forts at Jublains (fig. 30, p. 92), Lympne, Richborough and Kaiseraugst, and churches lying next to the fort walls, at Boppard (fig. 57, p. 149), Kaiseraugst, Zurzach (fig. 64, p. 162) and Richborough, and freestanding at the Lorenzberg. As far as barracks are concerned, observations at the fort of Deutz, (fig. 57, p. 149), although heavily built over, have made it possible to suggest a remarkably regular layout of 12 internal strip buildings, most or all of which may have been barracks. At Richborough traces of mortar floors suggested the presence of timber buildings, probably barracks, and at Portchester (fig. 79, p. 203), mortar floors, eavesdrip gullies and the overall layout suggested the presence of at least four small buildings of timber, dating from two separate periods in the life of the fort. Timber buildings also, of fairly extended plan, and also in two phases according to reassessment of the original excavation plans, existed at the Moosberg (fig. 89, p. 238), but here there is no evidence for Roman military involvement or occupation.

Rather more comprehensive plans of the total interior of late Roman forts come from two other sites in Raetia (Bavaria), at the Bürgle and at Isny. The Bürgle (fig. 67, p. 170)[28], excavated in 1925, was a diminutive fort of curiously irregular outline, a blockhouse containing space, even with two-storeyed portions as reconstruction drawings seem to suggest, only for a century of men or so, by no means a full auxilary unit of earlier days. The fort was long and narrow, with rooms arranged on either side of a central path or corridor. The stonework of the outer walls was heavily robbed, and the main gate was heavily recessed inside a flanking courtyard. The interior accommodation consisted of a series of timber barrack rooms facing each other across the central passage. At the south-west angle, there was a suite of five rooms, one containing a hypocaust which may have belonged to the officer in charge.

At Isny (fig. 67), on the other hand, a small trapezoidal bluff of land was wholly taken up with the fort, established in the third quarter of the third century. Its complete excavation has enabled its internal arrangements over the period from then until the late fourth century to be elucidated.[29] The area within the defences did not alter over the century or so of use of the fort. At all periods there was a stone building opposite the main entrance which may have been the commandant's quarters, incorporating large rooms, small offices, a courtyard and bath building, and there was a lean-to building with foundations at least of stone the length of the south wall. This may have been a store or a stable, since it seems to have been divided into compartments only 1.4m wide. Apart from these, the space within the walls was taken up by timber buildings of four separate periods, in which barrack blocks in two or three main locations were continually renewed and rebuilt. In the later periods, Constantinian and Valentinianic, they were provided with heating systems over all or part of their length. As far as space within the fort is concerned, provision seems to have been made at no time for more than about 15 rooms, of maximum dimensions of 7.5 by about 4m. If the garrison was of cavalry, this is enough to accommodate only two *turmae* of earlier days, and bespeaks a total garrison strength of only about 150 men. Garbsch suggests that the remainder of a garrison of more regular size was spaced out in the watchtowers known to east and west of the fort along the road from Bregenz to Kempten. This is possible, but no accommodation for troops has yet been found attached to the smaller watchtowers to compare, for example, with the living quarters found at Pilismarot-Malompatak (fig. 74, p. 187) or the slight indications of similar provisions at Mumpf or Sisseln on the Swiss frontier.[30]

Taken as a whole, the late Roman forts of the western empire have little to add at present to the largely theoretical debate over late Roman design and layout of interior buildings. The suggestion that Diocletian's palace at Split (fig. 91, p. 243) encapsulates the contemporary fort or fortress design[31] is not capable of support from any other western site. The argument must be carried forward by analysis and proper dating of the better preserved layouts of forts on the eastern frontiers such as Dionysias, El-Leggun (fig. 12, p. 28), or Da-ganiya.

3

Contemporary Records

It is crucial for the proper understanding of Roman defensive strategy to assign an accurate date to the construction of fort and town defences. The methods by which such structures are dated involve archaeological or historical evidence. Matching these is never easy, and often impossible. Archaeological evidence comes in the form of dated material found in association with the structures on the ground. The written historical record consists of occasional inscriptions, pictorial representations on dated coins, or chance descriptions of cities, towns or forts in literary works of many genres. This evidence will be discussed in some detail later.

The easiest datable archaeological material is the coin, which usually bears an inscription enabling at least the approximate date of minting to be determined. Other artifacts—pottery, metalwork or other manufactured articles—are dated more or less exactly either by a complex series of established links with similar material found at other sites, or on art-historical grounds from a known manufacturing source or style of decoration. Occasionally the date of manufacture of such material can be almost as exactly determined as that of a coin. Often, however, only a general date within 50 years can be given, and commoner, undecorated material can be even less precisely dated.

A further element of difficulty is added to any material from the position in which it is found. The construction-date of a wall can be determined archaeologically either by material sealed under the wall at the time of its construction, by material found in the actual levels of construction (in the wall itself or its foundation trenches), or by material associated with the primary occupation of buildings associated and contemporary with the walls or found in the earliest silting of the ditches.

Material actually within the wall or underneath it and covered by its construction provides the most obvious information. If it can be dated exactly, it furnishes what is known as a *'terminus post quem'*, a date before which the wall cannot have been built. There is no guarantee, however, that any material found in this position was not already of some age by the time the wall covered or contained it. Thus a *terminus post quem* is only a starting point for an overall view of the date of construction: it is not immediately a close and secure date itself.

The record kept of vital evidence for dating is sometimes infuriatingly imprecise. In the nineteenth and early twentieth century, the exact provenance of datable material is often glossed over. Coins found 'in the wall', for example, may come from the wall-core, and provide a *terminus post quem*. If, however, they came merely from disturbed earth and debris dug from round the wall during excavation, they are of little relevance at all, and, if anything, help date only the period at which the wall was quarried for stone by later builders.

Surer ground is provided by reused masonry and occasional inscriptions found in considerable numbers in the lower portions of many late Roman walls. These originally came from buildings and monuments—temples, public buildings, tombs—dismantled to provide stone for the city defences. The latest datable material of this type found incorporated within a city wall will of course provide a *terminus post quem*. It is to be expected, however, that some of this material will have belonged to buildings already of some respectable antiquity at the time of their demolition, and that any date they suggest will be substantially earlier than the date of their incorporation within the defences.

Material other than coins and inscriptions can of course be found in similar critical positions, but this can rarely provide a close date unless it is found in some quantity, rather than as isolated fragments. If sufficient excavation has been done to examine pre-existing layers, the structure of the wall itself, and the earliest occupation levels within the walled area, the closest date will be obtained. This is a counsel of excellence, however, which has been achieved at very few sites.

In the absence of this kind of information, it is clear that in most cases archaeological dating is bound to be approximate. In an effort to counterbalance this problem, it might be thought that such features as the constructional style, ground plan, or, even more important, tower and gate-types provide a ready typology which can be used to date similar sites as contemporary. It is becoming clear, however, from detailed archaeological evidence, that this is a dangerous assumption. It appears that at no particular time in the late Roman period was any particular shape of tower or gate in use. In general therefore typological dating of buildings is only meaningful where there is clear evidence that the use of a similar style was imposed as a result of an overall decision by, for example, a military authority.

By contrast with this archaeological evidence, contemporary or near-contemporary historical records are often far more precise. Roman practice was to commemorate the construction of a new building, including city walls and gates, with a dedicatory plaque usually giving the full names and titles of the reigning emperor. Where these survive in full, they can normally be dated to the exact year. Inscriptions of this nature were normally long, and the emperor's titles would take up the major part of the text. Thus, if only a portion of the inscription survives, it will often not refer to the building to which it once belonged: alternatively, if it had originally been placed at some ostentatious position, it may never have mentioned the building it once commemorated at all. It is rare for an inscription to survive actually in position on the building to which it refers:

perhaps the best known examples are those at Nicaea (Iznik, in modern Turkey), where several Greek inscriptions record the construction of city walls and gates under the emperor Claudius II.[1] If the findspot of the inscription is not certain, it is often difficult to determine to which Roman building, or to which phase of construction or reconstruction of the building, it should belong.

Equally rare, and mainly a phenomenon of the eastern part of the empire, is the commemoration of the construction of city walls on the reverse type of a coin issue. A series of Thracian towns walled under the Antonine emperors was marked in this way.[2] The issue of coins or medallions can, however, be as equivocal as any other piece of evidence. Coins and inscriptions had great propaganda value in antiquity, and for the archaeologist seeking absolute dating, the claim that a certain circuit of walls was built under a certain emperor may have glossed over the fact that the walls were rebuilt or only partially repaired at the time.

Inscriptional or coin evidence of this type is but one part of the whole gamut of written historical evidence from antiquity which may have a bearing on the subject of defensive construction. Contemporary writers sometimes mention defences in the course of their description of events. This can happen directly, where an author describes the construction of a military camp or the walls of a town, or by implication, where a writer mentions a siege or a city gate.

It is probably a fair assumption that when a historical writer records a specific fort or town as under construction or occupied he speaks with a certain amount of first-hand knowledge, whether gained from personal sources or from official reports. Even when they are conspicuously partisan, like the Panegyricists,[3] or the severest critics of the emperor like Lactantius,[4] there is little reason to suppose that in mentioning specific facts and mentioning places by name the surviving sources were deliberately perpetrating falsehoods. Even their most ardent assertions which might be suspected of distorting the facts are probably not without a grain of truth.

The main sources for the period 260–400 are a mixed assortment. Prime among them is Ammianus Marcellinus, who accompanied the emperor Julian in his campaigns in Gaul and therefore his detailed knowledge is probably at first hand.[5] His history is a chronicle of events on several fronts presented in chronological order, with each year's events as a rule kept separate from the next. Thus it is episodic, with summaries of the strengths and weaknesses of the reigns and personalities of the various emperors at the ends of their reigns. Although his history originally continued Tacitus' *Annales*, only later, more detailed, books remain, covering the period from 353 to 378.

Unfortunately, there is no mainstream source similar to Ammianus for the period from 260 to the reign of Constantius II. Several works more or less loosely based on a postulated lost 'Imperial history' survive. Of these the most important are the Scriptores Historiae Augustae and the two lesser works, Sextus Aurelius Victor[6] and Eutropius[7], on whom later writing is based. Two Byzantine Greek writers who occasionally contribute new facts, probably gleaned from the same source, are Zosimus[8] and Zonaras.[9]

The Scriptores Historiae Augustae, a collection of imperial biographies ostensibly by six different authors, has now been established as the work of one single man who wrote, probably in the years around AD 390, about events from the mid second to the mid-third century. It has been shown to be thoroughly unreliable in some of its statements. Although some are genuine, some of the authorities and documents it purports to quote are complete fabrication.[10] This has led to the view that the whole work is suspect, even though it must be admitted that even the most obvious historical spoof must retain sufficient proximity to the facts to prevent it from being instantly dismissed. At a certain low level, the facts which the *Historia Augusta* uses are probably reliable: it is the political interpretation of them in the main which is suspect. There is therefore little reason why the author of these biographies should have falsified details of frontier campaigns, when the broadest outlines of these would be known to contemporary readers of Eutropius and Sextus Aurelius Victor. In the broadest measure, all three authors' accounts betray a parallel or common source for such information.

Further source material comes from contemporary writings of genres other than the historical, whether the self-consciously literary letters and verses of Ausonius,[11] the letters of Sidonius Apollinaris,[12] or the poems of Avienus.[13] Best of all for the military historian are the records of emperors' campaigns as related by the emperors themselves (like Julian)[14] or by eye-witnesses of their achievements.[15] Some sources have to be treated with caution: official biographers and eye-witnesses of imperial campaigns are prone to over embellishment, as are the panegyrists. Even Ausonius, acting as a sort of court poet, cannot always be relied upon to present the climate of the times as it really was. Yet for all that, these writers are truly contemporary, and may therefore be treated as basically good sources particularly when writing of the construction of a military fort or the provision of new defences, neither of which is likely to be fabricated by the author to gain for his imperial subject some political capital. A panegyrist may give a biased interpretation of past events, but it is far more difficult for him to claim that a building was the work of an emperor who did not build it.

Other writers are unashamedly specialist authors. Potentially the most useful for the study of the late Roman army and of Roman tactical thinking in the fourth century is Vegetius, who wrote a manual on military affairs in four books.[16] Vegetius, however, appears to be far from consistent about the period of military history to which he has confined himself, and despite the date of authorship, it contains fragments and snippets of information from all the imperial centuries.

Itineraries, road books and a late Roman list of garrisoned posts, the *Notitia Dignitatum*, provide other useful sources. Itineraries, whether actually in map form, like the Peutinger Table,[17] or in list form, like the Antonine Itinerary,[18] give names of towns, road posts and other establishments of note, with some indication of the distance between them. In both documents, routes travelling east-west across the empire made use of the frontier roads along the Rhine and Danube, and thus a list of named stations along that road can be reconstructed. Since it is a drawn map, the Peutinger table uses small vignettes to show major towns, with a number of variations of form. Only a very few have any indication of surrounding defences. Both these Itineraries are thought to date from the early years of the fourth century AD. They may have had their origins in Roman military travellers' handbooks, but both seem to incorporate some early imperial material.

This evidence for frontier dispositions is given greater weight by the *Notitia Dignitatum*.[19] In the *Notitia*, a document whose exact date is difficult to determine, all late Roman civil and military commands are listed. Both halves of the empire are represented, and therefore the document provides a comprehensive picture of the frontier dispositions at about the turn of the fourth to fifth centuries. Each frontier commander has a chapter, which contains a list of the garrisons under his disposition, each with a named place at which the garrison was stationed at the time of the *Notitia* record. Many problems remain, not least of which is the exact nature of the *Notitia*: were the troop commanders' lists kept up-to-date by recording the changes in troop locations which may have occurred, or were the lists compiled once for all at one particular date? Could the information stored in some of the *Notitia* chapters, therefore, have been out of date when others were still current? Such problems have a considerable bearing on arguments which seek to establish the length of occupation by the Roman military forces at any particular site.

Further evidence relative to the date at which forts or towns were occupied, if not built, comes from the two great bodies of Roman law-codes. These, the Theodosian[20] and the Justinianic,[21] are a compilation of rulings on many subjects under main headings, the results of imperial decisions on subjects which were referred to the emperor for a policy statement. As well as giving dated legislation which sometimes refers to construction work and frontier problems, they also give the places at which the emperor signed the laws. It is thus sometimes possible to work out where the emperor was on campaign at various times of the year, providing an accurate and useful cross-reference with other historical sources. Though the laws within the codes are dated at the time of issue, they, too, must be treated with a certain amount of caution, for the offences against which they legislate were obviously not always curtailed by their enactment and in some instances the laws will have only given official recognition to a state of affairs which had already been in existence for several years. Other sources, basically fragmentary, like the chronographer of the year 354[22] or the *Consularia Constantinopolitana*,[23] contain occasional information which refers, not to events in Rome, but to frontier affairs not included in any other source.

When one takes a close look at the accumulated source material of this type, a general picture of the sorts of administrative and political constraints imposed upon military and civilian builders can be built up. Of paramount importance was the cost of constructing city walls: it is hard to say in real terms how much it cost. Building costs are seldom expressed in terms of monetary expenditure, but it is reasonable to assume that the scale was so great that only those cities which could receive financial help from the emperor could normally build walls. Many of the larger provincial cities kept a notable ambassador at the imperial court to plead their case for special attention, and the prospect of financial help towards a major project like the construction of defences would merit sending a special representative.

To some extent, of course, costs were related to the time taken over construction. Even the amassing of materials sufficient for the job was a time-consuming operation. The re-use of masonry, sculpted blocks and monuments from former buildings on the site, perhaps destroyed or abandoned, would cut down the amount of material thus suddenly made necessary. On the actual speed of construction, there is unfortunately very little evidence. The circuit of walls built at Rome under Aurelian took five years to complete.[24] Where great haste was necessary, construction could be quicker: the walls of Constantinople, running some five miles, were built in 70 days.[25] Those of Verona, a similar distance, were built in eight months.[26] Military building was probably no quicker: a Valentinianic *burgus* was constructed in 48 days.[27]

The construction of fortifications on the frontiers was a matter of direct military concern. Each early imperial legion had its own technical staff of architects, joiners, carpenters and builders who were readily available for construction work where necessary and this was probably also the case with the later, post Diocletianic legions. In an age of increasing specialisation, the design of constructional works of a more permanent kind may also have been the province of one or more separate units of troops. In the early as well as the later periods, most troop units produced their own tiles, which in some cases had a wide distribution over several forts in a specific area of the frontiers, thus suggesting that there were regional depots and stockpiles of building materials on which troops could draw as the need arose. Towards the latter part of the fourth century, some frontiers had tileries controlled by the Dux of the frontier, whose products in many cases bore his name and thus provide valuable evidence for co-ordinated campaigns of fresh building or rebuilding along the frontier. In Germania particularly, there was large scope for private tileries in the later period, providing the materials needed for the construction of large buildings at Trier.

In general, the army provided the labour for its own fort building, though for exceptionally large projects help could be requisitioned from civilian labour. Whereas in the first two centuries fort construction had been largely the science of exact layout, in the later Roman period it grew to be a skill much more closely linked with the natural defensive resources of the location. Once the choice of site had been made, the military architect then had to decide the shape his fort should be to take best advantage from it.

There is little evidence about the choice of site, but on some frontiers, particularly in areas where the frontier had been static throughout Rome's presence, the selection of a new site was a relatively rare occurrence. Even new campaigns beyond the Rhine in German territory tended to make use of previous Roman encampments partly because they were already defended and partly because they already occupied the best sites.[28]

The main responsibility for tactical decisions about the siting of new forts lay with military commanders in the frontier areas. The series of Diocletianic *centenaria* on the African limes, all built to a standard plan, were ordered and positioned by the *vicarius praefectorum praetorio*, and by the *Praeses Numidiae*.[29] A formidable rescript dated 365 from the emperor Valentinian instructed the Duke of Dacia Ripensis to make sure that he built his quota of towers each year without fail, and saw to the upkeep of those already built.[30]

There are also several instances of the emperor himself taking the initiative in building, and it appears that constructional activity was at its keenest, on both civilian and military fronts, at the place where the emperor happened to be.[31] Since the majority of late Roman emperors came to power at the head of support from the army, the welfare of the troops and the strength of frontier arrangements was of great concern. Some sources record emperors who took a hand in the actual siting and construction of forts.

Historians record that Diocletian personally supervised the construction of new forts on the eastern frontier, in Egypt and in Mesopotamia. His activity is also attested by inscriptions at sites such as Kasr Bser in Syria, and in many other places.[32] Julian, besides rebuilding walled towns or forts devastated by barbarian incursion in Gaul, claimed to have rebuilt 45 cities, probably including forts on the German frontier as well as Gallic towns.[33] Valentinian's personal intervention on the Rhine and Danube frontiers included a check on all the fortifications on the two rivers as well as the construction of new forts at weak points.[34] These included a fortification called 'Robur' opposite Basel, on which he was engaged in 373,[35] and also a strongpoint by the Neckar river—presumably in barbarian territory—which encountered grave difficulties because foundations had to be placed actually within the stream bed. Ammianus relates that 'the emperor's more than usually strenuous attention to the work, and the patient toil of the soldiers, who had often to work up to their necks in the river, won through despite everything'.[36] But the most impressive eye-witness account of the construction of a fort by an emperor is Themistius' description of Valens building the fort of Charsovo, in Thrace, in 368–9. Here Valens took an active part in selecting the site, ordering the material and in insisting the work was promptly carried out.

> The emperor was not blind to the topographical advantages of the terrain. He spotted a thin strip of land running out into the shallows, and culminating in a high hill, from where all the surrounding area was visible. Here he established a new fort, following the faint traces of an old fort found to be of use by some previous emperor, but abandoned because of the difficulties. There was no stone nearby, no baked bricks: you could not easily achieve anything: everything had to be brought from miles away by thousands of pack animals. Who then would not excuse those who had formerly withdrawn from the site on the grounds that the design could not be carried out? ... You would have said, though, that the stone seemed to bring itself to the site, that bricks appeared by magic, and that the fort began to rise without the help of architects or masons. Such was the extent of the soldiers' discipline and such the imperial mastery of the site's problems.[37]

Such forts were normally built by the troops, legionary or auxiliary, infantry or cavalry, who were stationed in the neighbourhood.[38] But portions of some frontiers, particularly those in Africa, relied on civilian builders. In Tripolitania, there are *centenaria* and farm-cum-blockhouses which according to their inscriptions were built by the local landowners.[39] In these areas, by the second century, the established policy was to lease empty lands on the frontiers to farmer settlers, who, at any hint of danger, would be ready to become a militia force to protect themselves and the interior parts of the empire against trouble from outside. Such settlers, called *limitanei*, were usually discharged soldiers or members of barbarian tribes, who were given the privilege of land inside the empire in return for this service on its borders. That this practice, begun in the late second century, continued until the later years of the fourth, is shown by rescripts from the Theodosian code which reiterate that the privilege of settling in these frontier areas is to be regarded as belonging to the descendants of former *limitanei*, or, if not available, to men who would originally have qualified for such lands.[40]

The anonymous *De Rebus Bellicis*, with a surprising lack of originality, also suggests just such a system for use on the frontiers to defray the expenditure in keeping troops on permanent guard. By providing a series of towers which would be paid for by the landowners themselves, the army would escape the costs of upkeep and establishment of the *limes*.[41] This system, which clearly existed in fact, is reflected in the *Notitia Dignitatum*, where the *praefecti limitum* had the *limes* area under their control rather than specific forts, and seems to have been unique to Africa.[42] Though other frontiers had towers, and even small blockhouses akin to those found in Tripolitania and other areas of Africa, there is no trace of a similar organisation elsewhere.

In the second and early third centuries, there had been recognition of the value of civilian assistance towards policing and construction in the frontier areas. Civilians in frontier provinces were sometimes required to perform guard duties in part payment for their safety. There was also requisitioning of local people to serve as *burgarii*, and to provide goods and services for officers of the imperial post. The responsibilities of civilians, and in particular of decurions towards the upkeep of public services and buildings became progressively more burdensome.

Public buildings in most cities included temples, the forum and basilica, workshops and stables for the *cursus publicus*, and, above all, the city walls and gates. These in particular occupied a special place in the religious and civic life of a city: the space immediately around the walls, both inside and out, was originally left unoccupied by buildings and hallowed as the exact line originally set by the founder when he marked out the city's limits. Among cities which were building walls for the first time in the late third and fourth centuries, often among buildings already standing, these exact traditions could scarcely be maintained. City walls were *res sanctae*. They could not therefore belong to anyone, and could not really be dedicated to the emperor, since they belonged, in a sense, to the gods themselves.[43] The emperor, however, held overall control; he had to grant permission for walls to be built and often maintained an interest in projects of provincial cities who wished to extend, broaden, heighten, or alter their walls in any way.[44] No inscription has survived from the western Roman empire which specifically records the dedication of completely new city walls in the late Roman period. At Grenoble, the walls were given to the city by Diocletian and

Maximian. This occasioned a dedication of the main gates by naming them after the emperors' tutelary deities, Jupiter and Hercules.[45] Cities in other parts of the empire also had gates which bore these names, and may show that the benefactions of these two emperors were more widespread. The renaming of cities was another method by which an emperor, as patron, would record his patronage.[46] It is likely that emperors became increasingly reluctant to concentrate their gifts on any particular city unless it was the imperial capital. Most provincial cities were left to themselves to find the money to finance building schemes including city walls. From Nicaea, whose walls were built in 259, a pair of inscriptions show more subtly the amount of imperial help.[47] One records the emperor Claudius II in the nominative case, probably as the donor, whereas the other gives the emperor's name in the dative case, and records completion of a portion of the defences which the townspeople themselves had had to build from their own funds.

If direct imperial patronage was not forthcoming, the most important way in which the emperor could help was to put the army at the city's disposal. The army was clearly nearer at hand in the frontier provinces, and it was here that pressure from outside the empire was most concentrated. Inscriptional evidence for construction work carried out by the army shows that their tasks were weighted towards buildings in these provinces.[48] Some emperors, too, had more of a reputation for building than others: Probus was notorious for using his troops as a labour pool.[49] Egypt was the area in particular which benefitted from his activity, but he was also clearly busy in rebuilding parts of Gaul, for Julian, himself a great builder, admired Probus' achievement in putting the Gallic cities back on their feet after their catastrophe of 276.[50] Other emperors were praised for their keenness at building, while also being criticized for the expense to which the empire was put.[51] In rebuilding the Gallic cities again after the catastrophes of the mid fourth century, Julian's troops worked their hearts out for their leader.[52]

If troops were not called upon to do the construction work, then the building could be done by civilian builders (usually the city guilds) under the supervision of a military architect or overseer. Under Gallienus in the mid-third century, military advisers were sent round Thrace to strengthen the walls after a spate of destruction and raiding by tribes beyond the Danube.[53] Architects and *agrimensores* were regularly lent to provincial cities, usually to settle boundary disputes, and overseers were sent in many cases to organise work for civilians.[54] By the later years of the third century, however, there was a great shortage of skilled craftsmen, despite earlier imperial encouragement for the training of architects.[55] The Diocletianic Edict on Maximum Prices does not establish architects at the top of the tree of professional craftsmen[56] but measures taken by Constantine to re-establish the popularity of their training attempted to redress the balance.[57] In the later third century, it was still a mark of special favour from the emperor if craftsmen could be found for municipal schemes: Eumenius' panegyric *De Instaurandis Scholiis* at Autun echoed the city's gratitude both for the imperial help given in rebuilding the city and for the fact that foreign (British) craftsmen were made available for the work.[58]

According to the normal pattern of building, it was expected that the city had to find the cost of materials used.[59] This probably explains in part the frequent use of old statuary and inscriptions in the walls of the cities built after the period of barbarian invasions. The Theodosian code, however, in its section on architecture and buildings, seems in general to rule against the pillage of old monuments to build new ones, though one ruling did authorise the use of material from temples for the construction of other public works, such as roads, bridges or aqueducts.[60] There is very little information, however, about how the cities faced the main cost of the walls. The question of finance for both construction and upkeep of the defences was at all times a difficult one: local authorities were strained by their taxation budget to a very great degree without having extraordinary expenditure on city walls to cope with. From time to time, emperors made remission of tax for cities to use on their defences: in 358, Constantius II allowed a quarter of the poll tax against wall building.[61] In 374, the amount was increased to a third by Valentinian.[62] But earlier still, it had been possible to divert funds from other sources, with the permission of the emperor, to pay for putting the defences in order. An edict of Diocletianic date allowed the use of part of the city's entertainment allowance to repair the defences, on the argument that the citizens could not otherwise feel safe to enjoy themselves.[63] The precedent established, in 371 the praetorian prefect Probus used part of the allowance for the materials for Sirmium's theatre to strengthen the defences since the city was under immediate threat of attack from the Quadi.[64]

Rescripts in the Theodosian Code which deal with public works paint a depressing picture of the state of many important buildings in the later period of the empire in Rome and elsewhere. Although the Codes do not specifically mention them, city walls were no exception.[65] For example, one of the gates of Savaria (Szombáthely) actually fell down while Valentinian was in the city in 374.[66] In general there seems to have been a reluctance to repair old buildings, but an over-keenness to build new ones in honour of the governor, many of which seem never to have been finished. The only new buildings allowable unreservedly and without the emperor's permission were stables for the *cursus publicus* and storehouses for the *annona*, thus emphasising the importance of these two aspects of the communications and the supply system at this time.[67] Several rescripts emphasised that on no account must there be any new building, and that the completion of older, unfinished buildings was more important.[68] Private buildings had evidently proliferated in public places, and these were to be torn down.[69]

The code emphasises that the burden of normal expenditure and upkeep of the public buildings of a city lay with its decurions and its guilds.[70] These, bands of tradesmen who formed societies within a city, sometimes played an important part in the construction of city walls: at Rome in particular, for the construction of its walls in part in 270–275, the guilds were organised into a series of military style work gangs.[71] Maintenance of the walls was also continued on the same lines, under the control of a number of magistrates who held overall responsibility. Such evidence as there is from Rome in the fourth century fails to specify exactly what services all the tradesmen's guilds had to supply, but the requisitions included, lime, transport, sand and bricks.

Where the guilds were not organised in this way, construction work on the walls was organised as a public duty to which everyone was liable unless he could find, send, or pay a substitute. At Antioch where repairs were carried out on the walls in the late fourth century, the local writer Libanius protested that craftsmen and shopkeepers were being compelled to heave columns around, and that peasants who brought their produce into the city markets were forced to load up their sacks with builders' rubble to take out of the city on their way home.[72] This overloaded both their sacks and their donkeys' backs. Public works of all kinds, not just the city walls, were built by requisitioning materials and labour from the unsuspecting taxpayer. One must assume that such drastic measures were not a yearly event, but that normal repair and maintenance was an item on the city's budgets (against which the emperor would allow the tax relief mentioned earlier), but that an emergency or a large-scale scheme like the construction of city walls might entail the participation of all the citizens in the task of actually erecting the defences.[73] At Athens, where the walls were rebuilt after the invasions by the Heruli in 267, this work was done and paid for by the citizens themselves. Two verse inscriptions survive, and both record in rueful tones how fortunate Amphion in the foundation myth of the city of Thebes had been in his ability to move large stone blocks into place by playing his lyre.[74] These half-humorous grumblings about the difficulty of their task provide a welcome palliative to Themistius' rather sanguine view of the emperor Valens' building activities in Thrace (see p. 61).

Some idea of the complexity, the length of time that the construction work could take, and the different methods that could be used, comes from a series of inscriptions at Adraha in Syria, where the building of the city walls was carried out over a period of some fifteen years, 259/60–274/5.[75] The cost of the walls, or at least part of it, seems to have been borne by the emperors Valerian and Gallienus, but the provincial governor is described on the inscriptions as the founder (or as the governor changed, the instigator), and there was great competition among local worthies to have their names included too. The inscriptions even allow a hierarchy of labourers to be worked out: the architect and his clerk of the works had names in Greek—they were clearly educated men. The contractor and his foreman also appear on the record: they bore local names.

The burden of upkeep of public works and of city defences in general was thus spread among the populace in many ways. Few were the cities in the later period which could actually increase in splendour during the later Roman period: the more important and prosperous were Rome, Constantinople, Antioch and Trier,[76] all of which figured well up Ausonius's list in his poem, the *Ordo Nobilium Urbium*.[77] New building, for all the idealism and idyllic quietude of his poems, which attempted to give the impression that a new age of peace had arrived, was rarely mentioned. Ausonius, however, was perhaps typical in other ways of the decurion class in the later Roman period, and shared attitudes to local affairs with many others. He clearly felt a great deal of local pride in his city, which he liked to compare favourably with Rome itself; no doubt other local dignitaries were as keen on their own town, and ready when they could do so to improve its facilities. To them, the provision and repair of city walls, and the public duty of maintaining imperial services, were necessary tasks which they carried out to the best of their ability and, on occasion, with more than a hint of civic pride.

4

The Pressures on Rome: Barbarian Invasions and Tactics

The construction of late Roman defences has normally been seen as a direct consequence of barbarian inroads in the third century. It is unfortunately by no means easy to prove such a direct historical relationship between the two. Evidence for the occurrence of invasions comes from casual or specific reports by Roman historians: often these are brief, for our sources for third century events are notoriously weak. It may be impossible, therefore, to pin down the exact date of an invasion and it may often also be uncertain whether two independent sources, speaking of an invasion by a Germanic tribe, are referring to a single invasion or to two separate ones. It is fair to assume that the *barbari* are only going to merit record in a late Roman imperial history when they force the writer to focus his attention on them. It is possible, therefore, that only the most serious of the barbarian invasions were reported. Many others, not quite as serious, may also have occurred but not merit a mention.

It might be thought that the course of an invasion could be charted from archaeological finds. Attempts have been made to show the line that invading barbarians took from the distribution of coin hoards and from recorded destruction within the right date range at Roman sites.[1] It is impossible, however, from archaeological evidence alone, to know who caused the destruction for which there is evidence on the ground. The burning of a Roman site or the demolition of buildings could always be accidental, and not necessarily the work of barbarian invaders. The reasons for the burial of precious metals, coin-hoards, in antiquity are even more inscrutable: to attribute every coin-hoard discovered (and for every one discovered, how many still lie unfound?) to psychological pressure exerted on Roman provincials by the imminent arrival of barbarian invaders is clearly simplistic. Such hoards may have been buried for other reasons, economic, private or social. Even supposing that such hoards were buried as a response to barbarian invasion, there is no guarantee that the barbarians ever actually reached the area where the hoards have been found. Hoards are buried in the expectation of trouble, which may not ever have materialised. A further consideration, however, is that hoards buried in antiquity and discovered today have never been reclaimed by their owners. One can speculate further about the reasons for this: were the owners killed in the barbarian raids? Or were the treasures simply not worth finding again in the wake of the inflationary spiral which seems to have hit the third-century Roman world?

It is clear from this, that evidence for barbarian invasions rests on a set of rather sketchy correlations between historical testimony and archaeological evidence. Any firm conclusions are necessarily limited and qualified. All the same, some attempt must be made to assess the scale and scope of such invasions, and to gauge the Roman view of the threat posed by their assailants before considering the fortifications themselves. There is no late third-century writer who weighs for us the political arguments of the time, and who provides the commentator's role on the events of this period. Such comment as comes from contemporary or near-contemporary writers is incidental.

The final problem relates to the temporal link between the invasions and their response—the fortifications. The archaeological evidence from the fortifications—the coins and pottery discovered in crucial layers underneath the walls, as well as the general typological design of the defences themselves—must also be consonant with a date at or around the time of the invasions. It has already been pointed out that when it has to rely on such things as coins and pottery to provide a construction date, archaeological dating is necessarily normally imprecise. There are therefore occasions when it is not possible to date the walls of the city or town archaeologically, and thus its assignation to the later Roman period is often on rather dubious typological or stylistic grounds. All these problems of interpretation must be given due consideration in approaching an assessment of the deterrent value of later Roman defences.

In Europe, the first serious raids appear to have been those of the Marcomannic Wars, though there had been spasmodic trouble on the Raetian frontier since the time of Marcus Aurelius. The forts at Kumpfmühl and Lorch (Lauriacum) had with others at Albing and Ločica formed the initial answer to the problem.[2] Under Marcus Aurelius the stone fort at Regensburg itself was built. The date of the earliest stone buildings at Vindobona, in the centre of Wien, is not precisely established, but they may have been constructed at about this period.[3] The legionary fortresses at Lorch and Carnuntum were also defended with stone walls, in common with many other auxiliary forts along this stretch of the Danube.[4]

There were threats towards the end of the second century in other areas too. Traces of what may have been a series of raids by Germanic pirates along the east coast of Britain have been revealed in a series of destroyed sites in Essex.[5] If the deposits which have yielded a considerable amount of very closely datable pottery of the years 180–200 can be interpreted as the result of raids by Saxons, then the eventual Roman reaction of constructing forts at Brancaster and Reculver to guard the main sea-approaches can be well explained.[6] Northern Britain, too, saw constant, though not ultimately serious, inroads into the military areas across Hadrian's Wall.[7] In Spain, there were threats of pirate invasions from Moors in Africa, and this may account for the fortification in what appears to have been a second-century style of many of the towns in the south and along the eastern seaboard.[8] In the main, however, the frontier lines themselves held out, possibly as much by force of treaties or payments as by the force of arms. Such agreements, often reinforced by cash payments by Rome to tribes outside the Roman world for keeping the peace, were common along the frontiers of the whole empire.[9]

More serious socially than these pressures from outside the empire was the effect of two plagues which are recorded as having swept parts of Gaul and Spain in the later second and early third centuries.[10] These may have left some areas severely depopulated, though the reports of their effects are probably exaggerated: Orosius records that no family was left untouched by the disease. How seriously this affected the cultivation of land and prices in general cannot be known, but the dearth of manpower in country areas could seriously have raised the price of food for town dwellers, and thereby accelerated the financial crises which were to break in the mid- and late third century.[11]

During the course of the third century, the pressures on the empire from tribes outside it increased on all frontiers. Franks and Saxons menaced the provinces in the north-east—Britain, Germany and Gaul. Alemanni and Iuthungi threatened the central portions of Gaul, Germany and Raetia. In Pannonia, there were inroads by Sarmatians, and further east other tribes, among them the Goths, penetrated deeply into Dacia, Thrace and Moesia, even on occasion reaching Greece. The reasons for such invasions seem mainly to have been the need for ripe crops to pillage, but this hardly explains the vast distances some tribes seem to have covered. Tribal movements deep inside mainland Europe may have displaced and unsettled a chain of barbarian tribesmen, with a consequent 'domino effect' on the Roman frontier areas. One of the most vulnerable portions of the empire was the flatlands east of the Rhine, where large, fertile tracts of land were guarded by several hundred kilometres of artificial frontier formed by palisade or wall with watchtowers and forts. Evidence for the gradual infiltration of this area in the third century by continual raiding across the frontier comes from scattered sources.

In Germany and Gaul the first positive evidence for raiding was the occasion when Severus Alexander met his death in 233 at Mainz, where he had come in answer to an emergency summons.[12] His successor, Maximinus, had found little trouble,[13] but in 238–9 the Germans invaded again, since the Rhine frontier had been left sparsely guarded in the wars which followed Maximinus' claim to imperial power. Balbinus, who had been entrusted with the task of fighting the Germans, was murdered in Rome before reaching the frontier, and Gordian III, though there is no record of his success in any historical source, was soon afterwards celebrated as *Victor Germanorum*.[14] The unrest in this area is evident in a scatter of coin-hoards of this date. One inscription dated 245 which records the construction of what probably was a watchtower by the *iuniores vici* at Bitburg[15] suggests that the pressure on the frontiers was not yet severe enough to prompt more than the occasional local attempt to do something about defence.

Areas bordering on the River Rhine and the upper portions of the Danube were the victims of raiding, particularly from the Alemanni, in the second quarter of the third century onwards. In 233, the Raetian frontier was thoroughly wrecked by the Alemanni in a series of raids which not only destroyed forts on the frontier between the Rhine and the Danube, but also caused widespread damage throughout the province[16]. The insecurity of the times is marked by the number of coin-hoards of this date, some found deep within the province, and a remarkable cache of weapons and other military equipment from near Straubing testify to the widespread nature of this crisis and to its suddenness.[17] Repercussions were felt in Noricum (Austria) and in Pannonia.[18]

In the aftermath of this catastrophe, some frontier forts seem never to have been rebuilt. It may be that since the earlier campaigns of 213 against the Marcomanni, Roman military commanders had already begun to scale down their concentration of forces. Severus Alexander and Maximinus Thrax were quickly able to restore the situation, despite this withdrawal of manpower, for Maximinus was able to assume the title *Germanicus Maximus* in 236.[19] Despite this, there appears to have been intermittent pressure on this area for some years to come, for more coin-hoards suggest further insecurity around 242, and there may well have been a renewal of raiding in the early 250s.[20]

There seems to have been little serious invasion in the west from this point until the more complex political period of the breakaway Gallic empire, though in the jostling for power between Aemilianus and Valerian in 254, the Gallic provinces suffered through the withdrawal of troops from the German frontier.[21] Gallienus was then sent to look after Gaul, while Valerian went to deal with the eastern frontier, where he was soon captured and died while a prisoner.[22] Part of Gallienus' campaigns took him in a sortie across the Rhine into German territory and he took the honorific title of *restitutor Galliarum* for which his lieutenant Aurelian may in some measure have been responsible.[23]

In 258, it had appeared that the situation on the Gallic frontier, if not completely calm, was sufficiently stable for Gallienus to be able to turn his attention to the pressing affairs on the frontiers of the east. There is some evidence that building or rebuilding was ordered and carried out at Köln, and there was also further activity on the part of troops on the lower Rhine.[24] Accordingly, Gallienus set off for Illyricum, leaving his son Saloninus with his military commander Postumus in command at Köln. Almost as soon as he had gone, there was a serious invasion, and a grave crisis as the soldiers of the Rhine army proclaimed Postumus as their emperor.[25] Postumus held Saloninus under siege in Köln, and was probably therefore unable to devote all his energies to mopping up the bands of Alemanni who had invaded Gaul.

At its fullest extent, this 'Gallic' empire, as it is called, held sway over Gaul, Spain and Britain: the separation of this portion of the Roman world from the control of the legitimate emperor lasted from 258 until 273, when Postumus' successor Tetricus ceded power once again to the then emperor, Aurelian.

There is literary as well as archaeological evidence for the barbarian raid which occurred in 259–60 (fig. 23). From the distribution of coin-hoards ending with coins of Gallienus, it has been suggested that the barbarians advanced[26] through Gaul on three main fronts. The fullest literary source is Aurelius Victor, who wrote more than a century after the events he was describing. According to his

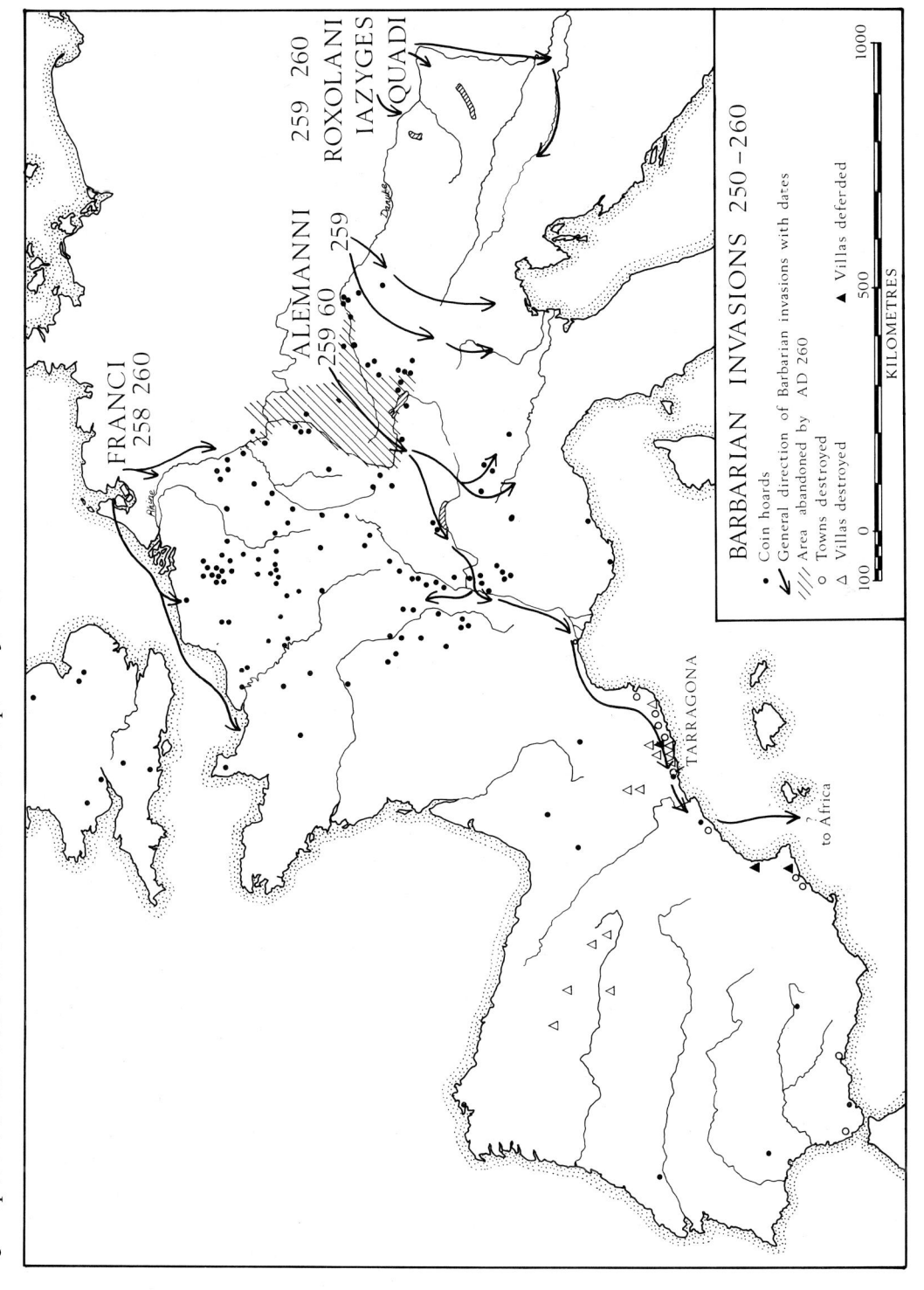

THE PRESSURES ON ROME: BARBARIAN INVASIONS AND TACTICS 71

23 Map to show main areas of barbarian invasion into Europe AD 250–260

record 'the Frankish nations ravaged Gaul and seized Spain. After devastating and almost obliterating the town of Tarraco (Tarragona) some of their number eventually got hold of boats and even reached Africa.'[27] Other sources add little to this sentence. The depth of penetration of this raid was particularly noteworthy. If the Franks crossed the Rhine in Lower Germany, created a trail of destruction through Gaul and Spain, and finally arrived in Africa, there is every reason to suppose both that their progress through the Empire, and the mopping-up exercise was a lengthy procedure, though it is doubtful whether the trail of buried hoards across Gaul can be really regarded as an accurate map of their progress. There is more archaeological evidence for destroyed or abandoned sites (normally villas) at this period from Spain than from France. Coin hoards of the period from Gallienus to Postumus (the late 250s) have been discovered in several places on the Spanish east coast. Taken with signs of villa destruction in the same area, this evidence suggests quite clearly the route taken by the invaders. Although Tarragona may have borne the brunt of the raids, many other villas and towns even as far west as Ilerda (Lérida) seem also to have been affected.[28]

This invasion has left little trace in Africa. There is some evidence of destruction at key sites in the interior of Mauretania Tingitana but there is no way of telling whether such damage, as well as the destruction of the profitable pickled fish industries, was caused by German invaders or by Moorish tribes who also periodically proved a nuisance.[29] One inscription, found at Tamuda (Tangier) mentions a military commander who defeated the *barbari* breaking into Tamuda.[30] The use of the word *barbari* in this inscription, unusual for Africa, may mean that these insurgents were not local tribesmen, but a band of the Franks who had crossed the Rhine in 259–60. The inscription is not closely datable, but is not inconsistent with a date in the late third century.

The Gallic Empire under the usurping emperor Postumus was not the only domestic problem which faced the legitimate emperor Gallienus at the time. The Pannonian armies also had rebelled, and were supporting their own champion, Ingenuus, as a candidate for imperial power.[31] Gallienus' struggle to combat this internal threat had denuded the Raetian frontier of troops, and an invasion by Alemanni in 259–60 reached deep into Raetia, destroying parts of the town of Kempten, and even causing concern for Italy.[32] Coin-hoards of this date have been found on the paths towards Italy in the valleys of the Rivers Lech and Iller. Pannonia, too, was affected by invasions at the same period, possibly from attacks by Goths which seriously threatened the peace of Dacia and the Balkans.[33]

There is reason to think that despite the existence of the Gallic Empire adequate measures were none the less taken for the defence of the western provinces. Trier and Tongres, along with other cities and towns on the frontier and nearby areas suffered some damage in the raids, but at Trier at least there was rebuilding under Postumus.[34] The Historia Augusta records that Lollianus (in 268–9) rebuilt several forts and cities which Postumus had placed 'on barbarian soil' during his seven years.[35] There is no clear archaeological evidence for any such rebuilding at this time, and it is possible that until about 300, when the first signs of free German settlement in the contested frontier areas manifested themselves, these areas were largely no-man's-land.[36] The small posts at Senon and St Laurent may be thought of as examples of Postumus' attempts to strengthen the road network by the provision of small patrol posts or extra defended strongpoints to protect stages of the Imperial postal system.[37]

Perhaps the best documented of the invasions at this period, though these affected only the Eastern part of the empire, were the Gothic raids into the Balkans and Greece.[38] The value of the account of this invasion, recorded in the *Historia Augusta*, is enhanced by the fact that it was taken from the historian Dexippus, who was not only an eye witness to the events described, but actually was instrumental in leading Athenian resistance to the barbarian raids.[39] The Gothic raid of 267 came as a concerted culmination to a series of inroads made by the Goths from about 238 onwards. Spasmodic pressure on the Danubian frontier and on Thrace had been met by a succession of third century emperors sometimes by cash subsidies (in effect payment to keep the peace) and sometimes by the withdrawal of this courtesy, and consequential further raids. Sustained resistance to the Gothic pressures was not assisted by the instability of the imperial office. All too often a Roman general who had triumphed over the Goths in a period of Roman superiority turned his back on them without properly completing the conquest, and marched on Rome to claim imperial power. There had been victories on both sides in the struggle, but the Goths had captured the city of Philippopolis in 250, and, at Abrittus in 251, had caused the death of the emperor Decius.[40]

After a lull of more than ten years after their last serious incursions into the province, two separate bands of Goths menaced the eastern parts of Europe in 267. One band crossed the Danube, and began to menace the Thracian cities, while a sea-borne force passed through the Bosphorus and began to attack Greece. The timing and the organisation of this raid suggests that it was planned by some war leader, but no name of an overall Gothic leader at this time is known.[41] The seaborne raiders concentrated their attacks on the Peloponnese and on Attica. Athens was one of their prime targets, and Dexippus, the Athenian statesman, was able to rally the citizens in the best traditions of Themistocles and Pericles to respond to the threat.[42] All was to no avail, however, since Athens eventually succumbed. Traces of destruction in many of the public buildings have been noted by excavators.[43] The libraries, however, were undamaged: the Athenian youth were less of a danger to the Goths if their leisure pursuits were left untouched. Athens and Corinth then fell to the Goths, who regrouped. Part of their band returned to Northern Greece, where they penetrated into Macedonia, and pushed as far west as Dyrrachium. Meanwhile the Danubian Goths were progressing into Thrace. They besieged the town of Philippopolis, and even employed a form of siege warfare, bringing machines up to the walls to attempt to storm them. These measures were relatively easily withstood by the townspeople, according to Dexippus' account, since they dropped large stones on the wooden siege mantles, and fired burning arrows into the timber constructions.[44] At last, however, the emperor Gallienus was able to mobilise a large army to combat them. The Goths, frustrated, gave up their siege of

Philippopolis, and turned to meet Gallienus. There is some confusion as to who gained the more decisive victory, but by the beginning of 268 the Goths had been soundly defeated. The victory could not be followed up, since Gallienus had immediately to contend with a new usurper at his back. Although this problem was dealt with, Gallienus fell victim to a conspiracy among his own army officers, and was succeeded by Claudius II. Although the main offensive against the Goths had been left in the hands of the army commander Marcianus, it was only on the resumption of hostilities by Claudius II that progress was again made, and even by the end of 269, the victory was not decisive and a peace had to be concluded. The Goths had been the victims of hunger, thirst and plague during the year and peace was desirable for both sides. The decisive assault came from Aurelian, who became emperor at the end of 269, and two years later dealt the Goths a blow from which they never recovered. From then onwards, the problems to be faced were those of resettling the areas devastated by raiding and of finding a use, as settler-farmers, for the Goths now decisively defeated.[45]

Gothic raids in Thrace and Greece threatened only one of several frontiers where pressure increased and decreased in measure with the activity or inaction of the Roman army or emperor. The most crucial problem engaging the attention of the emperors in the 250s and 260s was that of holding Dacia, at the best of times an awkward and cumbersome projection into barbarian territory. Despite its value to the Romans, Dacia appears to have been virtually lost in 257, a state of affairs finally affirmed by Aurelian in 270.[46] The extent of the problem can be seen from the assertion in a speech of 297–8 that 'under Gallienus' rule, Raetia was lost, and Noricum and Pannonia were devastated'.[47] By 260, the Alemanni had gone some way towards crushing the military and civilian administration of parts of the northern provinces: Avenches lay in ruins and the path towards Italy lay open.[48] An inscription of 260 records that there was some building in progress at Vindonissa.[49] More evidence survives, however, from the eastern part of the empire to show the wider activities of Gallienus at this period.[50]

The most disastrous of the raids into the Western provinces is usually reckoned to be the invasion of 275–6 (fig. 24).[51] It was the first emergency with which Probus had to deal after becoming emperor. But there are earlier signs that Aurelian, after winning back the Gallic provinces, had already taken some steps to secure at least some of the towns of Gaul before his sudden death. The walls of Dijon, as recorded in local verbal tradition handed down to Gregory of Tours, were built under Aurelian. Fortification of sites in this area may have had the dual purpose of security against threatened barbarian invasion and of control in the tussle for power between Aurelian and the leaders of the Gallic Empire.[52]

By far the most comprehensive account of Probus' reign and the raid of 276 is given by the *Historia Augusta*.[53] Several other sources agree that the Germans were in control of large parts of Gaul and Germany at this time, and that Probus had to regain 60–70 cities which had fallen into enemy hands. Possibly the most reliable source is the later emperor Julian himself, who was still able to write in admiration about Probus some 80 years after the raids. As ever, it is difficult to tell quite how extensive was the damage caused by this raid. Julian's comment about

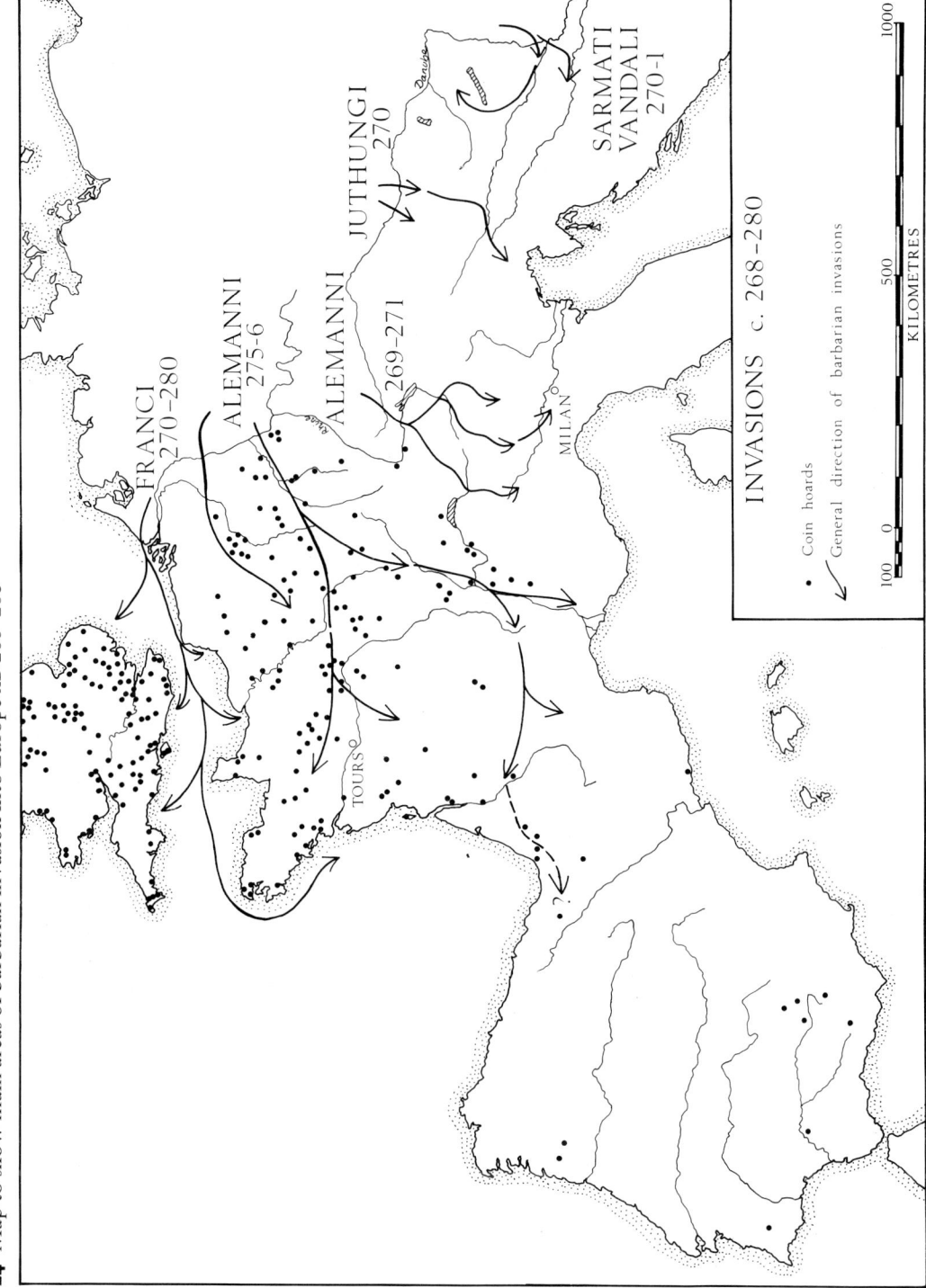

THE PRESSURES ON ROME: BARBARIAN INVASIONS AND TACTICS 75

24 Map to show main areas of barbarian invasion into Europe AD 268–280

Probus, that he set seventy cities on their feet, suggests that damage to unwalled Gallic cities was extensive.[54] One fragment of Dexippus' histories relates how a band of Germans had begun a siege of the walls of a city named as Tours. This city seems to have been within the path of the invaders, and the undated reference to this barbarian siege would most naturally fit within the context of the invasion of 276. However, to besiege a city, there must be walls at least partially built, and such evidence as there is does not suggest that by 276 Tours had any walls.[55]

It has been claimed that Spain was once more affected by this raid, despite the lack of literary testimony. A series of Spanish coin-hoards dated to the earlier part of Probus' reign suggests the progress of a band of raiders driven into Spain by the 'mopping-up' activities of Probus in Gaul as a result of the invasion of 276. The evidence, however, is more scanty than for the first invasion in 260. There is a concentration of coin-hoards in the south-west part of Aquitanica (the Bordeaux-Dax region), but this does not prove that the second invasion affected this area. The only evidence for military activity in Spain under Probus refers to the quelling by the emperor of several tyrant revolts in the provinces, among them Spain. Such punitive action, too, might produce a crop of buried coin-hoards.[56]

The maps of coin-hoards of the periods 260–85 suggest also that the English channel coasts were increasingly subject to hostile pressure. The build-up of this pressure on the coastal strip of southern Britain and Northern Gaul is not documented in any historical source. Traditionally, the Channel and the North Sea areas were policed by the *Classis Britannica* ('British Fleet'), and the Rhine mouth by the *Classis Germanica* ('German Fleet'), but by the third century there is almost no trace of either naval arm from tile-stamps, inscriptions, or historical references.[57]

Instead, by the 280s, there is historical reference to Frankish and Saxon pirates who had established some form of dominance over the channel area. Eutropius and Sextus Aurelius Victor who recorded this state of affairs, were both writing about a century after the events they describe.[58] Only Victor mentioned Franks and Saxons; Eutropius called them simply *Germani*. If these historians' accounts of the dominance already established by raiders on these coastal zones by the late third century can be relied upon, there is clear context for the coin-hoards found in coastal areas of southern Britain, northern and western Gaul, and even as far south as the Spanish coastline. The problems that this dominance posed to Channel trade and the relationship of Britain with the rest of the Roman world demanded strong solutions, and the appointment of a new naval commander Carausius in 286 was the first, and effective, step in re-establishing Roman control.

The barbarians' motives for such varied and deep incursions into Roman territories are hard to define. Prime among them was probably the desperation occasionally brought on by famine, consequent upon the failure of crops in their homelands. The prospect of rich cornfields across the Roman frontiers must have been a very attractive one, but a prospect hardly likely to encourage more than short term raiding with rapid strikes across the frontier and back again. Other reasons for the raids are equally intangible. The movement of tribes deep in central or eastern Europe is often claimed as having a 'domino effect' upon the tribes nearest the frontiers of the Roman world. By pressure from their immediate neighbours, such tribes were forced to look for space to settle within the somewhat underpopulated Roman provinces. The argument to prove that this was so is of necessity involved and tenuous. It is difficult enough to show by archaeological means that an invasion has occurred, still less a chain of such 'invasions' or 'displacements' of tribes stringing its way across central Europe. Even though the existence of such a chain of events may be impossible to prove, the motive of searching for land to settle could well account for some of the invasions.

Perhaps a more basic reason for this barbarian pressure was the attempt to secure, in some undefinable way, a portion of the obvious wealth available within the Roman provinces. This might take the form of winning the right to live within the Roman world, and gaining recognition by Roman emperors through service on the frontiers. This was a privilege sought after, and, once won, jealously guarded. From the third century onwards, and particularly from Diocletian's reign into the fourth century, it became an established practice.[59] It had been started as a method of regulating underused areas of the empire, normally the frontier zones, by giving plots of land to retired veterans. They thus became farmers who could usefully act also as a makeshift militia when their area came under threat from invasion. When this policy was widened to incorporate barbarian tribesmen rather than retired veterans, the benefits, too, were appreciably widened. As well as cultivating the under-used areas of frontier land, barbarians could be relied on to provide a fearsome warband to combat the effects of other German invaders.

Such a status was possibly the barbarian's ideal, but the available methods of achieving it were to create as much disturbance within the provinces as possible, and in the process, to grab as much plunder as they could. In this respect, too, the Roman world offered rich pickings. The driving motive for ever deeper raids, and the underlying reason for the provincial fear of barbarian arrival will have been the barbarian desire to collect plunder from the rich villas and settlements in the provinces.

The very effectiveness of the raids is surprising, not least because of the depth of penetration in the face of considerable difficulties of survival. As far as planning an invasion was concerned, the safeguarding of an adequate food supply to support a war-band was of prime importance, and there are very few instances in the literary record of raids by barbarians being planned to such a high degree.[60] The further a raiding party got from its home territory, the more dependent it would become on being able to commandeer supplies of food from the area in which it found itself. A single large group of barbarians therefore might gradually split up as the search for food became more hazardous and frustrating. Thus the Roman response to such raids might not be an initial attack on the participants (they would be notoriously difficult to meet in a pitched battle anyway) but to protect all possible sources of food from the warbands. This would mean safeguarding granaries, and protecting livestock by driving it into defended enclosures. In this way, the warbands could be starved into surrender or retreat, or possibly even given a safe passage out of the provinces by the promise of food on the frontier or on the route home.

Roman dealings with such barbarians, cornered and desperate within the empire, were liable to be completely unscrupulous and amoral.[61] An example of the sort of dealings to be expected, and a fair indication of the Roman authorities' view of the barbarian, comes from an episode recorded by the historian Ammianus Marcellinus.[62] In 370, a band of Saxons invaded Gaul. A count, named Nannennius, was sent to deal with the marauding band. The two sides lined up for battle, but the legions arrayed against them looked too impressive for the raiders. Thus they sued for peace, and, as a condition of the truce, they handed over many young men fit for military service, and were pledged a safe conduct out of the Empire. This was fulfilled, but a terse entry in a later chronicle and the tale told by Ammianus agree that a trap was set for the invaders. Just across the Rhine, the tribesmen were slaughtered to a man. Ammianus' comment on this is revealing of his disregard of the barbarians, and his classing of them as something less than human: 'and although some just judge will condemn this act as treacherous and hateful, yet on careful consideration of the matter, he will not think it improper that a destructive band of brigands was destroyed when the opportunity at last offered itself.'[63] Though there is no guarantee that this was the official attitude rather than the historian's own view, the number of instances of similar dealings throughout the Roman period suggests that control of the barbarians in frontier zones was more desirable than the preservation of a clear conscience on the part of the frontier commanders. If keeping barbarians hungry by promising subsidies which failed to arrive contributed towards keeping an uneasy peace, then such policies were pursued.[64]

The necessity for large-scale fortification in the late third century has sometimes been called into question. It has been supposed that the German tribes were no match for Roman warfare or siegecraft and that less powerful walls than those apparently built at this time would have been equally effective.[65] But that Germans in the west were not used to siege warfare or to storming cities is by no means established. Even as early as Tacitus' time, there are records of inter-Germanic sieges,[66] though the tribes' ineptitude at such tactics was a historian's commonplace. Greek historical writers tell of sieges in Thessaly at this period when Goths broke through the frontier in 269.[67] At the siege of Thessaloniki, the Goths were said to have used siege engines, but without success,[68] and they were relatively easily repulsed from several other eastern walled cities, including Philippopolis.[69] In context with such tales, the account of the siege at Tours must also be taken.[70]

It remained a commonplace among historical writers that the *barbari* were on the whole unprepared for and inept at siege warfare:[71] in his accounts of the campaigns of Julian, Ammianus Marcellinus records that the Germans attempted sieges when Julian was in winter quarters at Sens and Troyes.[72] On the latter occasion, they are said to have gone away disgruntled, thinking it foolish that they had ever even contemplated trying to take the city by force. Later on, Juthungi in Raetia were tempted in desperation to the 'unusual expedient' of besieging walled centres[73] and in Valentinian's time Ammianus, in describing Macrianus as a formidable German chieftain, records that he would even try taking the walls of cities.[74] By far the most effective way of taking walled cities, however, was by undercover negotiation. Successes which the Goths had in the east in their raids of the late third century seem to have been mainly achieved in this way.[75]

Despite this possibly biased view of the Germans' ability to storm cities, it is plain that walls were necessary to keep such raiders out. It was not to be expected that a city would undergo a long and difficult siege. If the Germans came, they came for quick plunder or for food, not to take over the centres of administration. But even so, these had to be well defended, as did granary stores in order to safeguard adequate supplies for troops on frontier duty. Not only were these walls now built, but a high degree of sophistication was lavished on their construction. The external towers, probably the newest feature of the western defences, were intended for the use of archers and *ballistae*, and often placed so as to enfilade the weakest parts of the wall circuit.[76] *Ballistae* of the later Roman period were light spring guns.[77] Their size and mobility would suggest that several could be mounted in a single tower. Wide windows were a prime necessity for this type of weapon, to give the widest possible field of fire. They had a range of some 400m, and therefore, if used carefully, could keep an enemy from coming in close to the walls.[78]

Whether *ballistae* were mounted on every tower and kept in full working order as a permanent feature is a matter of debate. One estimate of the life of a spring in use was only in the order of 8–10 years, and on such a basis the cost of keeping large numbers of such machines in working order was probably prohibitive.[79] No doubt, however, maintenance was but a small problem when one needed only to replace the sinews to ensure that the guns stayed in full working order. The *Notitia Dignitatum* mentions arms factories which could presumably supply ballistae mainly for army use.[80] Normally there would be little reason for all cities and towns to be in a continual state of alert. *Ballistae* could be brought and used if required, and legions had trained craftsmen who could make them if necessary.[81] As far as offensive machines were concerned, the fourth-century legions had none in permanent use. Ammianus' account of the siege of Amida in 359 shows how unfamiliar two Gallic legions were with artillery of any sort.[82] By the time of the *Notitia* there were special units of men trained as *ballistarii*.[83] These were created probably to guard against a shortage of men capable of shooting and looking after such equipment, such as occurred under Probus when there were brigand artillerymen under the command of Lysius.[84] Constantius II created a depot for such weapons at Amida to deal with his campaigns against Sapor[85] and they were also deployed on a re-used Trajanic ford some 20 stades from Mainz during the campaigns of Julian.[86] Other references show that the use of such machines against barbarian attackers was relatively common.[87]

There is no evidence that German weaponry or fighting skills had improved at all between the time of Tacitus and the third and fourth centuries.[88] The main weapons of the barbarians were still the sword and spear, though bows and arrows were also in use. Their mobility was not significantly improved, but the Roman view of them was, as ever, coloured by the accepted picture of barbarians as unscrupulous raiders who showed no pity and who certainly could expect no quarter if reduced to suing for peace.[89]

The view that the barbarians were really no match for the Roman army or their tactics was in direct contrast to the belief which saw the barbarians as an unstoppable destructive force. Traces of both views can be found in writings about the period. The historians on the whole favour the former, whereas Christian writers on the whole reflect a commonly held belief that the barbarian invasions were a scourge from the Almighty, associated with the end of the world. Sermons of men like Maximus, Bishop of Turin, returned occasionally to this theme at a period when Northern Italy was subject to raiding at the end of the fourth century.[90]

Examples of actual resistance offered to the invaders by Roman provincials are not easy to find, particularly in the western parts of the empire. The one outstanding example in the east was Dexippus, who not only organised a force of volunteers against the Goths, but also recorded their exploits. Among his writings he records a speech made to these volunteers after the capture of Athens in 267.

> 'I will run all risks to save all the things I hold most precious and to retain the respect the city has for me .... If after listening to others talk, any one of you is appalled at the depth of misery into which our city has fallen, he should remember that most cities have had the ill-fortune to be captured by an enemy catching them unprepared. One can see that Fortune will be with us: our cause is as just as can be. We fight against unjust aggressors, and usually the Gods decide on human affairs according to the justice of the cause ....'[91]

Men who, without the promise of imperial assistance, actually set about organising their own defence are rare, as in any society protected by a highly paid and, on paper at least, superior army. The records of such stirring deeds as those demanded from his followers by Dexippus are few in number—a couple of inscriptions refer to 'the enemy repulsed from the walls' of Saldae, in Africa,[92] and to an army commander, also in Africa, who 'forced his way into the province, and gave the barbari who had broken into Tamuda a sound thrashing'.[93] Perhaps the other star in the otherwise somewhat gloomy heavens is the early fifth-century exploits of an African philosopher and later bishop Synesius.[94] His letters, when arranged in their chronological order, show a life of activity against constant attacks from nomadic tribes in his native city of Cyrene. They paint an eloquent picture of the drive and energy required of a man who was to uphold the defence of his fellow citizens when all official sources of help were incapable or unwilling to give any assistance.

Even the troops, it seems, were prone to inactivity. Only when the emperor or one of his trusted adjutants personally supervised the resistance to barbarian raids could a firm response be expected. To keep his men in fighting trim at a period when they were often under attack, it was important to develop a taste for aggressive counter-attack to barbarian inroads. Lack of leadership or of initiative to take such a step could often lead to a laxity which many of the emperors found unattractive. A mercenary attitude among troops who drew large salaries and achieved little or nothing in times of crisis was also the cause of caustic comment from writers like Ammianus Marcellinus. He put into the mouth of Ursulus, contemplating the ruins of the eastern city of Amida after its siege and capture in 360, the words 'Just look at the so-called courage of our soldiers in their defence of the cities! And for this they are paid so much that the whole wealth of the empire can scarcely provide their reward.'[95]

By far the least complicated overall view of the state of defensive awareness in the late third century would be to suppose that the campaigns of wall-building then undertaken formed an immediate response to the unexpected savagery and depth of barbarian raiding in the years 260–80. Despite the lack of written sources which spell out the reasoning behind the provision of such an extensive series of costly and massive defences, the need for safety and the preservation of a Roman lifestyle in the face of prolonged banditry by marauding bands of raiders will have formed a strong impetus. This is not to discount other motives for building walls—civic pride, neighbourly rivalry or the successful securing of a government grant—all of which also may have played their part. The most fundamental shift, however, seems to have been in the minds of the provincials. No longer could they expect the frontier between Roman and barbarian to be securely guarded by well-paid troops from their military installations on some distant river bank. With the advent of the crisis of barbarian raiding, this frontier had now moved uncomfortably close to home for many provincial Romans. The provision of strong and secure city walls, until now unnecessary, was now an essential supplement to the traditional frontier defences.

5

Town and City Fortifications

The mass of evidence about late Roman fortifications from standing buildings, archaeological and old documentary sources presents a somewhat unwieldy bulk for examination and study. Accordingly the evidence will be examined by distinguishing, particularly in the frontier areas, between cities, small settlements or road posts (*vici*) and military establishments. The minutiae of such a division are open to considerable criticism, for while there are many walled sites which fall clearly and easily within these divisions, there are several which do not. The status of town or city was as much a technicality of termination within the Roman period as the appellation 'city' is today. Broadly the category will include all the *civitates*, *coloniae* and *municipia* whose status is known from Roman documentary sources, and it will also include other settlements of comparable size and apparent local importance but for whose exact status there is no evidence at present.[1]

Evidence as to which settlements had the status of *civitas*, in the later Roman period in Gaul at least, comes from a document called the *Notitia Galliarum*, 'the list of the Gallic Provinces', which is of early fifth century date, and lists all contemporary Gallic cities under their separate provinces (fig. 25).[2] Opinion is divided as to the original purpose of this document: it could be a list of the civil authorities,[3] or a record of the early bishoprics,[4] which, in the main, followed the mould of the civil structure. Whatever its origin, its existence and the information it gives about Gaul make it most convenient to consider this area first. Not all Gallic cities mentioned in this document have produced evidence for late Roman defences, but the study of selected groups of those which have is revealing.

In Lugdunensis Prima, the first of its provinces, the *Notitia Galliarum* lists five towns. These included the two large cities of Lyons and Autun, the town of Langres and the *castra* at Chalon-sur-Saône and Mâcon (fig. 26). At Lyon,[5] despite its position as capital of the three Gauls, and the preservation of other buildings from the Roman town, there is little trace of the Roman defences. The area that the walls enclosed was roughly triangular, and possibly as much as 65 hectares in extent. At one point on the north side the walls were found to be 1.45m thick. They were punctuated with projecting semicircular towers. It is possible

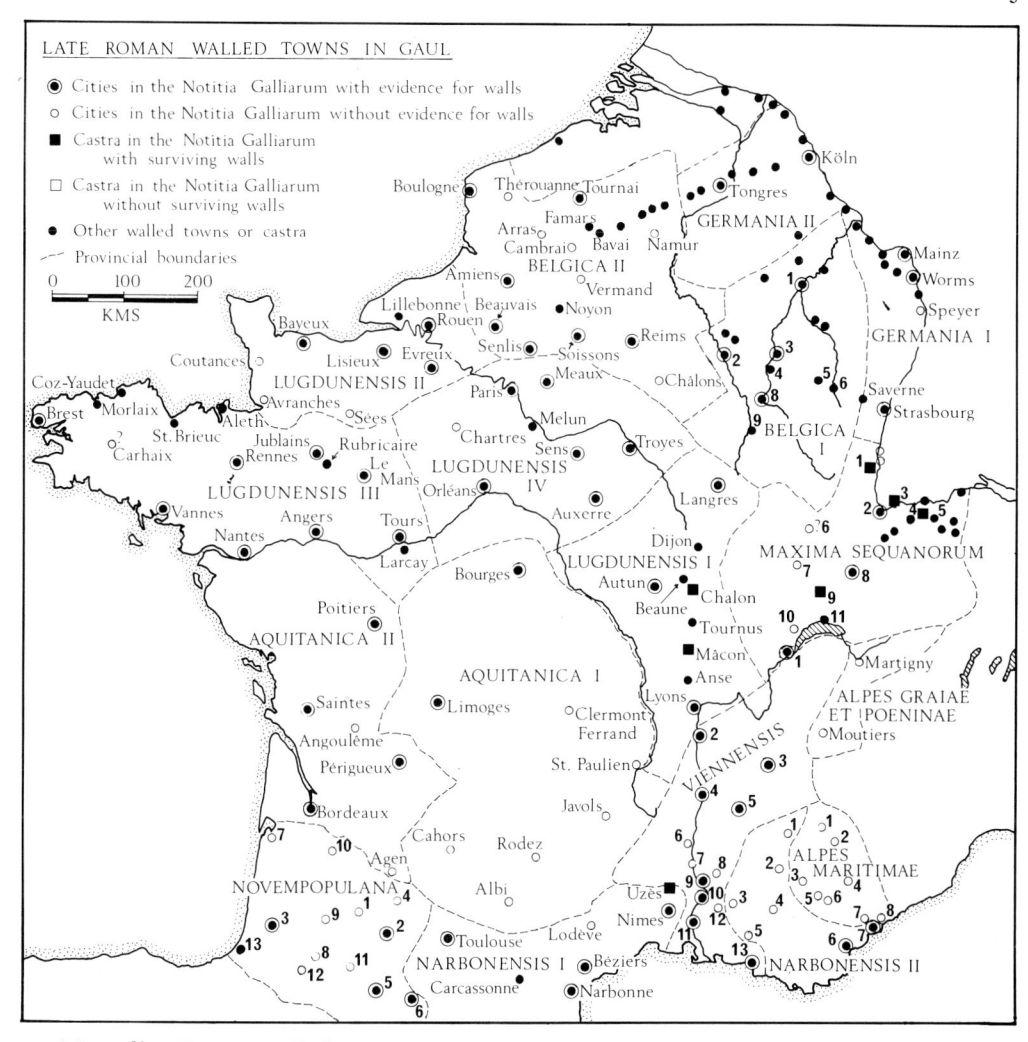

25 Map of late Roman walled sites in Gaul, the evidence from the *Notitia Dignitatum*. More details of the frontier provinces will be found in figs 51 and 62. Numbered sites as follows:

Novempopulana
1 Eauze
2 Auch
3 Dax
4 Lectoure
5 St Bertrand
6 St Lizier
7 La Teste de Buch
8 Pau
9 Aire
10 Bazas
11 Tarbes/Bigorre
12 Oloron

Belgica I
1 Trier
2 Verdun
3 Metz
4 Scarponne
5 Tarquimpol
6 Sarrebourg
7 Toul
8 Grand

Viennensis
1 Geneva
2 Vienne
3 Grenoble
4 Valence
5 Die
6 Alba
7 St Paul
8 Vaison
9 Orange
10 Avignon
11 Arles
12 Cavaillon
13 Marseilles

Narbonensis II
1 Gap
2 Sisteron
3 Apt
4 Riez
5 Aix en Provence
6 Fréjus
7 Antibes

Maxima Sequanorum
1 Horbourg
2 Basel
3 Kaiseraugst
4 Windisch
5 Baden
6 Portus Bucini
7 Besançon
8 Avenches
9 Yverdon
10 Nyon
11 Lausanne

Alpes Maritimae
1 Embrun
2 Barcelonnette
3 Digne
4 Glandève
5 Senez
6 Castelane
7 Vence
8 Cimiez

that at the time of the known disturbances of 68–9, and the subsequent attack on the city by the inhabitants of the nearby colony of Vienne, the city already had its defensive circuit, which remained unchanged into the later Roman period.

As capital of the Aedui, Autun[6] (figs. 2, 26) was famous in the third century for the schools to which Gallic noblemen would send their children. The walls of the colony date from their foundation under Augustus, and were faced with carefully squared blocks. There were projecting semicircular bastions at intervals. Two gates, each with a double portal, passenger archways and an arcaded gallery above, date from the original foundation. From re-used material in one of the two, the Porte St André, it seems likely that this gate at least underwent some rebuilding. After the raids of the later third century, the highest point of the town, on which the Cathedral now stands, was separated by a wall from the rest of the city thus forming a small defended 'Capitol' of 10ha. Parts of this wall consisted of lime-mortar and crushed tile, and reports speak of re-used stones being discovered in the blocking wall. Despite the presence of this late Roman wall, the full original circuit was still standing in 297, when Eumenius described the state of the city in his panegyric.[7] Little is known of the later period at Autun, though historians and writers mention the city at various times in the third and fourth centuries. After 296, rebuilding was taking place at the hands of British craftsmen,[8] and the walls were later described as 'a spacious circuit, but one weakened by time's decay'.[9] There was also an arms manufacturing depot at the city.[10]

Although the site of Langres[11] is an impressive one, occupying the top of a prominent hill, and at an important point for the crossing of main Roman roads, only one Roman monument, that of a Roman ornamental (possibly triumphal) arch, survives in the town. This may have been incorporated within the later circuit of walls, which have been located in places and are described as being 3.35m thick, and composed in major part of former Roman monuments. The only literary reference to Langres suggests that the walls were already built by *c*300. Constantius Chlorus was then involved in a narrow escape outside the city when being chased by Alemannic tribesmen and had to scale the walls with ropes in order to escape.[12]

Although described in the *Notitia Galliarum* as a *castrum*, the visible remains of the 'D'-shaped defended enclosure on the bank of the Saône at Chalon[13] presents an appearance archaeologically indistinguishable from *civitates* elsewhere. The walls enclosed about 11 hectares, were between 2.50 and 3.50m thick, and contained tile courses. Circular towers, some 6.80m in diameter, lay astride the wall. Chalon was an important commercial and naval centre,[14] and is mentioned as one of the principal towns in the province by Ammianus Marcellinus.[15]

Mâcon,[16] mentioned in the *Notitia Galliarum* as a *castrum*, was also of importance in the late Roman period. Like Autun, it housed an arms factory,[17] but little trace of the defences has survived. The late Roman town is known to have occupied about 10ha on a raised site slightly north of the present day nucleus. Only one fragment of the walls, faced in rough stone blocks, has been recorded.

Dijon is recorded in ancient sources as a *castrum*.[18] Gregory of Tours[19] also recorded that, according to local stories, the walls were constructed by 'Aurelius'.

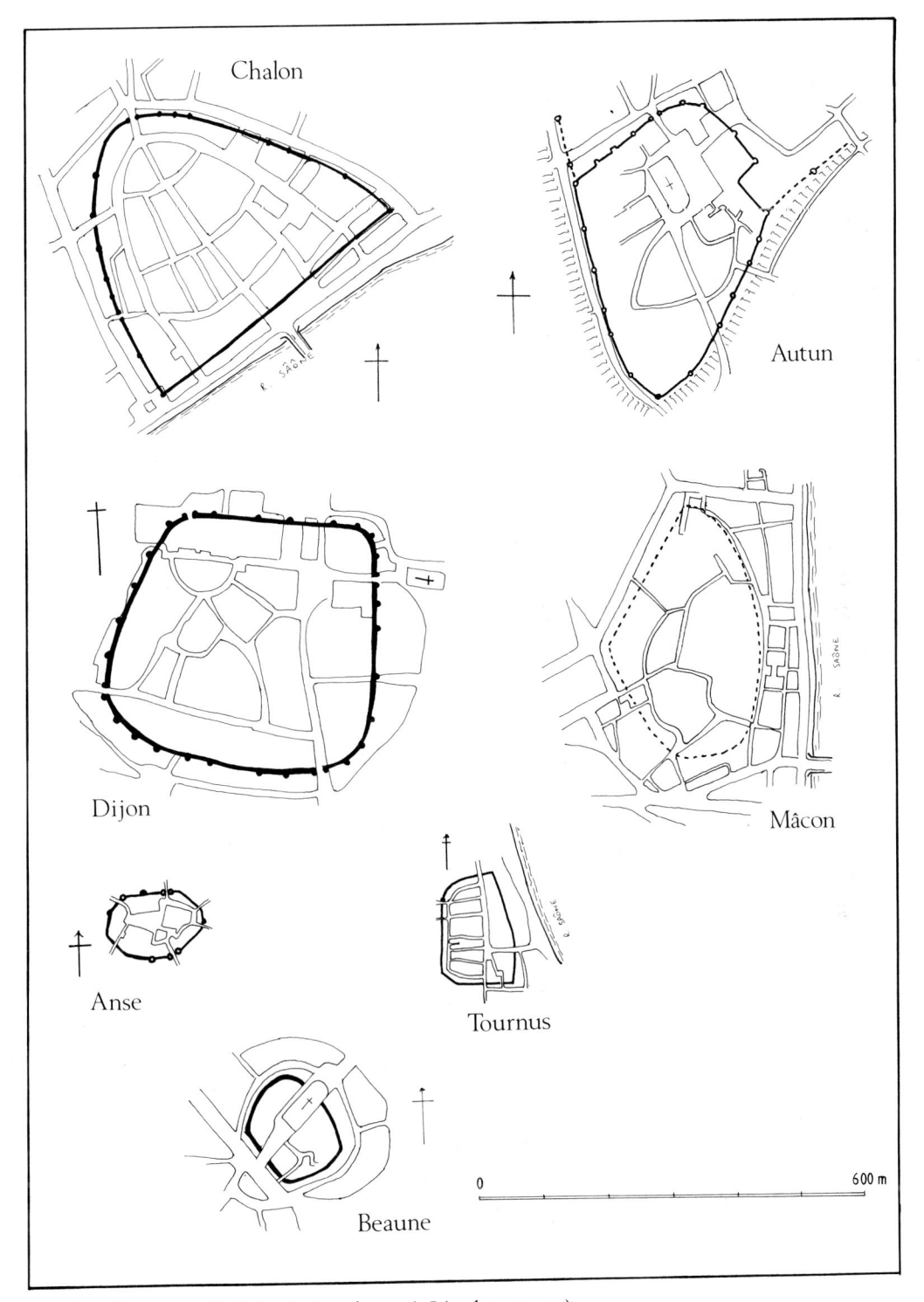

26 Late Roman walled sites in Lugdunensis I (*scale 1:10,000*)

He gave a long description of them, from which it is clear that, in his day, a stream ran through the centre of the defended area, that it had four gates facing the four points of the compass, and was surrounded with walls '30 feet high and 15 feet thick' fortified by 33 towers. Archaeological evidence[20] on the whole supports and confirms Gregory's description. The circuit was roughly circular, and occupied about 11ha. Large blocks of stone, often re-used material were used as foundation material. The walls, 2.40m thick on excavation, nowhere reach the breadth described by Gregory but the facing, where it is preserved, appears to have been in rough medium-sized worked stone, which accords well with his near contemporary eye-witness account. The gates have not been excavated, but two small posterns have been discovered, each protected by a tower flanking it. The interval towers were circular, and stood astride the walls.

The wall circuit at Beaune[21] was oval in plan, similar to Dijon, though at only 2ha, much smaller. The wall is recorded as 5m thick, and the wall facing was irregular, with large blocks of stone in the foundations. Twelve towers standing astride the wall ring the circuit at intervals of approximately 45m.

On the slight hill above the Saône at Tournus[22], there are traces of Roman walling, enclosing a roughly polygonal area of 1.5ha with only one known tower astride the wall. The date of the circuit is hard to determine. No re-used material is recorded from the walls, and a date in the second century has been suggested, but in the absence of excavations capable of establishing a secure date, it is preferable to see this small enclosure as one of the line of fortifications along the Saône valley in the later Roman period.

A roughly oval enclosure in the centre of the small village of Anse-sur-Saône[23] was of comparable size to Beaune and Tournus. It had strong surrounding walls, 3.55m thick, built of small blockwork with tile courses at irregular intervals. At least six towers stood astride the wall, and two main gates afforded entrance for the main road through the defences. Postern gates were defended by a single bastion. Re-used material found in the walls indicates that the walls are of late date.

This province, then, the first listed in the *Notitia Galliarum*, contains a number of late Roman centres of varying kinds. The main feature is the concentration on the line of posts—some large, some small—along the valley of the Rhone and Saône, north and south of Lyon. The slight remains also suggest that these defences were similar in style to each other. Normally they were small polygonal or near circular enclosures with walls of small blockwork, tile courses, and towers astride the walls. In many features of this design, all the walls follow the established pattern of the Augustan walls at Autun and those, as yet undated, of Lyon.

Comparison of this group of defences is thrown into sharper relief by consideration of other areas. In Lugdunensis II, for example, the *Notitia Galliarum* lists seven sites, in three of which, Avranches, Coutances and Sées, no trace of the Roman defences has been found. From the four in which remains do survive, and from a further site, Lillebonne, where late Roman defences remain although the site is not included in the *Notitia* list, some comparison with the group of sites in

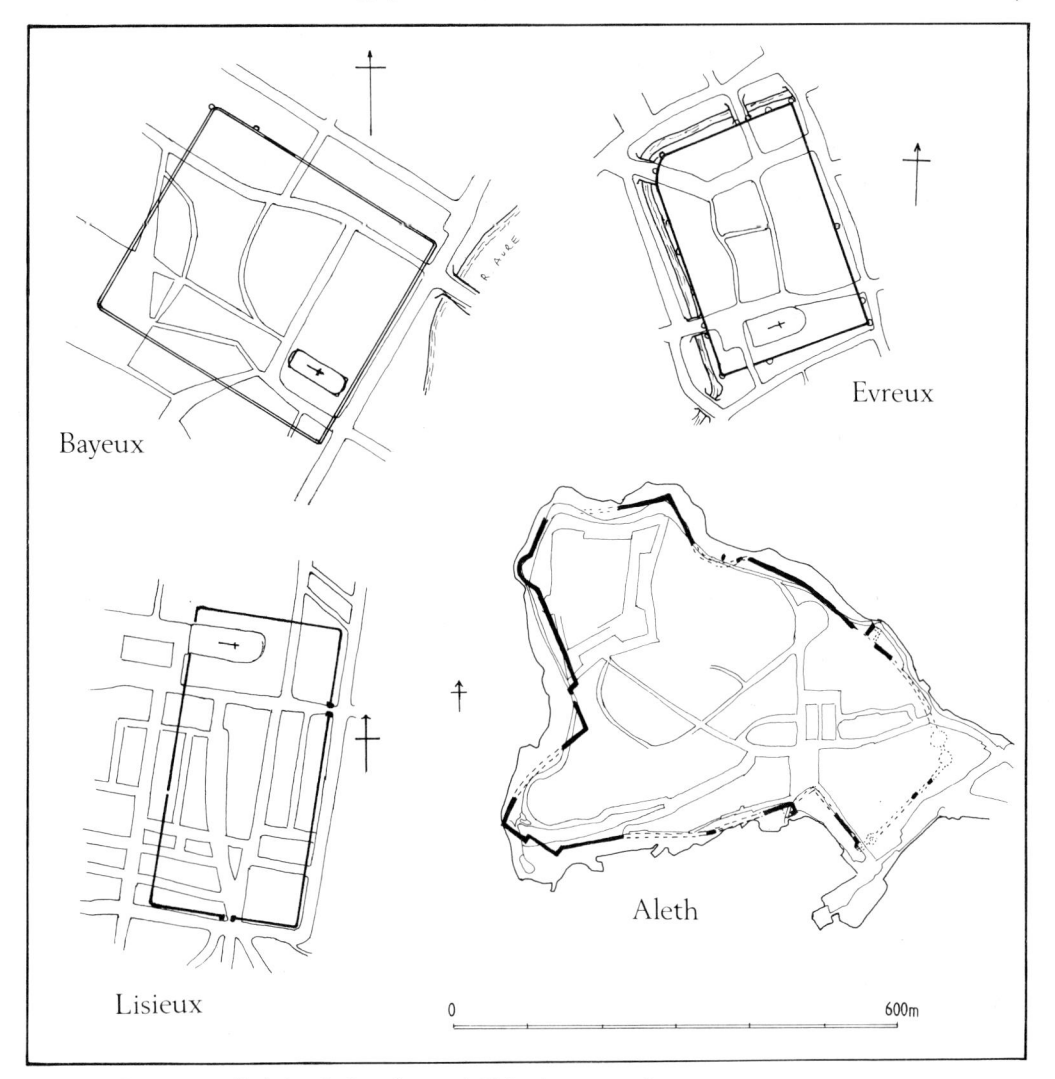

27 Late Roman walled sites in Lugdunensis II (*scale 1:10,000*)

Lugdunensis I can be made (fig. 27).

At Rouen, the capital of the province, the traces of the defences are fragmentary.[24] It has been thought in the past that the presence of walls enclosing a rectangular area was responsible for the grid pattern of the modern-day streets, but work on the river frontage has shown that there is much made-up ground there. Even if three sides of the Roman defences formed a rectangular shape, it is probable that a wall along the river frontage spoiled this symmetry. It has been claimed, on antiquarian evidence alone, that the city had four gates, and that its walls were defended by both square and round towers.

In the nineteenth century, much of the Roman circuit of walls at Bayeux[25] was still easily visible. Only part survives today, embedded within ostensibly medieval walls. The enclosed area, about 9ha in extent, was rectangular with rounded corners, and was set on the west bank of the river Aure. The Roman wall-facing was of small blockwork, rather coarsely dressed, interspersed with double tile courses. Re-used material is recorded from the wall foundations, among it a milestone of Septimus Severus. Roofing tiles too, occur in the tile courses. On the east wall, two U-shaped projecting towers have been incorporated into modern buildings. The north-east corner is rounded, and also sports a similar projecting tower.

The circuit of walls at Evreux[26] encloses the Cathedral and the heart of the town in a rather irregular rectangle with rounded corners. The foundations were of regular clay and flint construction. Above this were large stone blocks and a facing of small blockwork interspersed with double or triple tile-courses sometimes running right through the wall. The walls were on average 3.10m thick, and enclosed an area of about 10ha. U-shaped projecting towers have been recorded. No conclusive dating evidence has been found, though a substantial coin-hoard dating to the reign of Probus was found within the defended area. This does not, however, help to date the construction of the defences.

The site of the town walls at Lisieux[27] lies on a steeply sloping site on the east bank of the river. The enclosed area of 8ha is rectangular in shape, with rounded corners. Construction of the defences was clearly carried out without regard for existing Roman buildings on the site, which were sometimes used to support the walls. Finds of burnt stones and debris within the wall suggest that it was built after the city had suffered some destruction. The facing of the walls was of small blockwork and triple tile courses ran through the wall's full thickness. Two gates, not otherwise described, are shown on the published plan.

At Lillebonne,[28] the theatre of the Roman town was converted in the late Roman period into a fortified strongpoint. The southern entrance of the theatre was blocked with large masonry, and strong walls faced in small blockwork have been traced round a rectangular enclosure of 1.5ha which lay on the northern (stage) side. The date and purpose of this small enclosure are unknown.

Despite the poverty of remains of the Roman defences in the province, those that do survive display a common blueprint different from, but as distinctive as, those in Lugdunensis I. In plan, the walls all describe a rectangular shape with rounded corners. The details of facing, the design of towers, and the treatment of tile courses are all common features, and suggest that here a different designer was at work.

The same is also true for the third province in the large earlier Roman area of Lugdunensis. Here almost all of the ten sites named in the *Notitia Galliarum* have produced substantial remains of late Roman defences. As material for study, therefore, they form a large and complex group (figs. 28 and 30). Much of the evidence for the walls is so detailed, it is best presented in tabular form (Appendix I, p. 264–5).

The metropolis of the province was Tours,[29] where there is evidence for a siege in the accounts of late third century barbarian raiding (see page 76). The only known city wall, however, is within the mould of late Roman defences, and enclosed a quadrilateral area clasping tightly onto the amphitheatre.

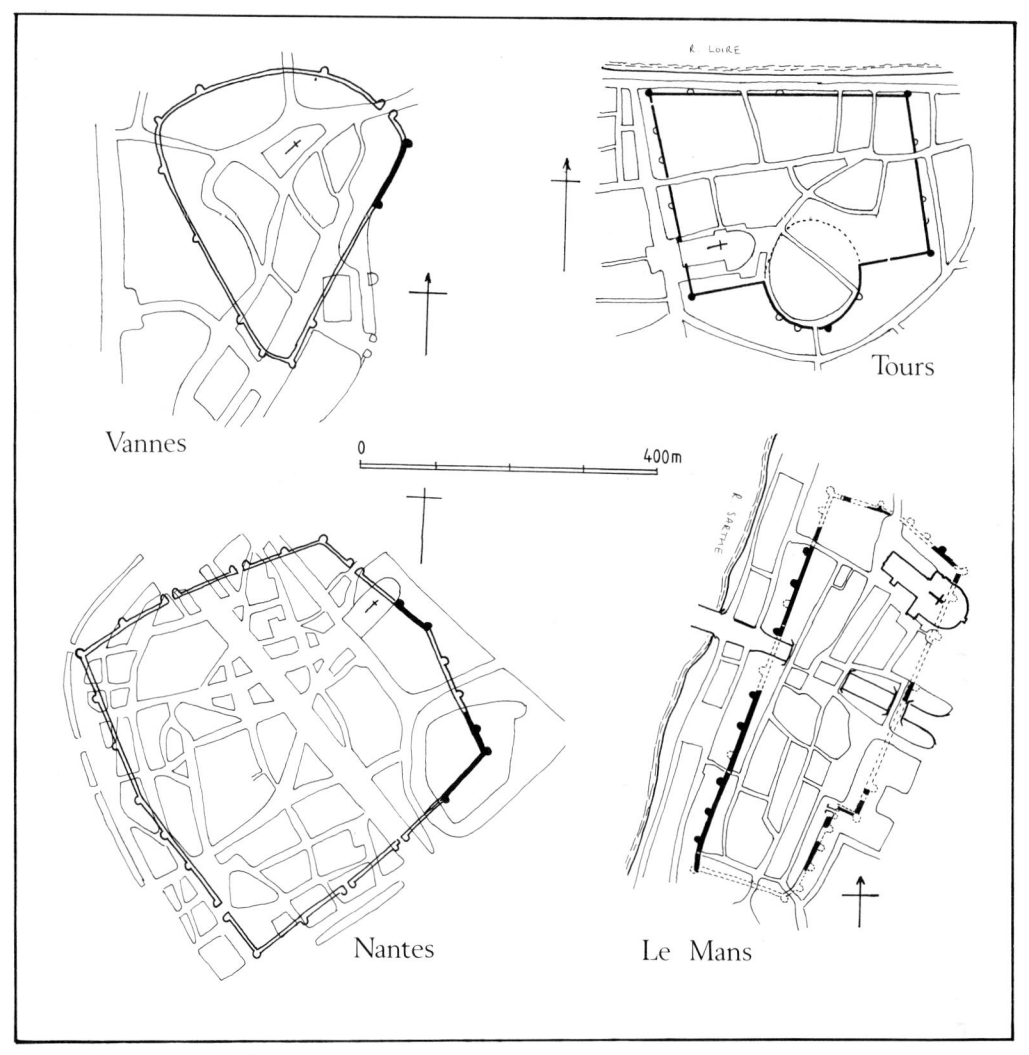

28 Late Roman walled sites in Lugdunensis III (*scale 1 : 10,000*)

The wall was built in the main of re-used materials. The lowest courses were formed of thick blocks of stone without mortar, and above this, in the substantial portions still visible, the facing was of small blockwork with double tile-courses. The wall continued round the outside of the amphitheatre to include it within the circuit as a massive semicircular bastion. Semicircular and three-quarter round bastions were built along the walls and at the corners of the circuit, and the only known gates are posterns.

At the south-west angle, the Tour de l'Archevêque, although largely restored, shows that the towers were solid up to first floor level at least, and also that they are not added on to the main walls of the enceinte. A postern gate in the south wall is still visible, its jambs made from blocks or stone, some showing decorations and obviously re-used from an earlier monument (fig. 29).

One of the best preserved examples of a circuit of late Roman walls is to be

29 Postern gate in the late Roman walls at Tours

found at Le Mans (pl. 1–3).³⁰ Basically rectangular in shape, it is nonetheless adapted to the contours of rising ground on the east bank of the Sarthe. The walls, though 4m thick at their base, decreased in width towards the top. They were founded on a base of large stone, some re-used, or on roughly coursed limestone rubble, and built of small blockwork neatly squared in local limestone, sandstone and ironstone. Triple tile-courses occur throughout, though in the upper and lower reaches of some of the towers, irregularities occur. Putlog holes, used in the construction, occur regularly in the middle of the tile courses. The mortar contains much crushed tile, giving the walls a pinkish hue.

Many of the U-shaped towers which project from the walls are still traceable, and two in particular, the Tour de la Magdeleine (pl. 1, fig. 15, p. 41) and the Tour du Vivier, still retain two stories of windows, at the height of the rampart walk and in the storey above. At rampart-walk level, a broad band of tiles, some six courses thick, runs right round walls and towers.

The site of the main gate lies in a re-entrant angle of the walls. It is buried in the cellars of a modern house, and was a single portal between two projecting semicircular or perhaps polygonal towers. Three other surviving gates on the west side are posterns. The best known is the Grande Pôterne (pl. 3), a wide single opening high up in the wall. The others lie at river level, at the base of the wall. One was below water level when found, and preserved the remains of waterlogged planks as its sill beams, and also the arrangements for barring the gate.

A notable feature of the walls of le Mans is the patterning in different coloured facing-blocks still easily visible near the emplacement of the demolished Tour Huyeau, on the Tour Magdeleine, and on a series of other stretches of curtain wall and towers on the west side (fig. 14, p. 39, pls. 1–2). Alterations to the patterning at various points suggest the work of different building gangs. One very clear break is seen at the site of the Tour Huyeau itself. North of this tower, a length of 60m, which included two towers, was built by men who used a consistent pattern.

Dating evidence from the walls is slight. Some sculpted stones but no inscriptions have been found in the wall, but these are not closely dated. At one point, the wall foundations, which are nowhere very deep, overlay a cremation burial in a rather characterless grey-black cooking pot, dated by its excavators to the late second or early third century.

The late Roman walls of Rennes[31] stand on the north bank of the river Vilaine, at the heart of the modern city (fig. 30). None of the circuit is visible today, but excavations have exposed various lengths of wall-foundation, particularly on the river bank. Above timber-piled foundations several courses of unmortared schist were laid, all originally below Roman ground level. On this base two courses of blocks of stone without mortar, containing re-used material— column drums, bases and milestones—were set. Above this, there ran a chamfered plinth, and a tile course which evens out any irregularities, then the small blockwork facing of the wall interspersed frequently with triple tile-courses. The close spacing of red tiles, together with the pink coloured mortar, earned the city the name 'Ville Rouge' in the Middle Ages. Putlog holes in the wall were usually located within the tile courses. There are also signs of patterning in the facing similar to that at Le Mans.

A postern gate, formed of large granite blocks including a substantial group of re-used milestones, was located in 1890 and re-excavated in 1968. In the group were milestones of several emperors of early to mid-third-century date, the latest of which are three stones which can be definitely assigned to Tetricus (270–3).

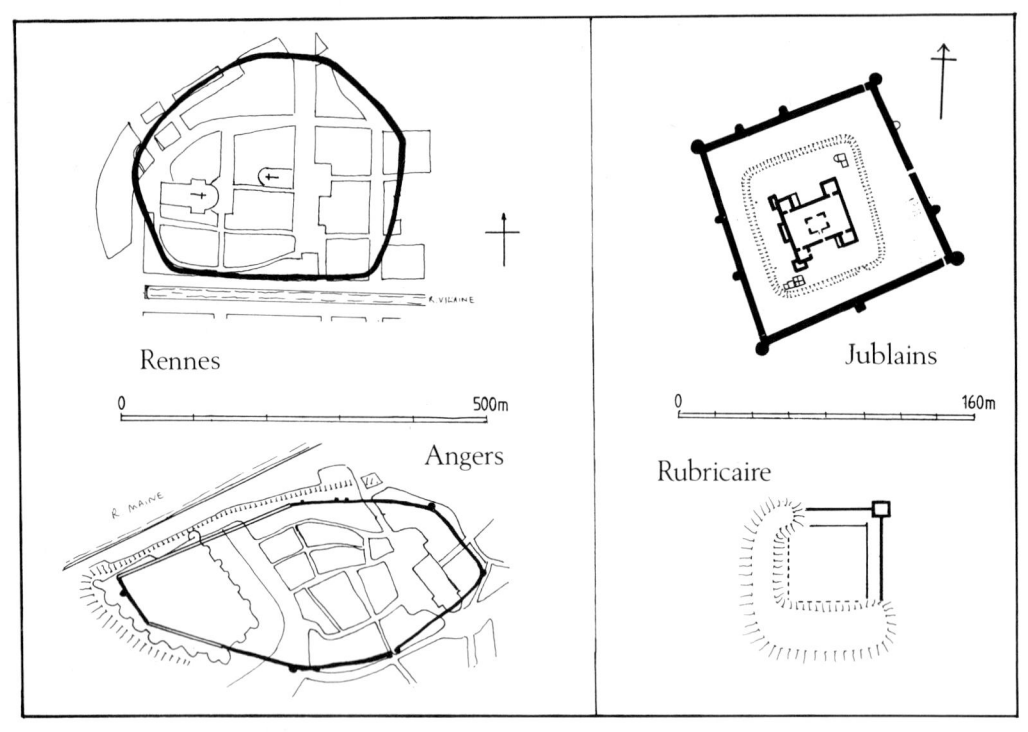

30 Late Roman walled sites in Lugdunensis III (*scale 1:10,000 and 1:4,000*)

This evidence strongly suggests that the walls were built in the years immediately following 273.

At Angers (fig. 30),[32] the Roman walls ring what is virtually a city acropolis, later occupied by the Château of the Ducs d'Anjou on the south and the Cathedral on the north. Where the walls were not founded directly on the rock there was a foundation course of large re-used blocks. The facing of the wall was of small blockwork, usually of three courses between double or triple tile courses. At the western end, the late Roman wall was built against the outer face of a pre-existing building. Semicircular towers, one 6.20m in diameter, project from the walls at intervals, and a series of small rectangular external towers have also been noted. The plan of the Porte de la Vieille Châtre, one of originally four gates, has been reconstructed by work in the cellars of houses on the south of the circuit. Its flanking towers were U-shaped and lay astride the wall. There is no close dating evidence from the walls themselves, although several undated inscriptions as well as other reused material have been found.

The walls of Nantes[33] enclose a pentagonal area, defended on three of its sides by the Rivers Loire and Erdre (fig. 28). Most of the circuit is now destroyed, but some indications of the original wall-structure have been revealed by excavations. A single course of large stones, including some re-used material, served as a foundation. The superstructure was slightly offset, and had a facing of small blockwork with triple tile-courses. On the site of one of the Roman gates now stands the medieval Porte St Pierre: the Roman gate was a single portal flanked by a pair of U-shaped towers astride the walls. Along the nearby curtain wall, three U-shaped projecting towers have been discovered; they were spaced at intervals of 30m. Evidence for dating the walls comes from re-used stones in the foundation course, of which two were milestones. These confirm a construction date for the walls after 275.

The kite-shaped circuit of walls at Vannes[34] was built in a position to dominate the harbour (fig. 28). The line of the walls is followed in part by the medieval circuit, and the site of the Roman harbour has silted up.

The Roman wall, of small blockwork and widely spaced triple tile-courses, was built on a foundation of large blocks levelled off to form a flat platform. Traces of projecting U-shaped towers have been recorded, and at least one of the towers of the medieval enceinte stood on Roman foundations. No dating evidence has been found from the walls, which contain surprisingly little re-used material.

The promontory of Aleth[35] at St Malo, has been the subject of considerable archaeological research in the last ten years. On the promontory itself lay the Roman and medieval towns (fig. 27). The medieval castle site, slightly south of the town, and perched on a separate small promontory, was in Roman times the site of a small *castellum*.

Fragmentary walling hugs the edge of the promontory. The towers appear to have been hollow, large and rectangular, possibly more like buttresses than defensive towers, and at least one round tower has been distinguished on the landward side. Here the main gate must have lain. There is no clear indication as to the date of the walls. The *castellum*, by contrast, was small and connected with the harbour to the south of the promontory. Portions of its walling have been distinguished within the medieval masonry of the Château walls. The Roman work has triple tile-courses. Pottery finds and coins in the main levelling layer inside the fort suggest that it was first occupied in the 340s.

Jublains,[36] recorded in the *Notitia Galliarum* as *civitas*, was in late Roman times the site of a small strongpoint, which was largely excavated in the last century (fig. 30). There were three separate elements of the defences. At the centre was a building of H-shaped plan, built of massive masonry with four square towers round a rectangular courtyard. This was protected by an earth bank. Outside this there ran the rectangular enclosure of walls, 4.5m thick, covering in all only just over 1ha. The wall was of rubble concrete faced with small blockwork interspersed with frequent tile-courses. The towers, except for one on the south side which was rectangular, were U-shaped and projected from the wall. The main gate was probably in the east wall. Other smaller gates, no broader than posterns, lay in the shadow of towers on the other walls. Modern replacement, in the cause of consolidation, of much of the Roman wall-facing has made detailed study of the constructional features impossible. There is no independent dating evidence.

The medieval castle at Brest[37] stands on a spit of land jutting into the harbour, and fragments of Roman masonry within the medieval walls suggest that the site was originally that of a Roman fort. Because of the present-day military importance of the site, access to the castle is severely restricted, but on the landward front there is an outer face of Roman-style walling with triple tile courses. The outer walls of the Château form a trapezoid shape, which was probably that of the Roman wall-circuit.

The *civitates* in Lugdunensis III, of which these, with the unidentified Civitas Coriosolitum and the Civitas Ossimorum, (if it is not Brest) form the total as listed by the *Notitia Galliarum*, provide a rich field for detailed study. It is by no means as easy to draw rapid conclusions from the study of this provincial group of defences as it has been in the case of the two former provinces. The defences do not show any marked similarity of plan, but there are common characteristics which when viewed against the backcloth of other regional traits, are clearly identifiable. The overall design of the walls, Tours and Vannes apart, is of note. Tile courses are almost as frequent as the cubes of small stone blockwork. In addition, the distinctive patterning has been noted at at least two of the sites. The scaffolders' putlog holes are normally sited, not as elsewhere in the small blockwork, but actually in the tile courses. These three points of detail are sufficiently distinctive to suggest that the defences in this province formed a group built to a standard pattern, given an individual interpretation by a common builder or designer.

The construction of defences at coastal sites within Lugdunensis III may also have been part of a military design.[38] The protection of these coastal areas was of considerable importance in the late third and fourth centuries, and it is possible that ports in this area were circled with defensive walls at an early stage to give back-up protection to areas of the Gallic mainland more immediately exposed to Frankish and Saxon raids. The more exposed parts of Gaul and Britain—the Channel straits and those areas on their immediate proximity—were protected from a surprisingly early date against such pressures, but the construction of a series of heavily defended bases to defend the straits and the approaches to them was a development of the late third century (see p. 201). As a back-up protection, coastal sites like Aleth, Vannes, Nantes and Brest were supplemented by a few further well chosen sites, Morlaix, le Coz Yaudet and St Brieuc. Outside Brittany, the coastal *civitates* of Avranches and Coutances were probably also fortified as part of these protective measures, though no trace of the late Roman defences of either has been located.

The diminutive scale of the defences at Jublains suggests that it can hardly have ever been considered as anything other than a fort. The question of the purpose of such small defended sites as this is one of considerable difficulty, and a further two sites in Lugdunensis III illustrate this to good purpose. At Rubricaire[39] near Jublains, lies a square *castellum*, apparently in style and plan something like a *quadriburgium* (see p. 27–9) with ranges of casemate buildings inside a wall of moderate thickness round a small square courtyard. Projecting towers completed the defences and a bath block lay nearby outside the walls. The reason for siting a small road fort at this point is not clear. In the present state of knowledge, however, it is not possible to postulate a series of such posts along all late imperial highways, and its original strategic purpose, like that of Jublains, must remain a mystery. Similarly at Larçay[40], on the southern bank of the River Loire near Tours, there is a small fort with thick solid walls and projecting corner and interval towers. It encloses a mere 0.3ha, though the northern wall is missing, and was clearly of diminutive size. It stands some 20m above the river bank on a small steep bluff, and its walls contain a substantial proportion of reused Roman material. The purpose of this small enclosure might have been to provide a private fortification for a local villa-owner or his tenants—or even a refuge for a small *vicus* or other community nearby.

The fourth group of Gallic fortifications which appear to exhibit certain common characteristics is to be found in Belgica Secunda.[41] In the Notitia Galliarum, 12 *civitates* are listed under this province: of these, five, Châlons-sur-Marne, Vermand, Arras, Cambrai and Thérouanne have revealed little or no information above their late Roman defences. The evidence from the others is patchy, but in the case of three of the sites, Beauvais, Senlis and Soissons, enough survives to enable some meaningful comparisons to be made.

At Beauvais[42] the walls form a polygon, enclosing an area of approximately 10ha around the Cathedral in the centre of the modern city (fig. 31). The site was enclosed by a ditch, part of which still channels a small stream, the Thérain, and preserves the distinctive outline of the wall circuit on its western side.

In detail, the walls were 2.5–2.6m thick, constructed carefully in a single building operation. The base consisted of three courses of soft limestone blocks, which include some sculpted and reused stones, above which, sometimes offset, the wall was built of rubble concrete poured in layers between an inner and an outer facing of carefully built small blockwork. Double tile courses occur regularly at intervals of every twelve courses of small blockwork. They normally run through the whole thickness of the wall. The putlog holes occurred exclusively in the row of small blockwork immediately above each double tile course (pl. 4). They were therefore 1.65m apart vertically, but their horizontal spacing varied considerably.

The towers were U-shaped apart from the square corner towers, and lay astride the wall, with a slight rectangular projection of the inside face. On the north side, the Tour de St Hilaire which contains a single window at the approximate height of the rampart-walk still stands. This window does not itself appear to be Roman, but the Roman window has been blocked, and was a little higher than the one at present visible. Rebuilding has obliterated any other trace of windows in the tower.

The oval plan of Senlis[43] differs completely from Beauvais (fig. 31). The wall enclosed approximately 6.5ha around the most elevated portion of the present-day town. On average, it was 7m high to the rampart walk, and 3.30m thick. The foundations were composed of several courses of large stones, placed without mortar. Above these was a wall of rubble concrete, poured in layers between a double facing of small, carefully shaped stones. There were triple tile-courses up to the height of the rampart-walk. Above this height they were exclusively double, as they were also on the outside face of a well preserved tower in the Château Gardens.

All the known towers were U-shaped and lay astride the wall. One of the towers has a complete frontage of two stories of three windows at rampart-walk

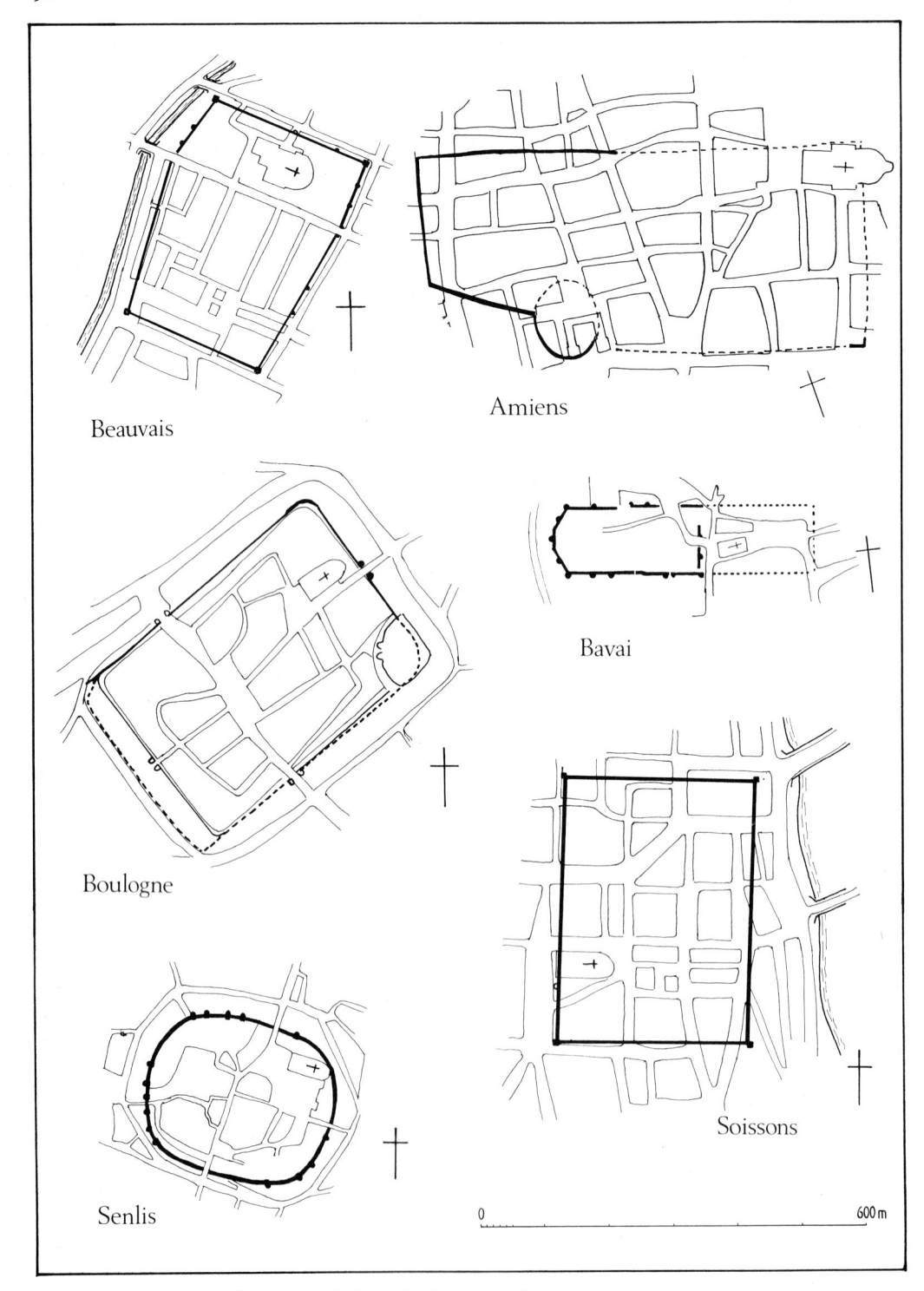

31 Late Roman walled sites in Belgica II (*scale 1:10,000*)

level and at the level above. A single window at a corresponding height in both upper stories looks inwards to the city and doors lead out of the sides of the towers at a slightly lower level to the rampart-walk. There were probably six gates. The Porte de Reims had two flanking towers; but another, the Porte St Rieul, facing northwards, had only narrow rectangular butresses which projected internally. The four gateways had a single tower defending a simple opening.

Little of the city wall at Soissons[44] has survived (fig. 31). The ground plan was rectangular, and the defended area seems to measure 400 by 300m, an area of 12ha. The thickness of the wall has not been recorded; its foundation was composed of three or four courses of large stones, some of which were re-used material from other monuments. Above this was the layered concrete core faced by small blockwork of square and rectangular shape. The tile courses in the visible portions are triple, but some sources record that they were also double. The towers at the corners of the wall-circuit were approximately 10m square. Projecting interval towers were probably U-shaped.

A number of similarities between the defences of these three cities are highlighted by these brief descriptions, despite the differences of plan, which spring from differing strategic usage of the land. The facing is uniformly of well dressed limestone with double or triple tile-courses: a distinctive feature is the use of rectangular dressed stone at the angles and to cap the putlog holes. Other common characteristics are the positioning of the putlog holes immediately above the tile courses, and the tooled finish given to the small blockwork. The square shape of the corner towers at Beauvais and Soissons is unparalleled elsewhere in city or town fortifications in the late Roman west. U-shaped towers astride the walls are not common elsewhere among late Roman walled sites: most of those in Gaul are semicircular or U-shaped, and project solely from the front face of the wall.

The combination of these similarities of style and design points so comprehensively to the suggestion that this group of late Roman defences was built by the same architect that the few points of difference, notably the thickness of the walls and the diversity of ground-plan, have little relevance. It is fortunate that so much of the walls of the three sites in question survives to enable such close comparison to be drawn. At Amiens,[45] for example, where the walls enclosed an irregular quadrilateral (fig. 30), based on the amphitheatre, extensive portions of the southern and western walls of the circuit have been excavated, but at no point has the exterior wall facing been found. No towers, therefore, are known, apart from one large square one reminiscent of the corner towers of Beauvais and Soissons which protected one of the main entrance passages. The constructional style of these walls seems very similar to that of the three cities already briefly described: for direct and detailed comparison, however, too little has survived.

At Boulogne and at Reims, the position is similar. The Roman walls of Boulogne[46] enclosed a rectangular area of about 12ha round the present Ville Haute (fig. 31), and occupied the site not only of an earlier Roman defensive circuit—possibly a fort of the *Classis Britannica* (British Fleet)—but also of the medieval walls. These have so completely encased the Roman walls that no trace of the exterior Roman wall-face has been recognised. Under the medieval castle, however, in the eastern angle, the inside face of the Roman circuit survives to a considerable height. Here too the walling is in a similar style but without tile courses. At Reims,[47] the late Roman circuit, long suspected to have enclosed a roughly oval area, and to have incorporated as gateways within it a number of earlier Roman monumental arches, has now been located and established. It, too, was of carefully finished small blockwork and tile-courses, exhibiting many of the characteristics of the other sites of Belgica Secunda.

At least three other walled sites apart from the 12 cities listed in the *Notitia Galliarum* are also known to have existed within the province. These are of smaller scale than the *civitates*. At Noyon,[48] between Beauvais and Soissons, a small *vicus* was enclosed with defences in the later Roman period (fig. 32). The circuit was roughly circular, and enclosed about 4ha. Although the fundamental principles of construction are similar to those of other cities in the area, the details of finish are far cruder. The stone is not the same crisp clean limestone used at Beauvais and Soissons: it is an inferior brownish sandstone. Its tile-courses are irregular, and incorporate *tegulae* or roofing tiles. This clearly suggests that the *civitates* of the area were given precedence in the use of materials and skilled labour.

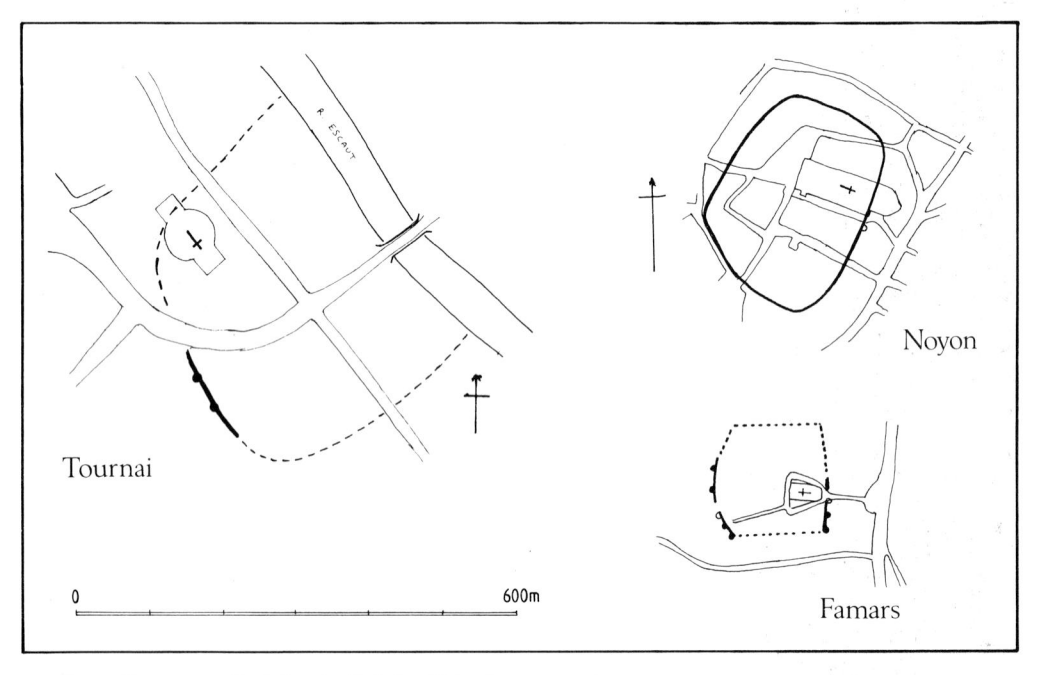

32 Later Roman walled sites in Belgica II (*scale 1:10,000*)

The late Roman walls of Bavai and Famars are more puzzling still. Bavai[49] was the site of an earlier Roman tribal capital, Bagacum Nerviorum (fig. 31). The extent of the earlier Roman town is not known, but the late Roman walls were extremely restricted, and enclosed no more than a small rectangular area of about 4ha round the former *forum*. In fact the walls leant on the forum and its substructure, the cryptoporticus, enclosing it completely. An additional walled area of similar dimensions lay to the east of this, giving the *castellum* (if that is what it was) the peculiar dimensions of 400 by 100m.

The defences of Bavai developed in three stages, as yet not clearly datable. In the earliest phase, the *forum* was used as a defensive structure in its own right. Only the open shop fronts to north and south were blocked off by a wall. A second phase clasped the whole of the forum within its ambit: it had large projecting semicircular towers, and on the eastern side formed an entrance. In the third phase, this wall was nearly doubled in thickness by the addition of an extra outer face and new, more frequently spaced, projecting semicircular towers. At some stage, the additional rectangular area to the east was also walled, but it is not clear to which phase of construction this belonged.

At Famars[50] a small fort with walls also of two periods was built around some of the structures of an existing *vicus* (fig. 32). The area enclosed was small, and the late Roman fortification leant in part against the walls of the former baths.

The problems of interpretation passed by the fortifications at Famars and Bavai are as intractable as those met with over the small sites of Larçay, Rubricaire and Jublains in Lugdunensis III. Famars and Bavai can hardly have been intended as anything other than small *castella*: in the case of Bavai in particular, the original fortification of the site seems to have been an immediate initial response to an emergency. The *forum*, the most solidly constructed building within the original town, was selected to act as a defended post, and given extra precautionary defences. It is unfortunate that dating of the phases of augmented defence in both cases is so imprecise. It would make best sense for them to be seen as a relatively speedy confirmation of the defensive usefulness of the posts at the two sites, rather than renewed defence in the Diocletianic and Constantinian periods as has sometimes been claimed.

The scope of operations of this group of engineers and architects in Belgica II may have extended also beyond that province. Of the sites whose remains can still be studied within Lugdunensis Quarta, (sometimes known as Senonia) at least one, at Sens, exhibits exactly parallel traits of building style. At Sens[51] the line of the Roman walls and ditch is followed by a tree-lined boulevard, making the small parts of the walls now visible very easy to trace (fig. 33). The wall was based on large blocks often of re-used stone, up to ten courses high. On this foundation, the superstructures faced with small blockwork and triple tile-courses, usually 12 wall-courses apart was built (pl. 5). One projecting semicircular tower still stands to a considerable height. The main gates have not survived. Dating evidence for the walls is provided by coins found embedded in them, which include those of the emperor Saloninus. This establishes that they were built after 268. The most recent inscriptions found in the re-used material at the base of the walls were of early third century date.

By contrast with Sens, the remainder of the late Roman defended sites in Lugdunensis Senonia were small and have left only slight trace. At Chartres,

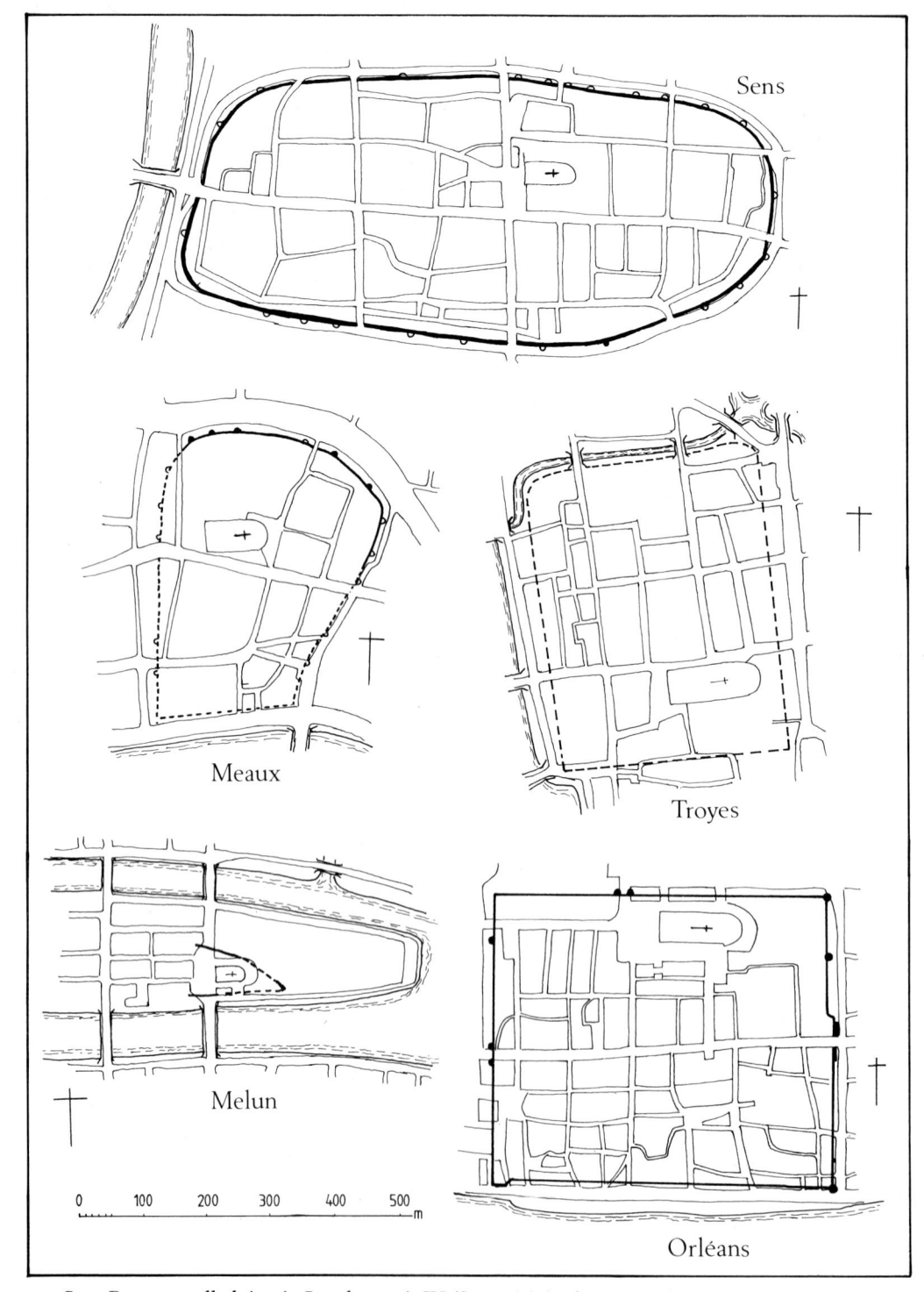

33 Late Roman walled sites in Lugdunensis IV (Senonia) *(scale 1:10,000)*

Troyes, Melun[52] and to some extent Paris (figs. 33, 35), the site of the walls is known or suspected, but no trace is now visible. At Paris,[53] the late Roman walls occupied only the Ile de la Cité within the Seine, leaving by far the greater part of the former large and expansive city without protection on the southern bank of the river (fig. 34). Contemporary descriptions show that this picture, confirmed by archaeology, was a fact in the late Roman period. The affection which the emperor Julian held for the city is therefore surprising.[54]

Only a little more survives at the other three *civitates* within the province. At Auxerre,[55] the site of the Roman enceinte occupies a roughly polygonal enclosure (fig. 35), inside which the Cathedral now stands, raised about 50m above the river Yonne. The walls were built on a foundation of large stone blocks above which was the normal facing of small blockwork and tile courses. The sites of four gates and two posterns are known. The projecting towers were probably semicircular in plan. Inscriptions found in the wall suggest a date in the later third century for the wall's construction, though a coin of Constantine was found in a layer of mortar near the wall and has been thought to date it to *c*325.

The walls of Orléans[56] form a regular rectangle (fig. 33), and though certain of the towers are still traceable within modern or medieval buildings, there is little sign of their Roman wall-facing. The walls, however, were in the usual blockwork facing. There is a tradition that the walls of Orléans were built by Aurelian, from whom the *civitas* received its name; originally it had been Cenabum, but the construction of the defences perhaps provided a likely context for the change. There is no accurate archaeological dating evidence, though many sculpted stones of second and third century date have been removed from the wall.

The Roman wall circuit at Meaux[57] was roughly D-shaped (fig. 33) with its flat side parallel to the river Marne. The facing of the walls now visible displays a high degree of modern cement-work. but the small blockwork and tiles appear in part to be original. Triple tile-courses are separated by seven rows of small regular blocks of stone, but this is not of the crisp limestone of the walls of Sens.

The walls of the *civitates* in this province are difficult to claim as members of a coherent group, and there is little or nothing among their various features which allows of any close comparison. The similarity between the group in Belgica II to Sens and possibly to Auxerre and Orléans may suggest that these defences, too, were the product of a single designer or builder. Study is hampered here by the lack of surviving remains.

The frontier provinces, Germania I and II with their chief cities at Köln and Mainz, and the province of Belgica I, based on Trier, contained a number of small fortifications as well as a few *civitates*. These will be better dealt with and discussed under the consideration of the Rhine frontier.[58] The other Gallic frontier province provides no coherent pattern. In Maxima Sequanorum (roughly present-day Switzerland), the *civitates* named within the *Notitia Galliarum*, include Besançon, Nyon and Avenches, in none of which have specifically late Roman defences been located. The *Notitia* also lists a series of *castra* within the province—at Windisch, Yverdon, Horbourg and Basel. These have revealed

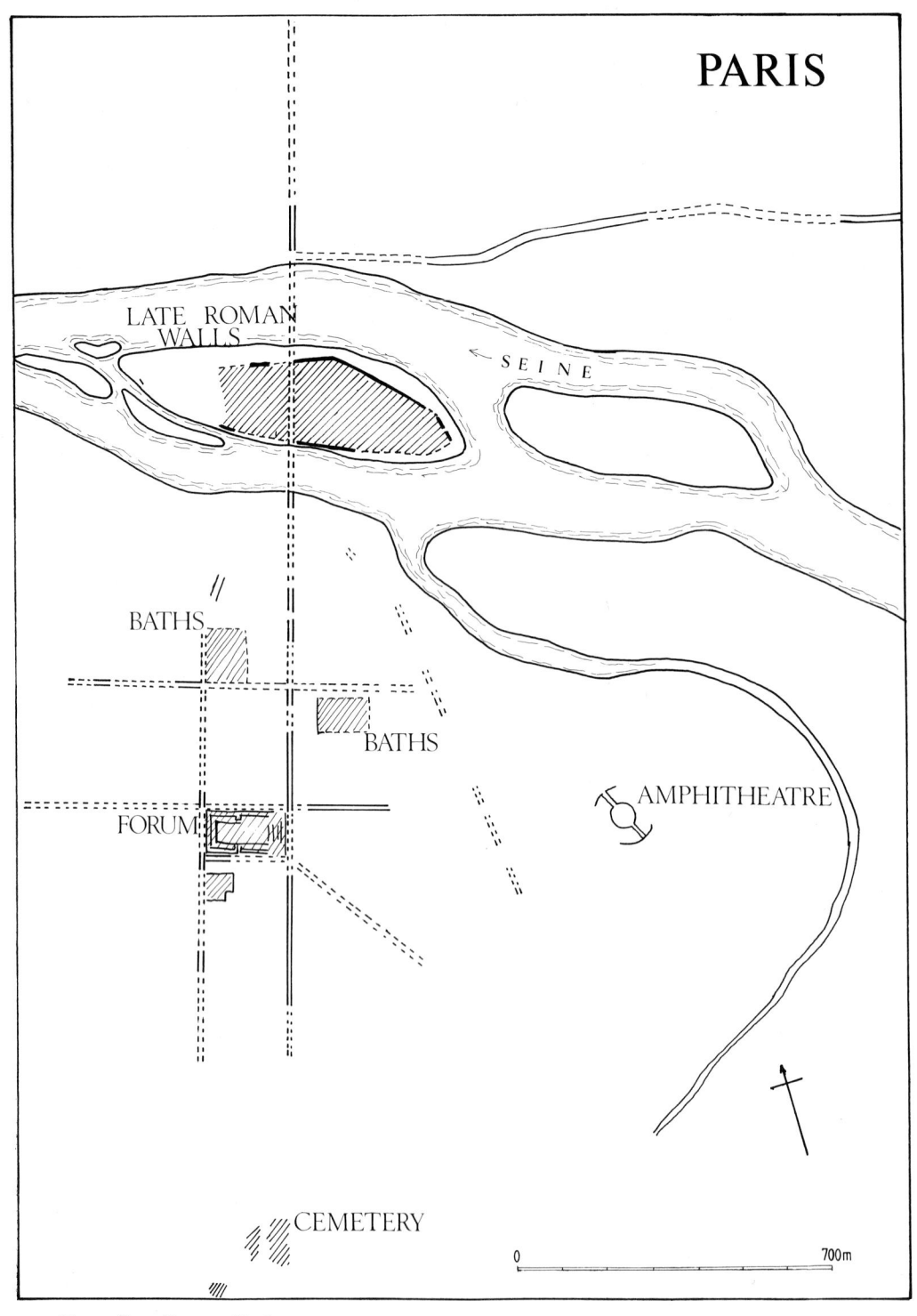

34 Plan of late Roman Paris

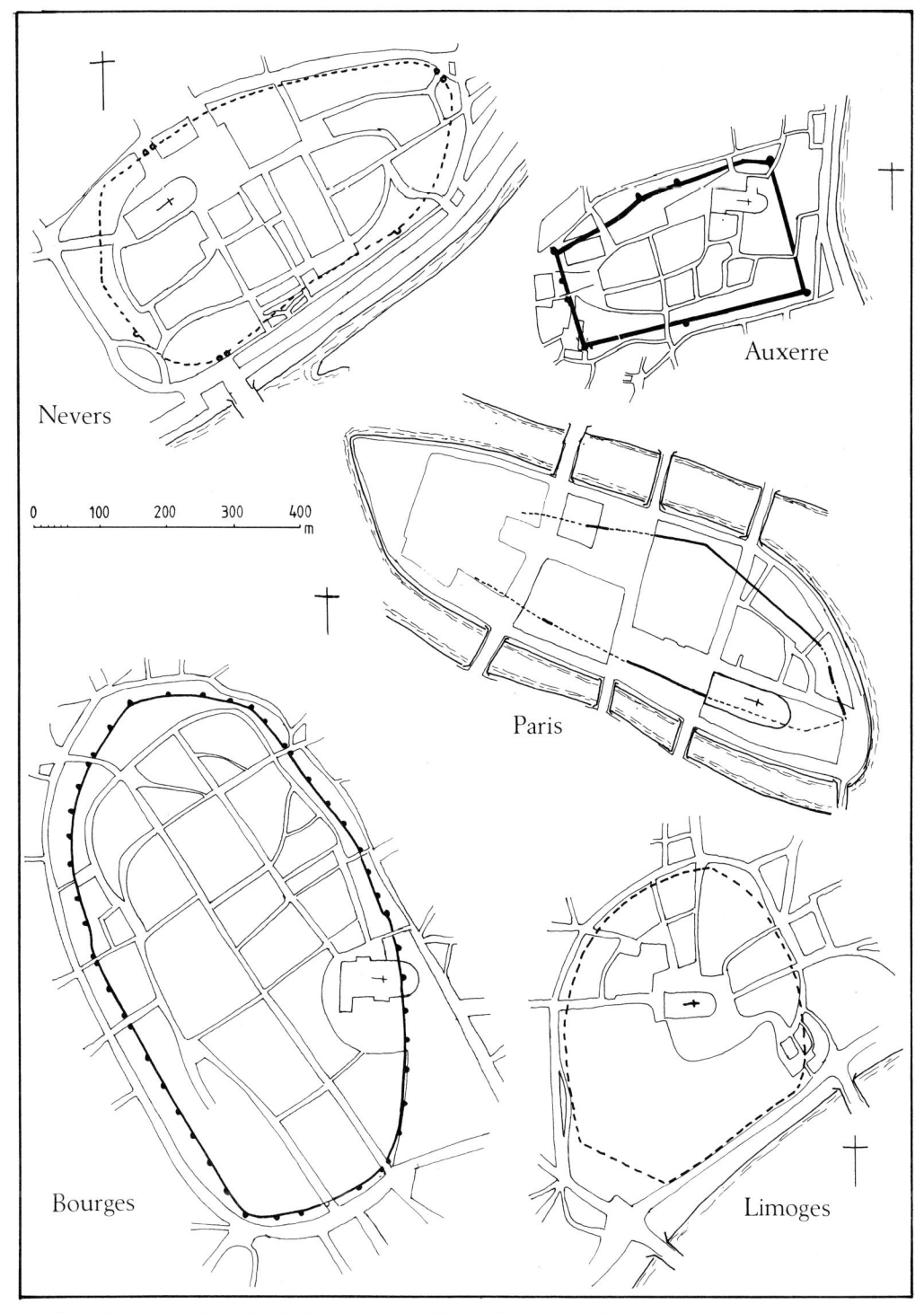

35 Late Roman walled sites in Lugdunensis IV and Aquitanica I (*scale 1:10,000*)

traces of late Roman fortifications but are also best dealt with as part of the consideration of the Upper Rhine frontier.[59]

Further inland in Gaul, a number of the southern provinces have produced virtually no trace of late Roman fortifications at any of the named sites in the *Notitia Galliarum*. This holds good for Graiae et Poeninae, Alpes Maritimae, and Narbonensis II. For Viennensis and Narbonensis I there is in each case an impressive list of sites. A great number of these, however, represent Roman towns of Provence, where defences were not constructed in the late Roman period, but in an earlier era. In Narbonensis I, for example, there is only very fragmentary evidence for rebuilding of the defences at Narbonne[60] and Béziers,[61] the wall circuit at Toulouse[62] is vast, and may not date from this period, at Nimes[63] the walls are definitely Augustan (fig. 2, p. 14), and the four other named sites have produced no evidence. Only at Carcassonne, ironically not mentioned in the *Notitia Galliarum*, is there a circuit of late Roman walls, and this stands to a considerable height (see p. 112).

There is a similar picture in Viennensis, where the walls of Vienne (fig. 2, p. 14)[64] and Arles[65] have produced some evidence for rebuilding and refortification in the late period (in the case of Vienne, of considerable complexity), but at all but three of the other named sites is there little or no evidence (fig. 36). The three exceptions are Geneva,[66] where the late wall circuit occupied a plateau of irregular shape whose full circuit is quite well known, Grenoble,[67] where the plan of the wall circuit describes a tight circle bustling with projecting towers, and whose gates are dated by inscriptions of the Tetrarchic period, and Die,[68] where in a surprisingly remote location an expansive circuit of late Roman walls enclosed the settlement.

Of the cities in Aquitanica I, few retain any trace of their late Roman walls. Slight remains have been located of Limoges and at Rodez, but the major site of importance is the capital of the province, Bourges.[69] In style and size, at 26 ha, the walls of Bourges compare very closely with those of Sens (fig. 35, p. 103). The regular cubes of small blockwork, ten or twelve between each tile course, form a distinctive architectural style. Research has shown that the walls lean against or incorporate portions of earlier monumental buildings belonging to the town—in particular a portico and a large fountain. Projecting towers, hollow, and provided with large U-shaped windows at first floor level where they survive to this height, are spaced at regular intervals round the circuit. No trace of gates now survives, but engravings of the Porte de Lyon at the south-eastern corner, drawn before its demolition in the ninteenth century, show flanking towers of large monumental blocks apparently decorated with engaged pilasters. This is remarkably similar to other surviving gates, notably at Périgueux (the Porte de Mars, fig. 39) and at Die (the Porte St Michel, pl. 8).

Dating evidence for the construction of the walls is meagre. Deep occupation layers piled up against the back of parts of the walls produced coins of the later third century and pottery of broadly similar date. This need have no connection, however, with the construction of the walls themselves, for which no direct dating evidence has been recorded.

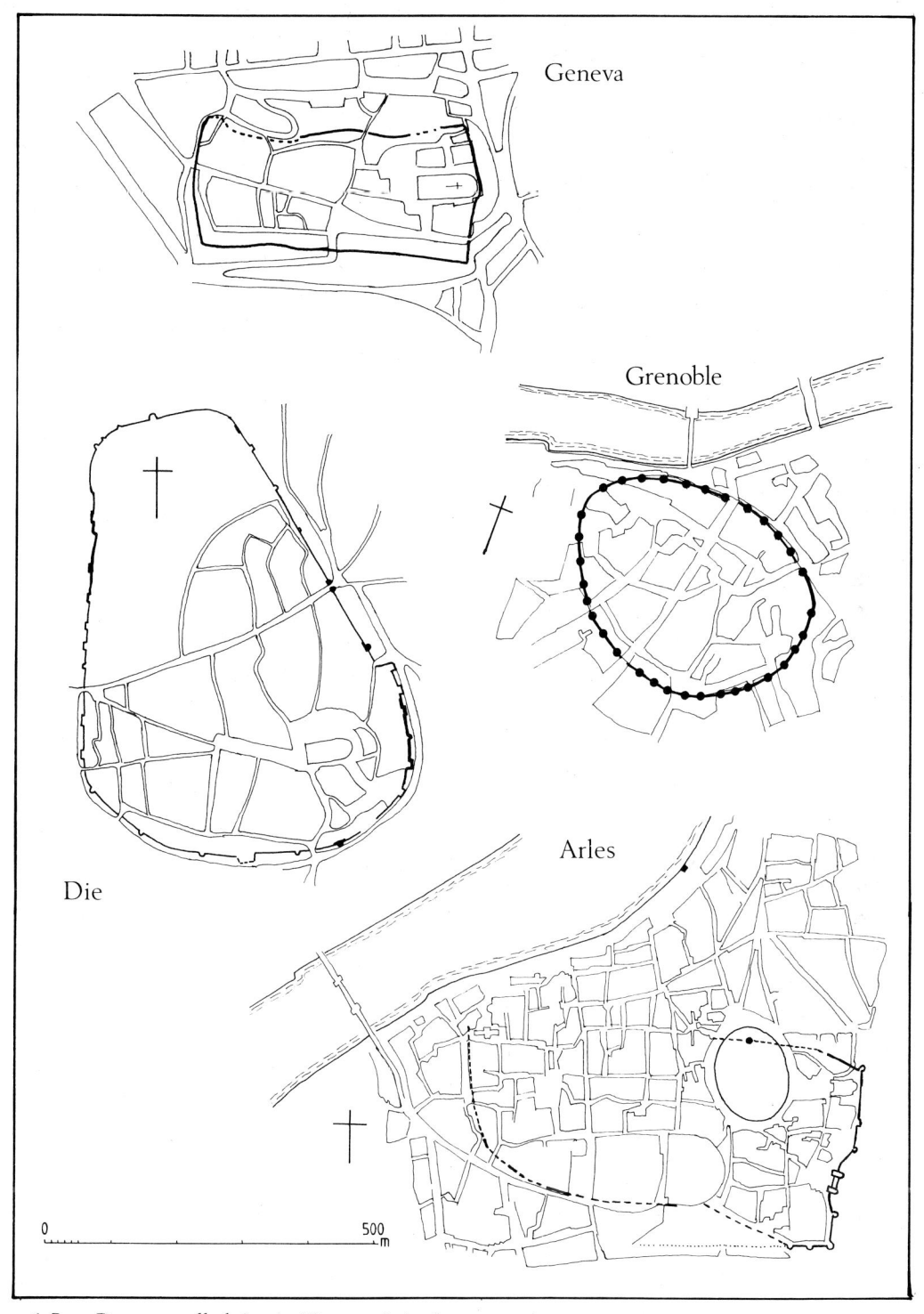

36 Late Roman walled sites in Viennensis (*scale 1:10,000*)

In only two further Gallic provinces do the sites preserve enough evidence to make some form of comparison viable. In Aquitanica II, a further group of defences with some common characteristics can be identified and recognised. The provincial capital was Bordeaux[70] where the late Roman walls are known (fig. 37). The circuit was large and rectangular, of normal rubble concrete construction with a wall-facing of small blockwork and widely spaced tile-courses. There is an unusual preponderance in the base courses of re-used large stone. In places this reached a height of 6m or so before the facing converted to small blockwork. The walls were 4–5m thick, and semicircular towers projected at regular intervals of about 50m. Ausonius portrayed Bordeaux as a city of foursquare walls, broad streets and spacious houses, gates set opposite each other in the circuit, and divided by a stretch of the river on which ships could easily sail.[71] Modern research has broadly substantiated this picture. A series of coins found in the city walls, the latest one a mint issue of Claudius Gothicus, suggests that they were built after 268–70. Re-used within the lower courses are many inscriptions, the latest of which dates to 258.

Though little remains of the wall-circuit at Saintes,[72] the site of the city of late Roman times lies on the bank of the River Aunis, enclosing part of the area known as the Capitol (fig. 37). The walls were founded on a basis of large blocks of stone, mostly re-used from earlier buildings. Finds from the wall-core include statues of first- or second-century date. No trace of wall-facing above these blocks has been found, and it seems clear that the large stones formed a substantial height of walling before small blockwork began above. One of the latest finds from the walls was a re-used inscription of early third-century date.

The late Roman walls of Périgueux[73] enclosed a relatively small but slightly elevated area of the city, incorporating the amphitheatre as a large defended bastion (fig. 37–8). Outside them lay all of the large temple complex dedicated to Vesunna, and much of the rest of the city. The walls were exceptionally thick and strong, and were composed of much re-used material, including a very high proportion of large stone blocks. On the west side, which incorporates the best preserved portion of the circuit, the walls stand to a height of 10m, of which all but the top 3m was composed of re-used stones placed without mortar (pl. 6).

The towers were semicircular and projected externally. Several gates have survived: the main gate was the so-called Porte de Mars, which faced east (fig. 39). This was a single passageway between two semicircular towers built of large blocks of stone shaped to include moulded pilasters and cornices. Other gates were built in similar style and defended by a tower.

Too little survives of other walls in the province to make meaningful comparison. At Agen, the late Roman site has not been identified. Fragmentary walls have been found at Angoulême,[74] and at Blaye, which was the site of a fort, the late Roman site has been thoroughly built over by a seventeenth century artillery fort. The only other known walled *civitas* within the province lay at Poitiers,[75] where the broad sweep of the wall-circuit of late Roman date has been identified. The published accounts of the finding of this wall lack precision, and are much confused by discussion of a so-called 'double wall'.

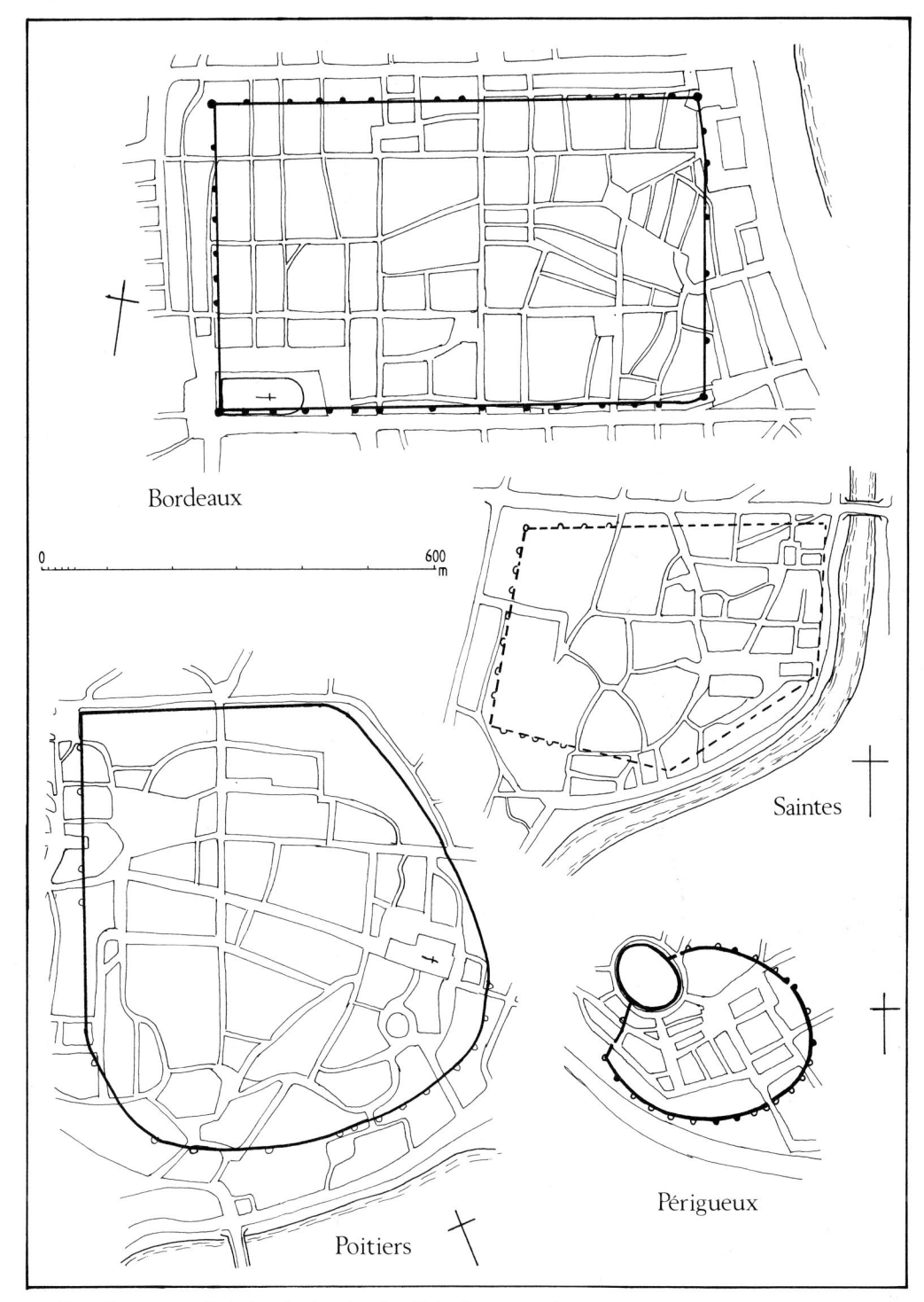

37 Late Roman walled sites in Aquitanica II (*scale 1 : 10,000*)

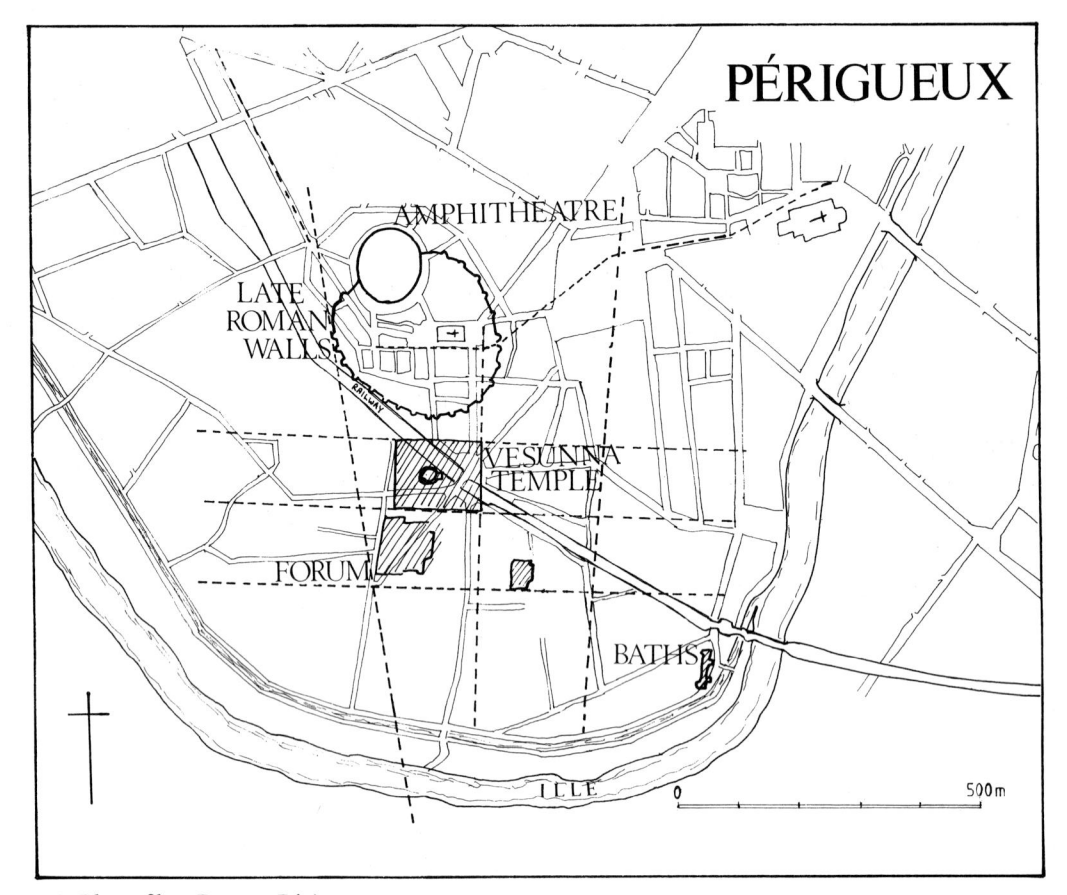

38 Plan of late Roman Périgueux

South of Aquitanica II lay the province of Novempopulana. Within this small but mountainous area within the foothills of the Pyrenees, there lay twelve Roman cities, according to the *Notitia Galliarum*. These were the chief cities of small tribal territories, and this status is reflected in the form of the late Roman defences. As in other provinces a number of the sites have revealed no trace of their late Roman walls: at Lescar, Aire, Bazas, Tarbes, Boas, Oloron and Eauze, the Roman walls are not known, though in some cases medieval defences have now enclosed the towns, probably on the site of Roman predecessors.[76]

The defences which can still be traced include those of Auch (fig. 40), the capital of the province.[77] The walls encircled a small but prominent hill: their facing was of small blockwork and widely spaced tile-courses. Semicircular towers projected at intervals from the walls, whose lowest foundation was in large blocks of stone. The full circuit has not been traced.

The late Roman walls of Dax[78] once enclosed a roughly rectangular area in the centre of the modern town (fig. 40). Substantial portions of them were still standing in the mid-nineteenth century, and their partial destruction caused a great furore at the time among archaeologists in France and Britain. Observers who saw the circuit before 1856 record that it was built of small blockwork and

39 The Porte de Mars at Périgueux

tile courses, with the tiles becoming a less frequent feature towards the top of the wall. The towers were U-shaped and projected from the walls. One of those visible today appears not to be bonded into the curtain. Other towers, now demolished, were hollow and contained doorways affording access from the interior of the town, and a postern gate built of large slabs of stone with a tile relieving arch is closely guarded by another nearby tower.

No details of the gates are known, but some rather unusual architectural features have been found in the walls: drainage outlets of worked stone, possibly re-used, have been recorded at various points, and there are also long rectangular stone blocks forming slits near the gates which might either be for drainage or for ballista fire, but whose purpose is still far from certain (fig. 41). The walls were constructed in usual fashion, with a concrete core between a double facing of small blockwork, and the whole wall rests on a base of timber beams and stakes. Little re-used material has been found in the walls, and there is a report, of dubious value, of the discovery of a coin of Magnentius from the wall mortar. There is an inscription from the walls which is very fragmentary, but may be that of a milestone, possibly of a late third century emperor.

The small town of Lectoure[79] lies on a hill-top encircled by the fragmentary remains of medieval walls which followed a roughly elongated oval. These defences are probably basically Roman since tile-courses and small blockwork are still visible at one or two isolated places. Re-used Roman material has also been found in the wall core.

The low lying city of St Bertrand de Comminges,[80] which was filled with large buildings in the first to third centuries AD, is dominated by a hill on which

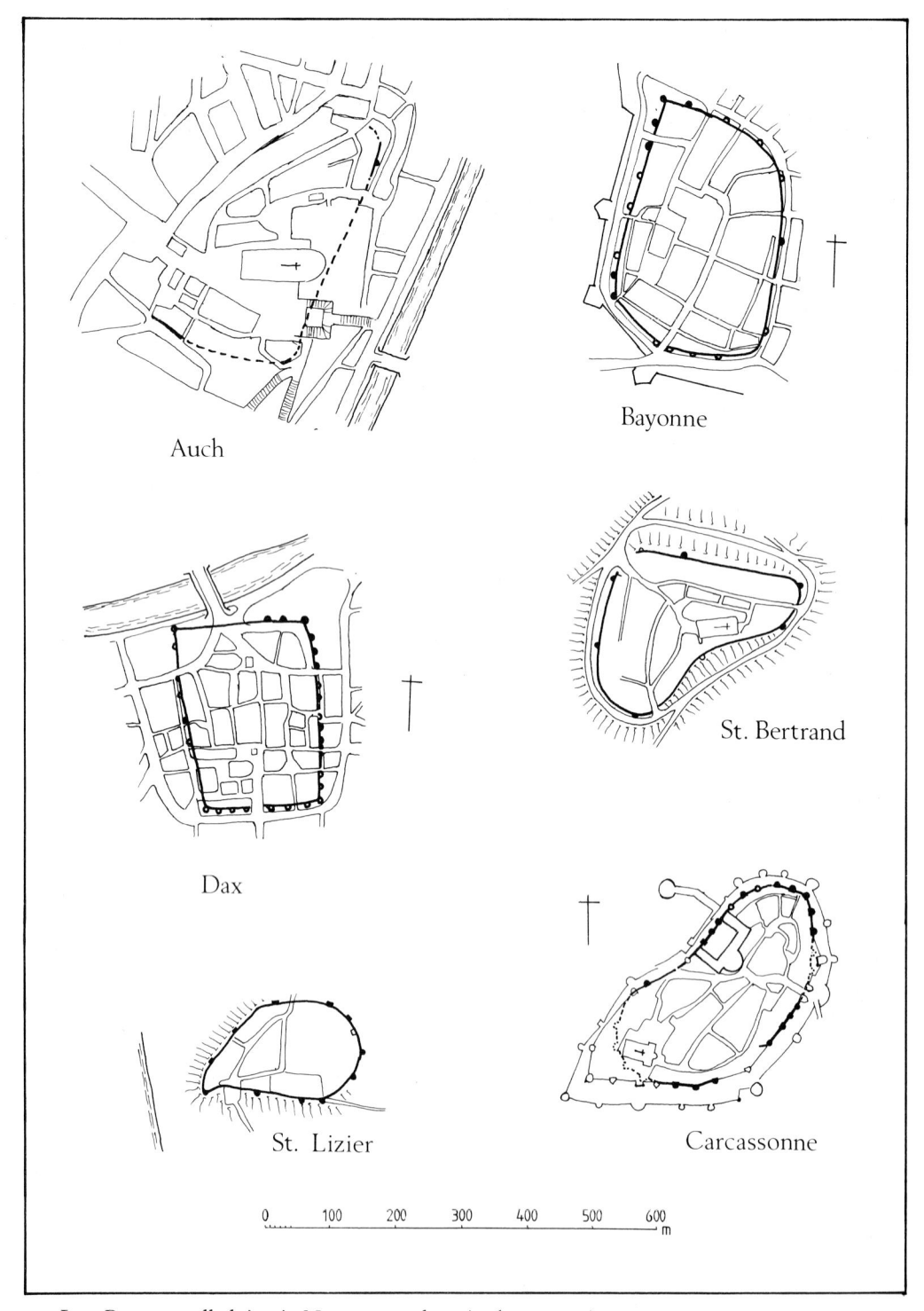

40 Late Roman walled sites in Novempopulana (*scale 1 : 10,000*)

Opening in the wall at Porte St. Vincent.

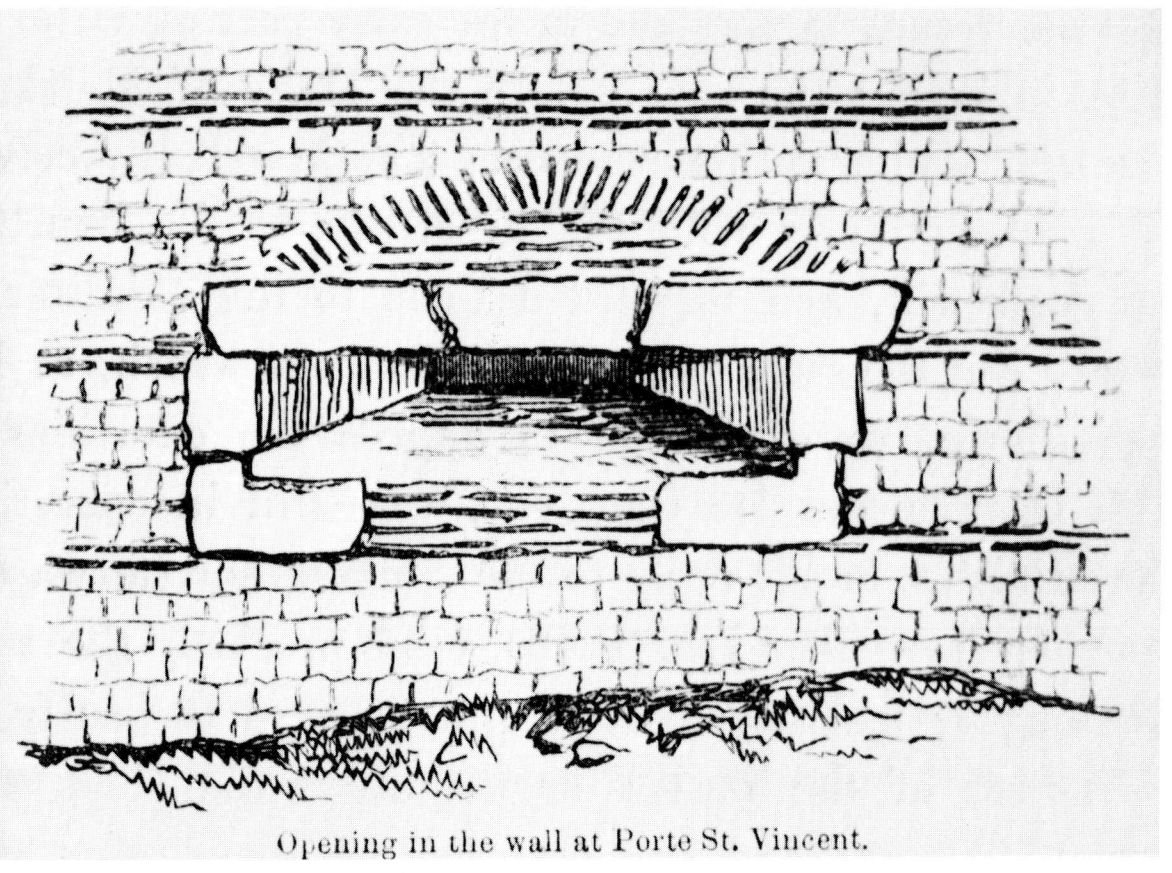

41 Dax, opening in the walls at the Porte St Vincent

stand the remains of a circuit of late Roman walls of typical construction (fig. 40). They are of grey granite, with irregular courses of tile at intervals, and enclose a small roughly triangular area at the very summit of the hill, precipitous in places. Semicircular towers project from the walls at irregular intervals, and there are three gates. The arrangement of the main gate is not known. Other gates face east and south, and seem to have been defended by a single tower. There is no dating evidence from the walls.

Like St Bertrand, St Lizier[81] is perched on the top of a steep hill overshadowing the modern small town (fig. 40). The late Roman walls, later rebuilt and refaced by medieval circuits, enclose a roughly pear-shaped area, with main gates to south and north. Little of the walls survives, but external towers of both semicircular and rectangular shape project from the walls.

One final site in Novempopulana, which does not appear within the *Notitia Galliarum* is at Bayonne.[82] There is no evidence that it was an important Roman centre before the later period, but because of its strategic importance, on the coastal road from Gaul to Spain, it soon surpassed other sites in the same region in terms of size and prosperity. The town was walled in the late Roman period with defences of typical construction (fig. 40). Little re-used material has been found in the walls, and they were all adapted to the terrain on which they were built. Built essentially of the local stone, and solid grey granite, with tile lacing courses, sometimes replaced by flat slabs of granite which bind the facing to the core, the walls describe an odd D-shaped enclosure on the bank of the river Adour.

This group of towns, the last Gallic group to be defined in this way, shares yet another set of characteristics which give their walls a distinctive stamp. The enclosed area was uniformly small: the circuit formed a kind of citadel, normally perched in a fine defensive position on the top of a hill and dominating its approaches, and in general terms the masonry techniques were similar—the walls normally of blockwork with widely spaced double or triple tile-courses. It is possible that this style of construction was also in use in the neighbouring province to the east of Novempopulana, for at Carcassonne, the surviving impressive wall-circuit fulfils all the criteria of the group.

The late Roman circuit at Carcassonne[83] has been encased within the medieval city which succeeded it (fig. 40). In places, medieval fortification has completely replaced the Roman, but the Roman work and its projecting towers occasionally survives to full height. The formation of the Lices, a level platform between an inner and an outer skin of medieval fortification, has had a curious effect on the Roman walls, whose line is taken by the inner circuit. In many places, the medieval builders have exposed the foundations of the Roman walls, and have had to underpin them with new masonry. This leaves the Roman work exposed, but isolated half-way up the inner wall circuit, with medieval masonry forming an underpinning, and also crowning its top.

The Roman facing is particularly distinguishable by the use of small blockwork and tile courses, which occur irregularly, but usually in double rows. The Roman towers were U-shaped, and lay astride the wall, presenting a uniformly semicircular external face. Some of these now rest on a square base, added to underpin them in the medieval period.

Some towers are well enough preserved to have their upper storey of windows extant, and at least two, the Tour de St Sernin and the Tour de la Marquière (pl. 10), show traces of two and three windows respectively at rampart walk height. In the latter tower, decorative semicircles of Roman tiles appear in one of the horizontal tile courses. The larger windows, approximately 1m wide by 1.50m high, have arches of voussoirs of stone interspersed with double tiles. Several postern gates of Roman date are known: the best preserved is that next to the Tour du Moulin d'Avar (pl. 9). Its jambs and lintel are formed of large stone blocks, and there is a relieving arch of tiles over the door lintel. The position of the main gate is not known, but this probably lay where the medieval Porte Narbonnaise now stands. Dating evidence from the walls is completely lacking, and since later arrangements on the site have caused so much disruption to the Roman work, it is unlikely that any such conclusive dating will now be found.

This discussion of the building styles of late Roman town and city defences in those of the Gallic provinces where some degree of archaeological evidence survives highlights a number of points of interest. First and foremost, it suggests that various Gallic provinces exhibited their own individual building styles,

1 Le Mans, the Tour Magdeleine

2 Le Mans, the Tour du Vivier, showing a detail of the patterning in the wall-face

3 Le Mans, The Grande Pôterne. The semicircular tiled arch is Roman, the pointed arch within it a medieval addition

4 Beauvais, the interior face of the late Roman wall

5 Sens, the external wall-face, showing a typical Gallic foundation of large stone blocks, capped by a wall faced in regular small blockwork and tile courses

6 Périgueux, the foundation courses of the wall, showing the large amount of re-used stone

7 Périgueux, the Porte Normande from the exterior

8 Die, the Porte St Marcel from the exterior

9 Carcassonne, the postern next to the Tour du Moulin d'Avar

10 Carcassonne, detail of facing on the Tour de la Marquière

11 Barcelona, externally projecting tower showing windows at first and second storey level

12 Gerona, exterior face of the walls, showing a projecting semicircular tower with windows at the height of the rampart walk

13 Zaragossa, towers projecting from the late Roman walls near the Torre de la Zuda

14 Sopron, exterior face of the city wall, with a projecting tower

15 The Porta Savoia at Susa

16 Burgh Castle, one of the externally projecting towers, showing the straight joint between the tower and the lower part of the fort wall

markedly different in their interpretation of a certain standardised design from towns or cities in neighbouring provinces. This might be perfectly natural, and a consequence of localised supplies of stone or other building materials being made available to a group of provincial sites. The actual standardisation of techniques of building rather than of materials at groups of sites within a region suggests however, that it was more significant than this.

If these provincial groups of defences were indeed built by individual groups of builders, a consequence of importance follows. Any group of defences built and designed in this way must be contemporaneous, or very nearly so. This means that any piece of dating evidence can be used to date not only the town defences to which it refers, but also to give an approximation of date for other defences within the group. Perhaps more importantly, the assorted fragments of dating evidence from all the members of a group can be pooled to give a general dating horizon within which all the group's defences were constructed. Thus, in Belgica II, the walls of Amiens cap a filled well dating from 278, and those of Beauvais produced a coin of Diocletian from the wall foundation (286). Taken together these fragmentary dating indications have a little more weight than if considered separately: they suggest that the Belgica II group of defences was being built from about 280 onwards, possibly taking some years over the programme.

Elsewhere the pooling of dating material gives further precision, and is at its most precise in Lugdunensis III, where, as already noted, the defences of Rennes contain a series of milestones, the latest of which is datable to 273, but all of which fall within the third quarter of the third century. At Nantes, within the same province, and possibly part of the same group, two further re-used milestones of 275 have been found within the walls. In the round-up of useable material for city defences, milestones of now disgraced emperors were collected and quickly consigned to the useful oblivion of the foundation courses. Thus this group of defences was under construction relatively soon after 275.

Sadly, other fragments of dating evidence are so isolated as to assist little in this overall pattern: at Sens, coins of 268 are recorded from the wall-core and show that the walls were built after that date. At Geneva, in Viennensis, an inscription found within the walls proves their construction after 269. In Aquitanica II, a coin of 268 has been recorded from the walls of Bordeaux. It is worth remembering, in this context, that the literary testimony of Gregory of Tours suggests that the walls of Dijon were built under Aurelian (270–5). The evidence for the major period of construction falling within the range 275–90 is hardly contradicted, though re-used inscriptions or sculptures from the walls of any of the towns or cities rarely belong to a date later than the early third century, and the great majority of such re-used material is even earlier in date.

Some evidence can be set against this preponderance suggesting a late third century date for the walls. Excavations at Tours have revealed mid to late fourth century material in layers associated with wall-construction, and there are similar indications from Saintes as well. Apart from this more recent information, at Auxerre, (Lugdunensis Senonia) in a mortar layer near the wall, a coin of Constantine was recovered. This could be interpreted as dating the wall to *c*325, though inscriptions found in the wall suggest an earlier date is possible. It remains to be definitely established that this mortar layer was in fact associated with the original wall-construction, and not with rebuilding or repointing. More evidence came from Dax in Novempopulana, where a coin of Magnentius (351–3) was reported to have been found in the walls. This was dismissed early this century as a piece of romancing,[84] and there is therefore no reason to treat it seriously.

Quite apart from the wider implications of dating which the pooling of this evidence can present, there remains a historical consequence of the discovery that groups of the Gallic cities were constructed by individual gangs of builders. This suggests a regional, if not a provincial or diocesan, plan for defensive construction. According to this plan, the major *civitates* were provided with walls, or were encouraged to build them, with help or advice given by imperial agents touring the provinces.[85]

The identification of these regional groups of defences (fig. 42) is not the only argument in favour of imperial help or encouragement in their construction. In the first place, all city walls needed sanction from the emperor, though in the insecure climate of the third century, this sanction was probably a mere formality.[86] Second, the great scale and the surprising numbers of Gallic cities which were ringed with walls for the first time at the end of the third century suggests something greater than a mere haphazard reaction to the pressures of the time. Hardly an important town in Northern Gaul seems to have been omitted from the defensive programme: if this had been left entirely to the decurions of individual cities, the construction of defences may not have been so strictly contemporaneous, nor so uniform.

What the presence of regional groups does suggest, however, is that the selection of the sites for special treatment was a matter of imperial rather than local priority. In Belgica II, for example, the major *civitates*, like Soissons, Senlis, Beauvais, Reims and Amiens were walled in a style appreciably different from that of Noyon. Whereas the larger *civitates* formed part of the grander defensive design, it is clear that smaller centres such as Noyon did not, and may have been left to provide their own defences as they were able to afford them. It would be interesting, of course, to know a little more about how the provision of these groups of Gallic defences was organised and paid for: did imperial funds rise to paying for the protection of these major centres, or were the walls built by help in kind from troops stationed on the frontiers?

The dating evidence outlined above, and the view that the provision of walls was a part of overall strategy, rather than a haphazard response to barbarian or other pressures, makes it more imperative to find an historical context for the programme of defensive construction. The dating evidence is crucial here, for, if the period 275–90 saw the major construction of the defensive walls, the context is clear. It was in 276 that Gaul suffered what appears to be the last of a major series of barbarian raids. In the aftermath of that raid, literary sources—one of them the later emperor Julian himself—testify to the restoration programme which the then emperor Probus had to institute. Probus (276–82) was a strong and energetic military figure: he was regarded by his soldiers as a hard taskmaster, and the

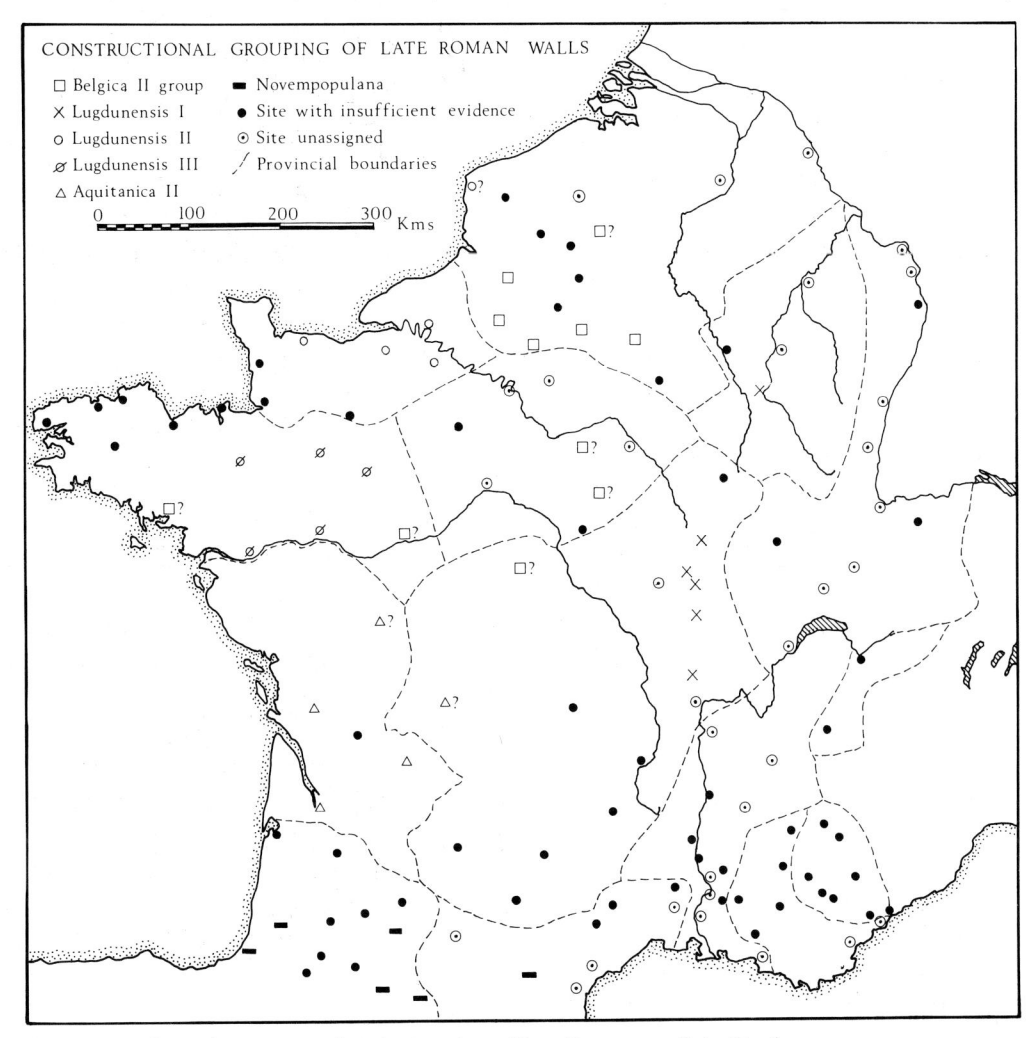

42 Map to show the constructional grouping of late Roman walls in Gaul

esteem with which Julian, a man of parallel endeavour and energy, regarded him speaks highly of his qualities and of his reputation.[87] It would suit the evidence well if within his reign, perhaps towards its end, after the strengthening of the frontier zone along the Rhine itself, the cities of maximum importance were selected for protective treatment. Probus' death at his troops' hands in 282 and the problems of succession which ensued until the emergence of Diocletian in 285–6 may have resulted in a lapse of effort, renewed again when Diocletian came to the throne. The grand design, however, as far as the Gallic cities were concerned, was probably that of Probus, and echoed in the admiring words of Julian some 70 years later when he described 'Probus, who set seventy cities on their feet again in less than seven full years . . .'.[88]

By contrast with earlier days, defences in the late Roman period were no longer symbolic either of the status of a town or of the pretensions to which it aspired.

They were a realistic necessity, and their very size and solidity suggests that the threat they answered was a very real one. The area now enclosed was extremely restricted: in Gaul towns often were now reduced in size to merely a quarter or so of their original extent. Where there was prominent ground which could be used to give extra defence, the late third century walls enclosed it. At Paris and Melun, for example, defence was achieved by siting the walled city on a small island within the channel of the Seine. In the case of Paris, this meant that the great expanse of the town of earlier imperial days was left without defences. There is every indication that this did not in fact mean the total abandonment of the earlier town site, but doubtless the palace quarters used by Julian when he stayed there were within the citadel.[89]

A similar emphasis on the use of defensible ground in the late Roman period often meant the reoccupation of a hilltop site which may have been the site of the Iron Age precursor of the Roman town. Sometimes this meant the abandonment of a low-lying site altogether and the use of a new site some distance away, implying the transference of a whole urban population. This seems to have happened, for example, in Belgica I at St Quentin, where the hill-top site of Vermand superseded the lower settlement. The return to hill-top sites of this type is more marked in some areas than others: in Raetia and Maxima Sequanorum there are several instances—at Lausanne, Kempten, and Geneva, and in this area too, there is a series of forts which also occupy similar high ground. Elsewhere, Carcassonne may be one of the best examples, where the colony lying in the plain was supplemented by the immensely strong circuit of walls on the nearby hill-top. Other towns, such as Autun, Narbonne, Béziers, Vienne and probably Avenches, made use of a portion of their former walled area as a kind of acropolis.

Conversely, the study of those towns which failed to survive the barbarian invasions is perhaps potentially as revealing as the study of those which rose again from the destruction. A small number of Gallic towns seems to have suffered so badly that they never again gained their full status after the invasions: Bavai, of course, is a case in point, Jublains (though there is a strongly walled *castrum* there), Carhaix, and Corseul in Britanny. The reason for a city's failure to survive seems to have been mainly its disadvantageous position. it is unlikely that the passage of barbarian raiders, whatever that entailed precisely, would have been so thoroughly destructive as to leave no trace at all of a former town.

It has sometimes been considered that the construction of the walls was carried out in a rapid fashion, in haste occasioned by the proximity of barbarian invaders. Despite the presence of reused material of all kinds within the walls of Gallic cities, there is rarely any clear archaeological sign of haste. Nor is it entirely true to say that the walls were built 'on the cheap'. Their construction must have been a fantastically expensive operation and might therefore naturally have incorporated any convenient building which happened to be on the chosen course, thus lessening the task. Occasionally the later walls were concentrated round one specific building, as at Bavai, where the walls clasped the former *forum/cryptoporticus* only. A list of other towns which incorporated standing buildings within their circuit or which leant against earlier structures can be compiled, and reinforces the view that this was a relatively common practice. Sometimes, however, surprisingly important and apparently dominant buildings were omitted from the wall circuit—for example the temple of Vesunna at Périgueux, or the monument to the goddess Tutela at Bordeaux. Perhaps the most unusual use of an existing building at first sight comes in those towns where the defences incorporated the amphitheatre or theatre within their circuit. At Arles, for example, the late Roman walls used both theatre and amphitheatre for part of their length, but at Périgueux, Tours, and Amiens the amphitheatre, and at Lillebonne the theatre, became a kind of advanced bastion of gigantic proportions. At Tours and Périgueux this was provided with towers and carefully adapted for defensive purposes. At Trier (fig. 6, p. 19), too, the amphitheatre had been incorporated within the defensive circuit, but here it became a kind of triumphal entrance, with its southern gate giving onto the exterior of the city and the northern one giving internal access. Given their size and impressiveness, the amphitheatres made a stronghold in time of need: perhaps they provided the only building within a Gallic town inside which a large proportion of the population could be easily transferred in times of danger. They thus served as a natural refuge within which people could cluster if the need arose, until the danger had passed.

This brings a wider question into focus. What were the city walls designed to protect? One of the prime aims was to provide security for the administrative and legal systems of the empire. Thus one might expect the walled centres to enclose official buildings—*forum*, temples and basilica, together with whatever other main administrative buildings may have been thought necessary. These might include a hostel for official visitors, grain-stores, a billet for troops garrisoned at the site, and any other buildings which, as a matter of state security, needed to be protected—arms' factories, official clothes' manufacturers, and the like. There would clearly be some room left over within the defended areas for housing, shopkeepers and markets. Reference is occasionally made to houses *intra muros* (inside the walls), showing that the defended areas formed an inner town which was not merely a citadel. The fact that reference to these dwellings has to be so specific, however, suggests that it was as common to live outside the city's enclosure as within it.

Thus far, discussion has centred on the Gallic towns alone. In Italy, too, although surviving evidence is slighter, many of the cities, particularly in the north, were provided with new defensive walls or had their existing defences refurbished. In the Danube provinces, the few Roman cities that existed seem to have adopted similar measures. There is a sizeable surviving group of late Roman defences in Spain, and in Britain, too, where most of the Roman towns had defences which date from the late second or mid-third century, there are some sites which were walled for the first time in the late third or fourth century.

One of the earliest major cities of the western provinces to receive new defences as a result of third century barbarian invasions was Rome.[90] Rome's ancient wall-circuit, the 'Servian' Wall, dating from the early years of the fourth century BC, had been long outgrown even by early imperial times, and in the insecure times of the late third century AD it seemed imperative to invest the city with a new defence which would mark it out as the strong capital of the world. Doubtless there was a mixture of motives for the building of the wall, not least the fact that if Rome built herself a circuit of walls, as a showpiece, other cities would be ready to imitate. The new circuit, which archaeology and literary testimony combine to place in the reign of Aurelian (270–5), was 18km in length, built of a layered concrete core and faced with tiles (fig. 43). According to one estimate, nearly one tenth of its length was composed of previously existing monuments which were incorporated within its circuit. These included the walls of the Praetorian camp and the Amphitheatrum Castrense.

Substantial portions of Rome's Aurelianic Wall still stand. Their complexity for study is compounded by several periods of rebuilding and improvements

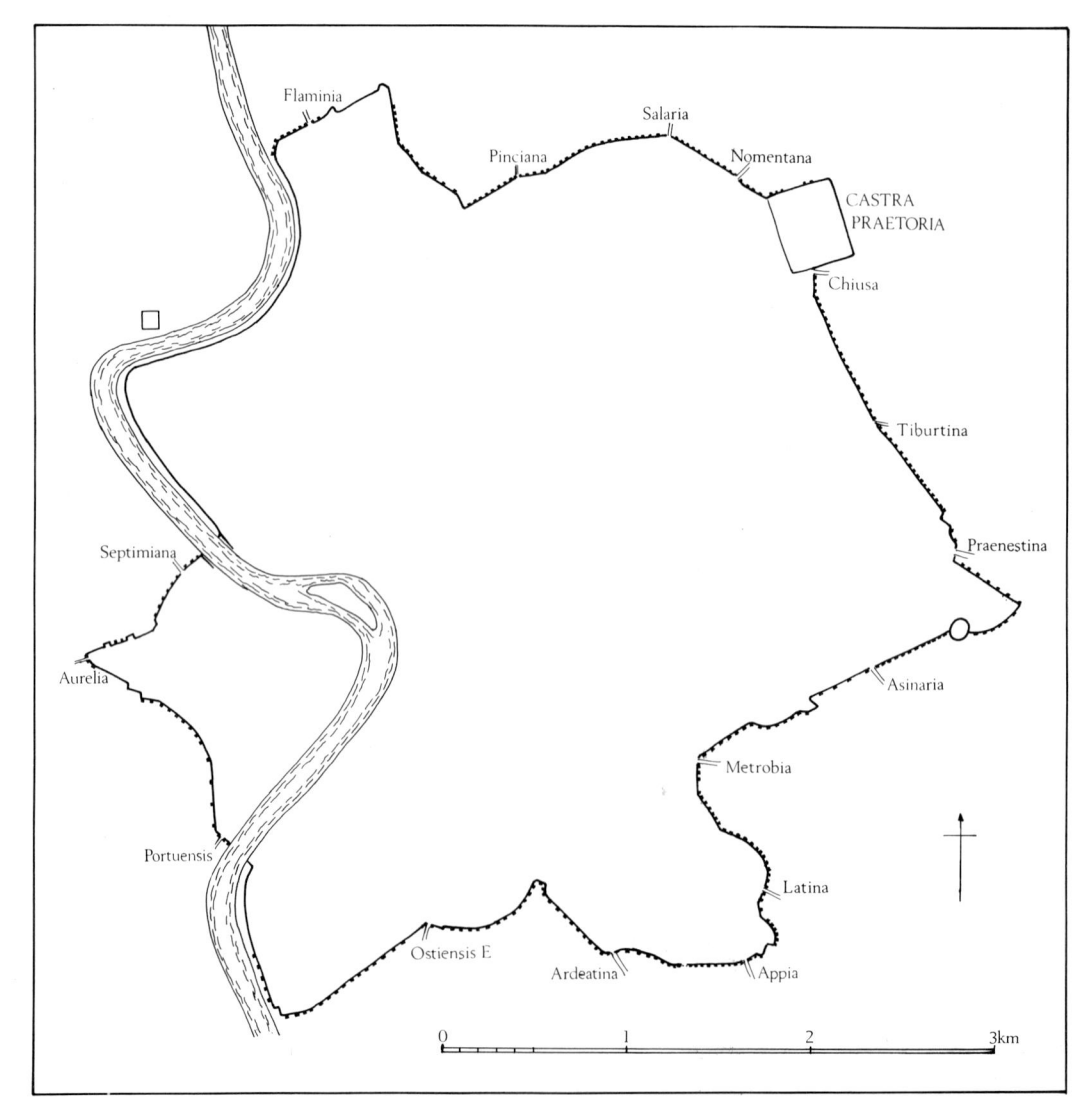

43 Plan of the Aurelianic Walls of Rome

made to the walls after their initial construction. Although the construction of the walls began under Aurelian, their completion was only achieved under Probus. Aurelian's wall, built by the city guildsmen, was of relatively modest height. It was 7.8m high and 3.6m wide, built in a massive rather squat style. Only occasionally does it have in it traces of the tradition of eastern fortification with interior walkways and arcaded galleries which are the hallmarks of a Hellenistic style. The gates were either double or single portals: they were protected usually by a pair of semicircular towers, and this appears to have been a plan which was almost standard (fig. 18, p. 45). The wall was also provided with external rectangular towers. These were spaced at fairly regular intervals of 30m: they normally had a pair of U-shaped windows in the front face at the height of the rampart-walk, and a single window in the side faces.

Among the more important additions to the wall were those of Maxentius in 306–12, who is said in the literary sources to have dug a ditch, which remained unfinished. Archaeological evidence seems to show that he did much more. The walls were heightened by another storey, which was added in the form of an arcaded gallery, and some of the towers were heightened to adapt to the new structure. The Porta Appia and the Porta Asinaria were now altered, too, on a far more grandiose scale. The Porta Appia, originally a gate of Aurelianic type— probably double portalled with semicircular flanking towers—was made more impressive by making the towers more pronounced, and by taking them up four storeys high. At the Porta Asinaria, one of the simple Aurelianic gates with no flanking towers set between normally spaced rectangular wall-towers, new semicircular flanking towers were added. These improvements were brought to a sudden end by the death of Maxentius outside the walls of Rome at the hands of Constantine and Galerius.

No more work was then done on the walls until Honorius substantially strengthened the whole defensive curtain in 401–3, and repaired those parts of it which needed attention. The windows of the towers, until now the typical large rectangular openings suited for *ballista* fire, were replaced where convenient by small slit windows. All the gates were now made single portalled only, and two of the smaller ones were blocked completely. Most of the gates were completely refurbished, unsually with added rectangular towers built round the base of the older semicircular ones. There is some evidence that this scheme of improvement was never carried through to its completion. Further restorations occurred under Valentinian III, Theoderic, and Belisarius.

Whether as a result of Aurelian's building programme in the capital or not, North Italian towns seem quickly to have accepted this lead. In fact, Verona had already rebuilt its walls under Gallienus.[91] Some, like Turin, Como, Aosta and Milan already had circuits of walls dating in most cases to Augustan times. Others had walls dating from even earlier still and it is probable that all repaired and rebuilt their defences where this was necessary. There is archaeological evidence for reconstruction at Como and Aosta, while outside the Alps, but still within Italy, the old colony at Emona (Ljubljana) was provided with new external towers to strengthen her first century walls. Milan, an important city which

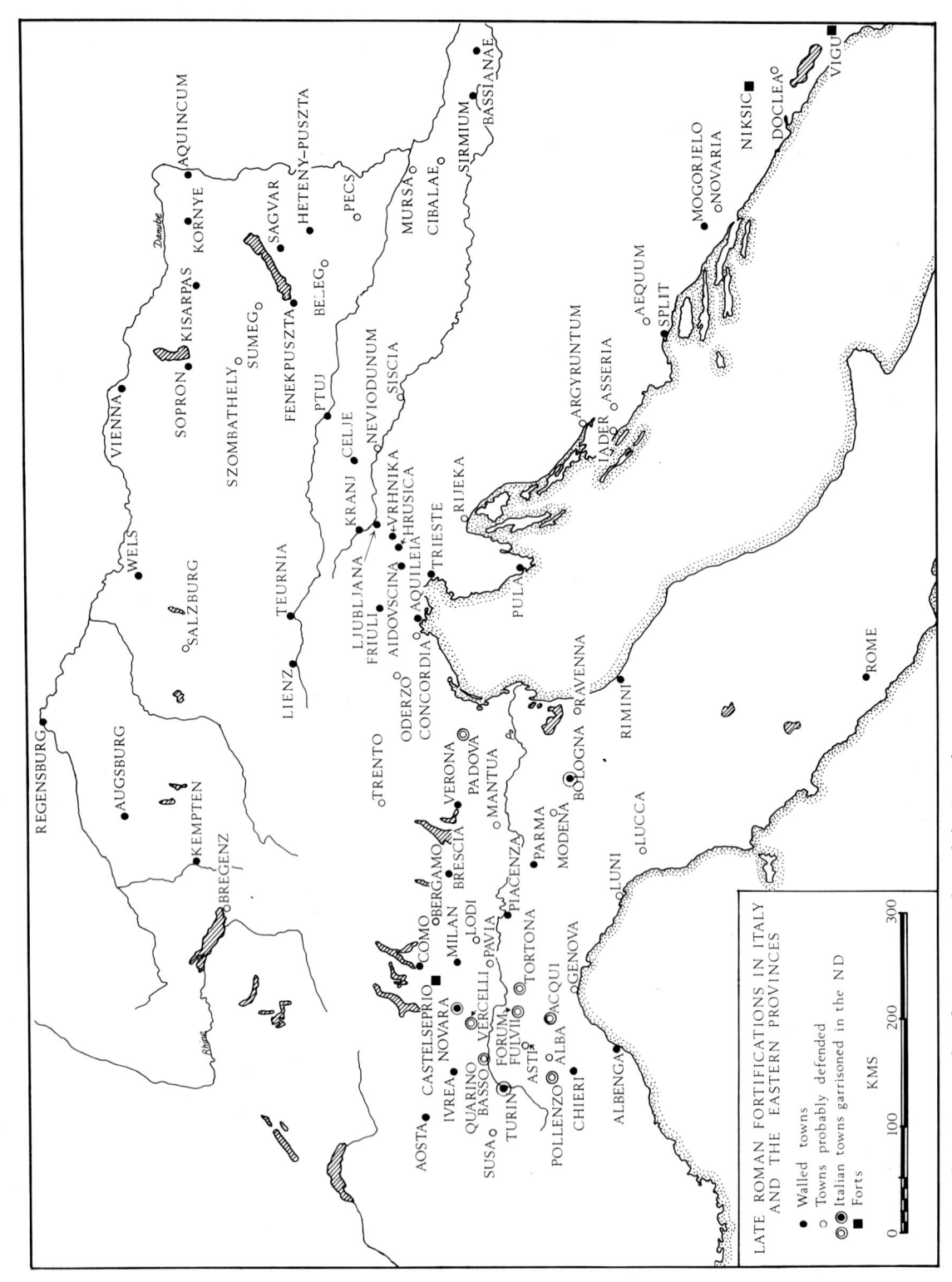

44 Map of late Roman fortifications in Italy and the Danube areas

figures constantly in the military history of this period, had a wall-circuit which was beseiged in the 270s. Some 20 years later, under Diocletian, the city was rebuilt and the walls raised again to make the city more glorious than ever.

Other Italian cities, until now probably unwalled, took measures for their defence. In no case is there any objective criterion for dating the defences to the late Roman period, but the relatively small area enclosed by the defences, the externally projecting towers, and the provision of gates with a single portal between a pair of flanking towers strongly suggest that these defences date from the later third century. For Susa there is evidence from the 320s that it possessed walls, since then its gates were stormed without the city being taken.[92] Brixia was also besieged at the same date but no details of its Roman wall circuit are known.

Unlike neighbouring cities in Gaul, most Northern Italian walled cities were not normally reduced in size in the later period. There appears to have been less haste about the defensive preparations, for, well back from the frontier line, they could construct defences in a measured fashion. Here there was no destruction caused by bands of marauding Germanic tribesmen, and so the walls contain less re-used material than those of comparable sites in Gaul. Consequently, there is less evidence for precise dating at those few cities whose walls can be tentatively diagnosed as late Roman. Even so, despite the fact that the cities of Northern Italy must have been better defended at earlier periods of Roman history, it is surprising that there is such a scarcity of well authenticated late Roman walled cities in the area.

Outside Italy, too, there was a similar lack of late Roman defended sites in Dalmatia and the Eastern Provinces. The road from Aquileia through Emona to Ptuj (Poetovio) had defended posts and cities along it. Ptuj, like other cities, was originally an open site spanning the River Sava; its late walls enclosed a relatively tiny area on one bank of the river where the topography allowed the siting of a citadel.[93] The large, open, low-lying part of the city to the south-west was abandoned. Celje (Celeia), now perched on a rocky hillock, was a similar defended site,[94] but other cities further down the Sava seem to have been poorly defended. Other cities in the neighbourhood but lying in the neighbouring province, Dalmatia, were similarly neglected, though Stremska Mitrovica (Sirmium), as the capital of Pannonia II and the birthplace of Maximian, received special treatment.[95] Its walls were strong and kept in good repair. Cities which lay nearer the frontiers, like Bassianae, may have been fortified earlier, possibly at the end of the second or the beginning of the third century (fig. 6, p. 19).

Further north from this area, in the frontier zones of Pannonia, Raetia and Noricum, the number of Roman cities was sparse, and those which had not already received walls by the early third century was small. At Kempten and Augsburg research has revealed that the late Roman towns withdrew into a protective circuit of walls set on available high ground.[96] At Sopron,[97] a rough oval of walls enclosed about 9ha within a wall built largely of large stone blocks and protected by U-shaped projecting towers. In the wall-core excavators discovered a coin of Diocletian.

Sopron was clearly not the only walled city in Pannonia, for Ammianus

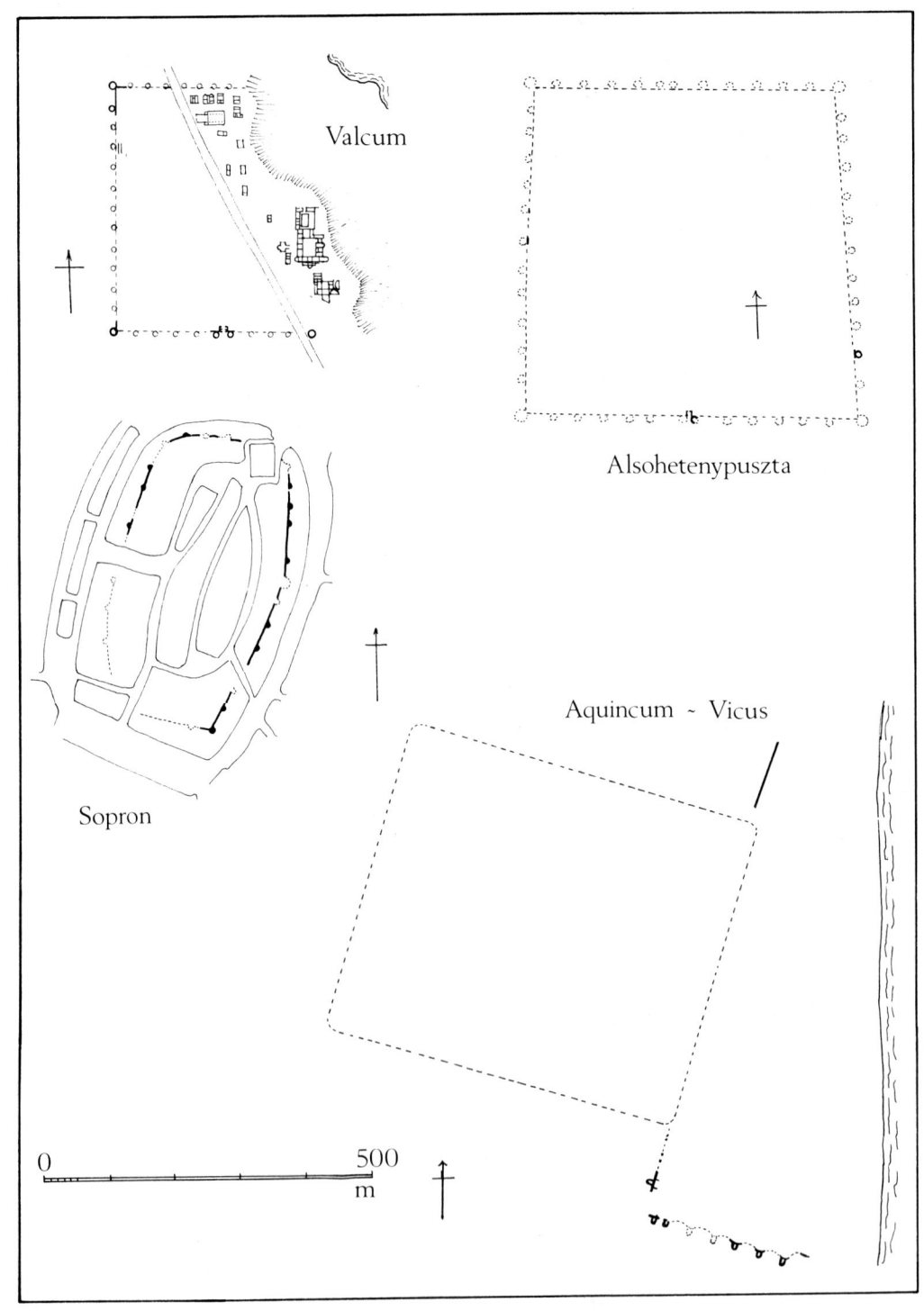

45 Late Roman walled sites in Hungary (*scale 1 : 10,000*)

Marcellinus records specifically that one of the gates of Savaria (Szombáthely) fell down while the emperor Valentinian was staying there.[98] No trace of the Roman walls is known at this site. At the capital of the province, too, Aquincum, there were late Roman defensive developments.[99] The settlement clustering round the eastern (riverside) front of the legionary fortress in the late Roman period was enclosed by a defensive wall of most unusual form (fig. 45). A series of large U-shaped towers were linked by a thick wall which looped between each tower, to form, in plan, a series of bowed curves, as it were pegged in a straight line by the towers. This construction, unparalleled elsewhere, must belong to the late Roman period, though it has not been closely dated.

One other group of late Roman settlements in the Pannonian provinces calls for attention. Well spaced out along the major Roman roads in the province, there lie a number of foursquare walled enclosures whose purpose may be military or civilian. Perhaps the best known of these is Valcum (Fenékpuszta) at the south-western end of Lake Balaton (fig. 45).[100] A powerful rectangular circuit of walls, festooned with circular projecting towers, with a similar pair guarding the gates, enclosed an area of some 15ha. Within this area was an assortment of buildings of at least two phases, including what looks like a large villa building, a basilical hall, and a number of small separate store-buildings. The overall dating of the walls and of the interior buildings has not been determined. The interpretation of the site is thus very perplexing indeed—is it a defended villa, small town, or is it some form of military establishment?

Similar sites, though far more fragmentary, are known at Környe,[101] Kisárpás[102] (where there was an associated cemetery, dating from the Constantinian to the Valentinianic period), Ságvár[103] (where a cemetery came into use in the period of Constantius II) and Alsóheténypuszta.[104] Only at the last have excavations taken place on the defences and within the defended area (fig. 45). It was the largest of the defended sites, and, though of quadrangular form, was not well suited to the terrain on which it lay. It enclosed approximately 22ha, and the walls, 2.6m thick, have been shown to reveal traces of at least three periods of use and alteration. The earliest phase of defences comprised a wall with projecting U-shaped towers, and fan-shaped towers at the corners. At a later stage these were superseded by large round towers. In a third building phase, these too were destroyed and rebuilt, and an extra courtyard added at the gates. On typological grounds, the construction of this fortification can be assigned to the reign of Contantine, followed by rebuilding with round towers under Constantius II, and further rebuilding under Valentinian. The name Iovia assigned to the site, however, would accord better with construction under Diocletian who claimed a special affinity with Jupiter. Coin finds from within the site suggest that the first main occupation came in the early years of Constantine's reign.

The nature of these walled centres needs some clarification: they lay spaced on the road network so as to complement existing towns and cities like Szombáthely, Sopron and Sopianae (Pécs), and they clearly formed convenient refuges for the local populace. The lack of firm evidence for town buildings within the walled areas suggests that they were not in themselves civilian settlements, but since the evidence of barrack blocks for troops has so far not been forthcoming, the claim that they are support bases for the frontiers or bases for the field army still needs considerable proof.

The main concentration of cities in the Spanish province walled in the late Roman period lay in the north.[105] There is no inscriptional evidence from Spain recording the construction of a late Roman town wall. Many Spanish cities have circuits of medieval walls which have more massive, probably Roman, foundations (fig. 46). The identification of such wall-circuits as Roman rests on the assumption that large well-squared blocks of masonry (*opus quadratum*) are indicative of a Roman date. Since there is usually little dating evidence, and usually no associated finds, the assignation of a date to Spanish defences is a matter solely for subjective assessment.

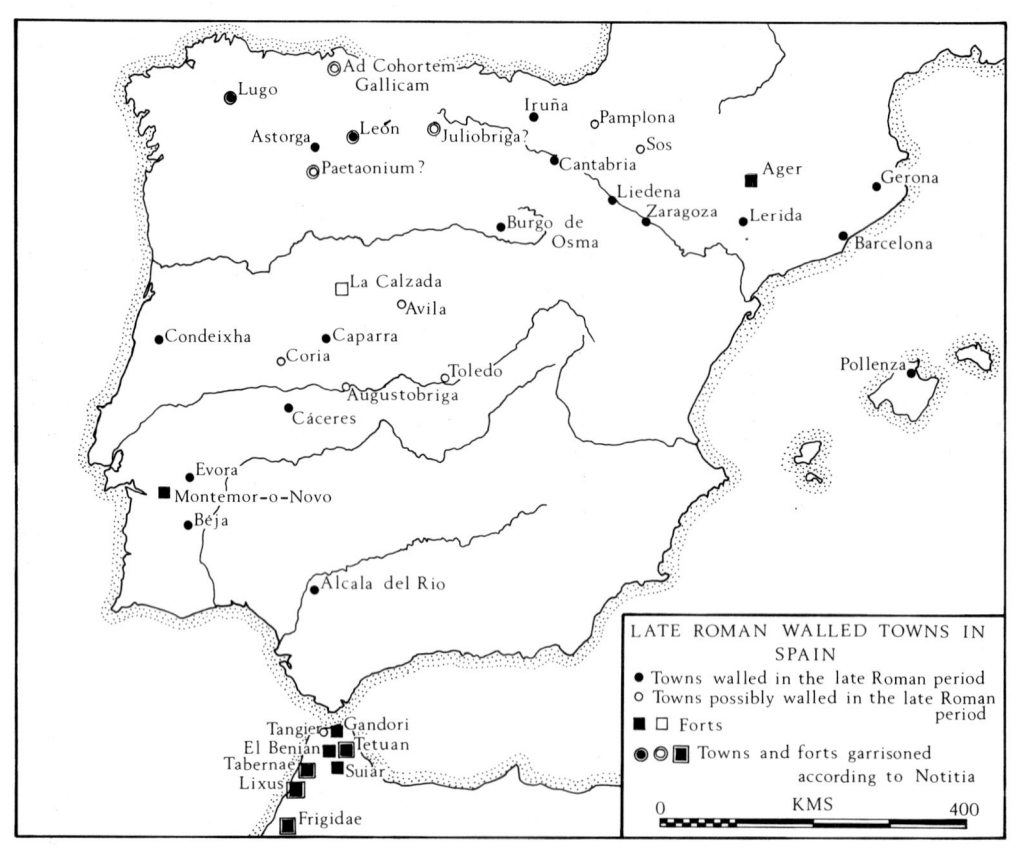

46 Map of late Roman fortifications in Spain

No document comparable to the *Notitia Galliarum* exists for the Spanish provinces, so there is no comprehensive and consistent list of Spanish towns which formed the late Roman provincial cities. Evidence for the late Roman defences of those Spanish towns known to have had Roman origins comes in various forms. Literary sources provide some indications of sites whose walls have not survived: the most detailed is a late fourth century description of the walls of Pamplona which records the circumference of the walled area as over two miles long. In this length there were 67 towers.[106] There was probably also a circuit at Lérida, for seventeenth-century drawings show a gate with polygonal towers which appears to be late Roman.[107]

The major evidence for late Roman walls in Spain, however, is archaeological. The general design of the Spanish walled cities was not dissimilar from those in Gaul. They had thick, solid freestanding walls, punctuated by frequent towers. Instead of small blockwork, the Spanish sites had a facing of large stonework in regular courses (*opus quadratum*). Tile courses and other distinctive features of Gallic walls do not appear in Spain. The most significant group of these towns, was recognised in northern Spain more than 50 years ago.[108]

The walls of Astorga[109] occupied a long narrow area of about 27ha in the centre of the modern town (fig. 47). The number of Roman inscriptions found re-used in the walls, the latest datable to the period of Severus Alexander, confirms their Roman date. The walls were about 4m thick, and largely obscured with a medieval facing of roughly coursed, but very neatly jointed stones. They were protected by frequent towers all of semicircular plan and often spaced less than 20m apart. No dating evidence has been found.

The late Roman walls of Barcelona,[110] which enclosed a small area of only about nine hectares, were sited on a small bluff overlooking the harbour (fig. 48). On the whole, these defences were of uniform construction. Facing in large blockwork, with some headers and stretchers, continues up the wall to rampart-walk height, 9.19m high, where there was a string-course and parapet. Above the height of the rampart-walk, the facing changes to a smaller style, and traces of rectangular battlements have been noted in the wall curtain, where medieval heightening of the wall has preserved the Roman curtain substantially intact. The lowest courses sometimes had a small offset plinth, sometimes moulded. The corner towers were round or polygonal. Rectangular towers project from the wall face at frequent intervals round the circuit. Like those at Rome, they bear round headed windows in their front and side faces at rampart walk height. All of the four known gates were single passageways flanked by a pair of projecting towers, of varying ground-plan. Re-used material, particularly inscriptions, has been frequently found within the walls, and suggests that they were built after the mid third century.

Burgo de Osma, the Roman town of Uxama Argaela,[111] lay in an irregular sprawl over the tops and slopes of two hills. In all, the fortifications, about 3m thick, and provided at intervals with rectangular projecting towers, enclosed about 28ha. The towers were concentrated particularly on the gentler access slopes of the town. The walls formed an extensive circuit, and the defended site lay in an unlikely position, perched high above the modern town. Little trace of the defences now survives on the ground.

Cáceres,[112] despite its southerly location, has a circuit of walls which has much in common with other sites in the northern and eastern parts of Spain (fig. 48). There can be little doubt that under the layers of medieval masonry of several periods, the walls of Cáceres date in origin from the late Roman period. A circular

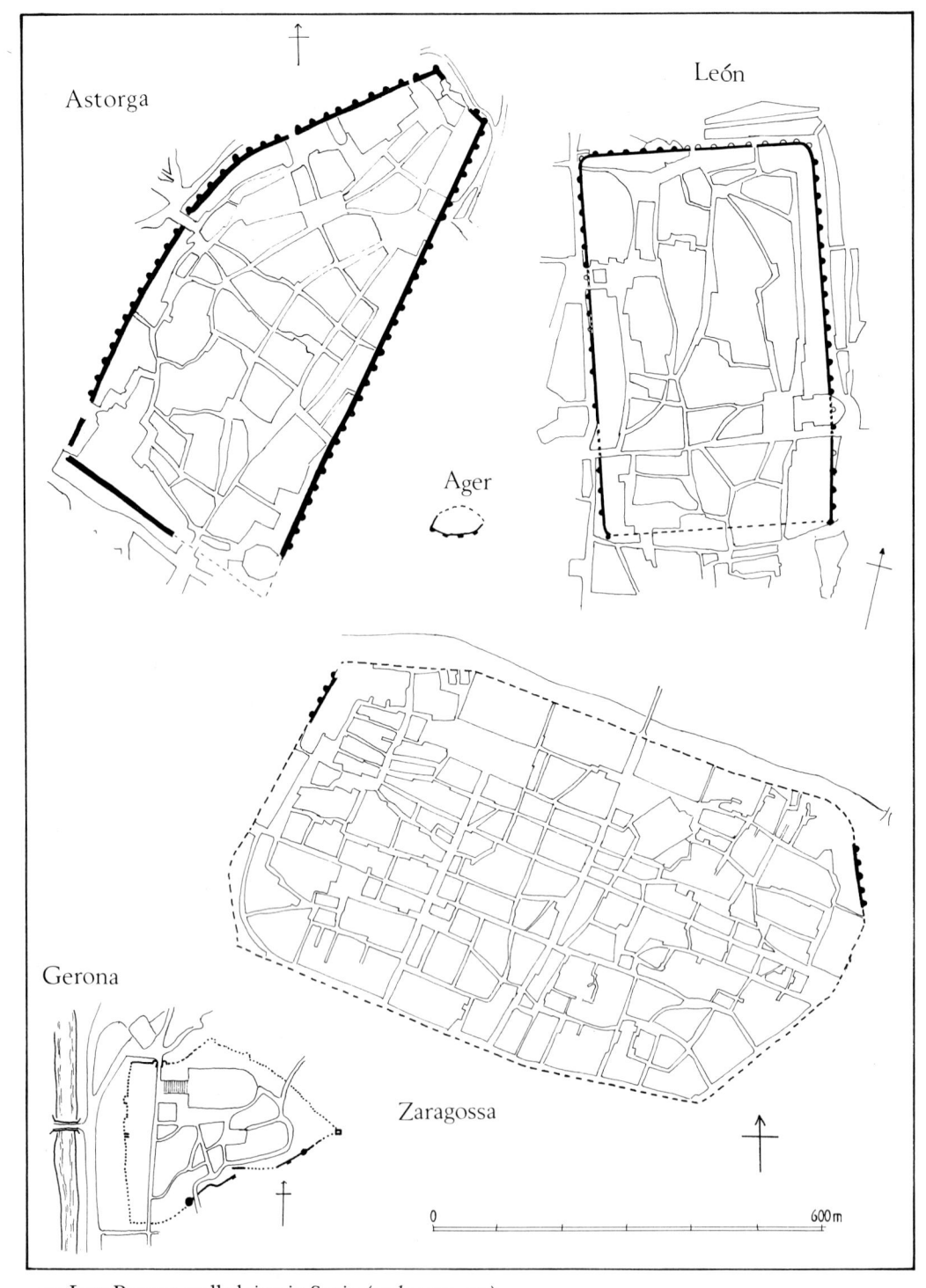

47 Late Roman walled sites in Spain (*scale 1:10,000*)

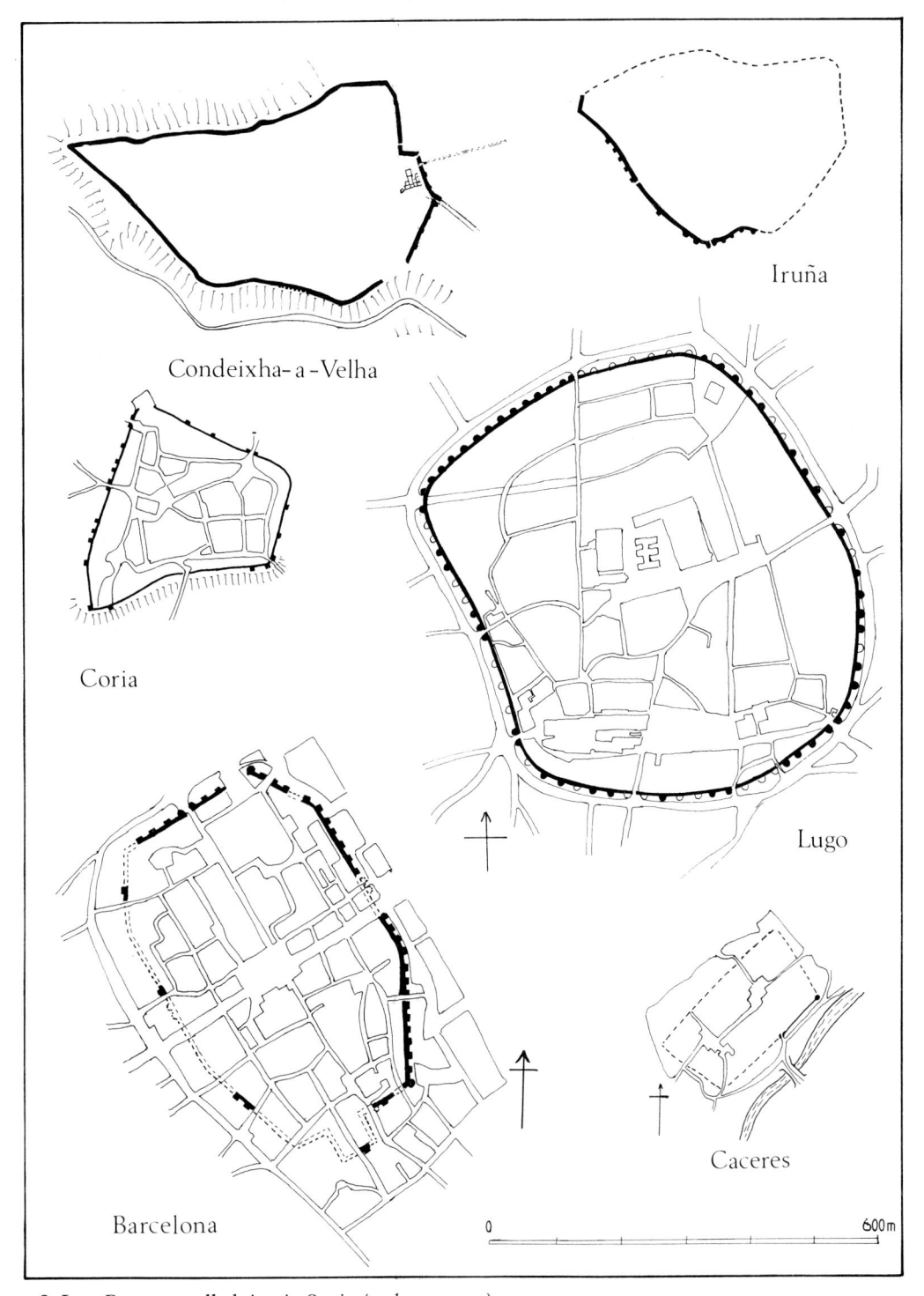

TOWN AND CITY FORTIFICATIONS 127

48 Late Roman walled sites in Spain (*scale 1:10,000*)

corner bastion, and projecting rectangular towers, all built in the typical large blockwork common to many of the Spanish sites is the hallmark of this work. Portions of one gate, the Puerta del Rio, set at a slight angle through the walls, its foundation formed of large slabs of granite and clearly Roman in date, can still be seen. No objective dating for the walls, however, has been found.

The site of Condeixa-a-Velha[113] occupied a triangular plateau, some 600m long and 300m wide at its fullest extent, which had sheer sides down to small rivers in deep ravines on two sides (fig. 48). The town flourished in the first and second centuries AD, but in the late third century it was walled for the first time. Portions of the wall were faced in medium stone blockwork, though larger stone is sometimes visible at ground level, and in the lower courses. Only on the western side were the walls defended by projecting towers, though at one point on the circuit, there was a series of nine small buttresses, presumably to strengthen the wall so that it did not collapse into the ravine. The towers were rectangular, and spaced at intervals of about 10–15m. The main single-portalled gate was enfiladed by a pair of rectangular towers. A subsidiary gate not far away was a simple opening through the wall without protecting towers. The wall itself was 2–3m thick, but at its base as much as 3.5m wide. It was constructed so as to thread its way between houses which were already on the site and standing when the wall was built. No dating evidence has been published.

The town of Coria,[114] near Cáceres, retains a very restricted circuit of walls, bustling with projecting rectangular towers, which must be of Roman date (fig. 48). The walls, built of large stone blocks, enclose a strategically well-chosen site of only 1ha. Projecting towers, often as little as 5–6m apart, are very prominent, and several gates, single passageways with flanking towers or with towers nearby, are still in use. There has been some medieval rebuilding, but some of the gateways, notably the Puerta de San Pedro, retains a grand chamber in one of the towers, and the beginning of steps leading up to the rampart walk. The construction date of the walls is in some dispute, but the general style, the tightness of the defended area, and the weight of the defences suggest that these walls are late Roman rather than earlier.

The walled area at Gerona[115] occupies a triangular site on a slight bluff between two rivers (fig. 47). Depsite heavy refacing in medieval times, portions of characteristically Roman work still remain, partly in the normal large blockwork, and partly in 'Cyclopean' style, with large irregular stone blocks laid together in dry stone wall technique. In the portions of wall which can be clearly identified as Roman there are projecting rectangular towers and circular towers sitting astride the walls (pl. 12). Two gates, one apparently no more than an angled passage through the walls, the other a single portal flanked by a pair of semicircular towers, still survive. There is a distinct possibility that the walls of Gerona incorporate work of pre- or early Roman date: walls of 'Cyclopean' style have not elsewhere been defined as late Roman work, and the combination of rectangular and round towers is unusual.

Iruña,[116] commonly thought to be the Veleia of the *Notitia Dignitatum*, is now a virtually deserted hilltop near Vitoria (fig. 48). The walls, built of large blockwork and containing re-used material, encircle about 10ha in a good defensive location, and stand 4.5m thick. Semicircular and rectangular towers project from the walls, and near the gateway which survives, the walls were of noticeably more careful construction. The gate was a single passageway flanked by a pair of bastions semicircular in intention if not in execution (fig. 19, p. 46). The approach to the town clearly commanded greater defensive attention, for here the projecting towers were spaced at intervals of less than 20m, as compared with their much more infrequent spacing, if not their absence, at other points round the circuit.

León,[117] the *Castra Legionis Septimae Geminae*, still preserves some of the finest Roman fortifications in Spain (fig. 47). It is of almost regular legionary plan, a rectangle 385 × 550m, with one corner slightly tapered, giving a perimeter of 1,400m, and an area of 19ha. The Legion was still based here in the late Roman period. At this period, earlier fortifications of Flavian date, which consisted of a wall faced in small blockwork, were augmented by the addition of an outer skin of masonry. This late Roman wall, which gave an overall thickness of 5.25m to the defences, was flanked by U-shaped towers 8.25m in diameter spaced on average 14m apart. Re-used material of all sorts, including many inscriptions, has been found in the walls, though there is no accurate date for the late Roman wall construction.

Lugo[118] still preserves an almost complete circuit of Roman walls enclosing 34 ha (fig. 48). Their facing is partly in the large *opus quadratum* blocks and partly in thinner slate-like stones. The walls are 6m thick, and still stand 11–14m high. Semicircular towers flank the walls, and are set about 17m apart. They vary in diameter between 13.9m and 10m, which is perhaps a sign of the work of different building gangs. In a tower which still survives above the level of the rampart walk, there are four large window-arches of inverted U-shape. The gates are single passageways set deep between large semicircular bastions. Re-used material in the wall-core and the overall style of the wall suggests that it was built after 250.

The late Roman walls of Zaragossa[119] depend to a large extent, like those of Barcelona and León, on a set of earlier Roman walls (fig. 47). The Augustan colony at Zaragossa was founded around 24 BC and soon provided with defences which incorporated monumental gates, destroyed in the course of the nineteenth century. The line followed by the walls is visible in the modern streets, basically rectangular in plan, and enclosing about 60ha. The late Roman walls provided an added strengthening only to this circuit, already 3.2m thick. In the portion best known and studied, near the Torre de la Zuda (pl. 13), the late Roman walls had been added to the external face to make a total wall-thickness of 6m, and in the process had completely demolished all but the foundations of a projecting Augustan tower. Three new projecting towers, of near circular plan and spaced 14m apart, protected this short length of wall. The added thickness of wall was built of re-used material, with an exterior facing of large blockwork, in complete contrast with the neater, smaller facing of the Augustan wall.

The late Roman walls of Spanish cities provide a great deal of variety. The defences at Barcelona, Zaragossa and León are similar in that they all have a double wall where earlier imperial walls had been strengthened and provided with new or extra towers. The diversity of tower types in general and the lack of datable material within even a fairly broad time-span make the attribution of the defences of any Spanish city to any particular imperial period or campaign very much a subjective exercise.

One regional group of fortifications, however, displays a set of common architectural characteristics. These sites, León, Lugo and Astorga, lay in the mining district of Lusitania. All three were defended areas of some size surrounded by thick walls with closely spaced semicircular towers. The mining areas of Spain were among the most important for the Roman world, and no effort was spared to keep them safe from attack. Braga, also in the mining district, was one of the cities mentioned by Ausonius as most illustrious in the Roman world.[120] Such Spanish garrisons as are mentioned in the *Notitia* lay in a heavy concentration round this area.[121]

This group of fortifications can be regarded as a series of defended sites built and strategically placed in the mining area to provide regional defence. The slight dating evidence suggests the later third century as a more likely date for the construction of this group than the fourth.[122] The close spacing of the towers, a feature which appears to be typically Spanish, is also found at Iruña and at Dax, in Novempopulana; and the chain of sites from Lugo through to Dax and Bayonne may represent the consolidation of a pass through the Pyrenees over which bullion from the mines had to pass. It may not therefore be accidental that the only garrison known in the south-western corner of Gaul was stationed at Bayonne, strategically placed to watch over this traffic.[123] Bayonne itself was not listed as a *civitas* in the *Notitia Galliarum*, and it was an oddity among other Gallic sites of a similar appearance in that it was built on what appears to have been a new site. In addition, the route from the mining areas of Braga and León westwards was of great importance to successive Roman governments, as is shown by the wealth of milestones of differing periods discovered along it.[124] It was probably not until the political position stabilised under Diocletian and the Tetrarchy that there was a reasoned enough calm in the Spanish provinces to set about protecting the internal wealth of the empire in this thorough and methodical fashion.

The German raids into Spain in the 260s appear to have taken the province completely by surprise.[125] Ampurias and Badalona seem never to have recovered. Tarragona, even though defended by its massive Cyclopean walls, also fell victim. Portions of Barcelona, though it was then protected by its Augustan defences (perhaps by then allowed to fall into some degree of decay), were damaged. It was natural therefore that, as an immediate reaction, Barcelona and Zaragossa, both uncomfortably close to what are natural through routes into the interior of Spain, should have been strengthened, and their defences brought up to date. The walls of Barcelona in particular, with their rectangular towers, appear to be closely modelled on those of Rome, and may therefore have been begun under Aurelian or Probus. At Zaragossa, the Augustan walls already had semicircular towers: the new defences were modelled on similar, though not exactly parallel lines.

In Spain, as elsewhere, there is a sharp contrast between the cities which survived the raids of the late third century and those which went under.[126] Those which survived were normally those which had a firmly established acropolis, on which the new defended area could be placed. This is strikingly obvious in the cases of Gerona, Burgo de Osma and Iruña. The cities which did not survive are those like Badalona, low-lying and situated on the sea coast—easy prey to marauders and raiders, even though encircled with walls.

The threat was not only posed by invaders by land. A scatter of coin-hoards from the latest years of the third century similar to that occurring down the northern and western coasts of Gaul is also recorded in Spain.[127] This may indicate the presence of pirates in this region in the last decades of the century. The very presence of the defended city of Conimbriga suggests that on the Atlantic coast, as on the Mediterranean earlier, piracy and sudden raids from seaborne enemies were a serious threat. On the south-facing coasts too, there was defensive reaction in Alicante and Elche, possible from Moors.

In the British provinces, by contrast, the majority of the major cities were provided with defences of earth or stone soon after the mid-second century (fig. 7, p. 21). Often, an earthwork defence of second century date was improved during the course of the third by the addition of a stone wall to front it. Although attempts have been made to assign the construction of these walls to known events within the (rather patchy) known history of Britain in the second century,[128] the overall pattern of town wall building such as is known from archaeology suggests that the picture is considerably complicated and that probably no single historical event provoked the construction of walls round *civitas* capitals and some other towns.

This complexity continues into the later period: the impetus of late third-century raids which elsewhere forced cities to look to their defences seems to have passed the British cities by, though not without exception. Such evidence as there is suggests that the town of Venta Icenorum (Caistor St Edmund)[129] was walled in the late third century with new stone defences with projecting square and polygonal bastions and an earthern rampart. These replaced an earlier, more extensive defensive system which has not been examined or dated. At Canterbury,[130] too, a town-wall with a rampart bank behind it can be dated to the late third century by the discovery of coins of Postumus in the rampart. Despite this construction-date, the style of Canterbury's walls showed no sign of new developments in Roman defensive architecture displayed elsewhere. They had no projecting bastions, and retained interval turrets and the earthen rampart. The gates were traditional in style, and within the mainstream of city gate construction in Britain and other provinces.

Another important site which had a new circuit of walls in the late third century was Brough-on-Humber,[131] where turf and timber defences round a possible military post or stores base were replaced in stone. Later additions to this wall included the provision of externally projecting towers at the gates and at other places on the circuit. These developments have not been accurately dated.

The provision of defences with earth ramparts in what was increasingly becoming an archaic style can in several cases be dated to the late third and possibly early fourth century. Among those sites where walls were added at this date to existing ramparts are Dorchester-on-Thames[132] and probably Ilchester,[133] Kenchester[134] and Rochester.[135] At Ancaster,[136] Cambridge (fig. 49)[137] and Godmanchester (fig. 49),[138] there are indications that an old-style rampart with a contemporary wall and bank was constructed from new at this period.

Only five of the known walled urban centres within Britain, as far as present knowledge goes, seem to have been provided with fortifications even approaching the continental model—with thick, freestanding walls and projecting bastions. Of these, three, Caistor,[139] Horncastle (fig. 80, p. 205)[140] and Bitterne (fig. 49),[141] lay in the southern and eastern coastal zone, and might be interpreted, not as urban centres themselves, but as stores or military establishments connected with the late Roman defensive system round Britain's south eastern coastline. The other two are Mildenhall (Wilts),[142] where a rectangular wall circuit appears to have superseded an earlier defended area on a slightly different line. Examination of the site has not been extensive, but projecting towers contemporary with the construction of the walls have been noted. A coin of 360 was found in the fill of a ditch overlain by the town-wall, thus demonstrating a late fourth century date for its construction. At Great Chesterford,[143] the walls described a regular polygon, and were freestanding (fig. 49). No projecting towers have been located, and a broad and shallow ditch has been traced outside the walls at several points. The debris of buildings overlain by the wall foundation provided a *terminus post quern* of the early fourth century for the wall construction.

A number of other sites dotted throughout the British provinces were also walled in this later period: a series of small posting-stations lying on Watling Street between Wroxeter and London have produced evidence of four-square ramparts or walls and ditches of this date.[144] There seems also to have been a move at many military sites in the late third and fourth centuries to defend the civilian settlements with a wall, perhaps signifying their status as communities independent of the military command, and able to defend themselves in their own right. Perhaps the best examples of this development are to be seen at York,[145] where the civilian town opposite the fortress was probably walled at this period, and at Catterick,[146] where there is no doubt that what may have begun as a *vicus* had become a fully-fledged town in the later Roman period (fig. 49). The defences of Corbridge[147] may also come in the same category.

The search for an overall pattern in the study of Romano-British town defences is liable therefore to be frustrating, and the imposition of over-rigorous historical explanations for their construction may well hamper progress in their understanding. There is one feature, however, which a substantial number of British defences share. This is the updating, or partial updating, of the walls at some time in the later Roman period by the provision of externally projecting bastions coupled with a wide and shallow ditch. This should rightly be viewed as an attempt to modernise the rather out-of-date defences which were to be found

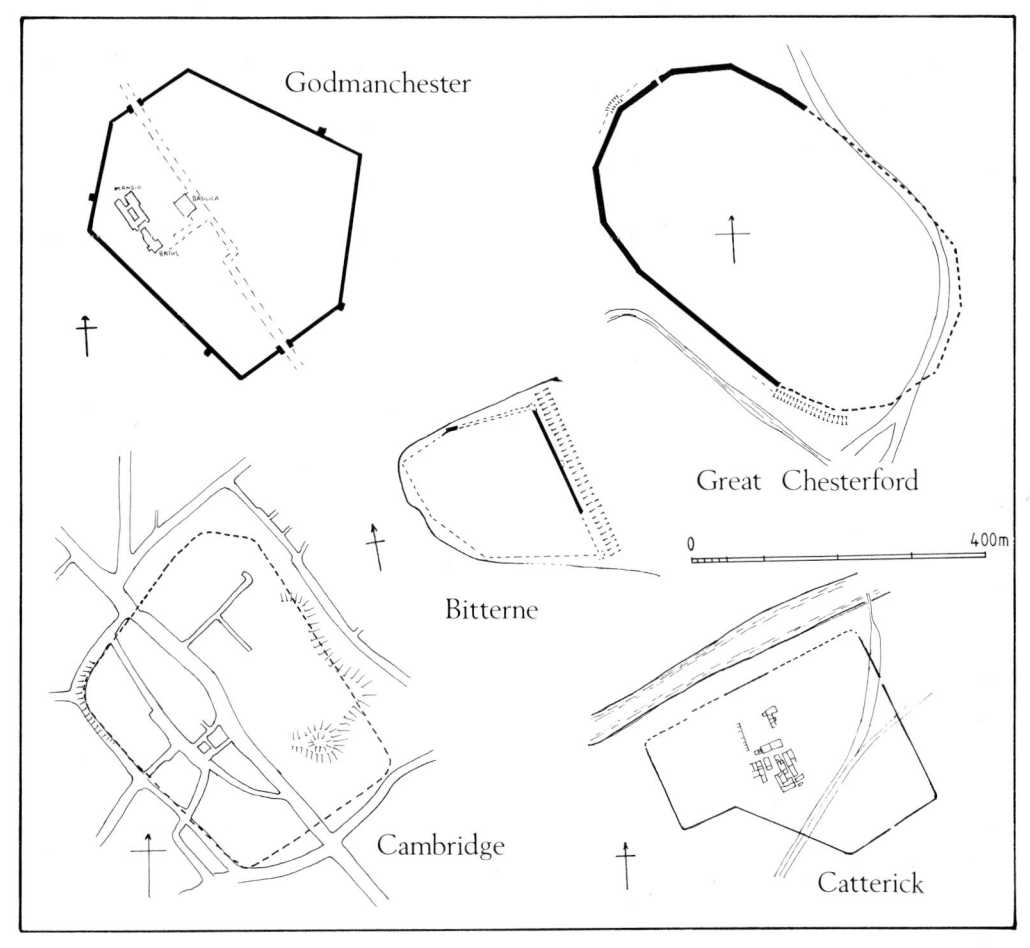

49 Late Roman walled sites in Britain (*scale 1 : 10,000*)

in the province. The dating of these has been claimed to belong to a single historical period—that of the restoration of British cities under Theodosius in 368-9,[148] but this is probably too simple a view to take. The earliest bastions to be found on a British town are probably those of Caistor St Edmund, which take their lead from the walls of the near contemporary forts of the south-eastern coastal defence system. Thus projecting bastions were known in Britain from the late third century onwards: the polygonal water-front towers at the legionary fortress of York have been claimed as Constantinian,[149] and the unusual fan-shaped projecting corner bastions at Ancaster and Godmanchester might also be claimed as of a similar date. An assortment of archaeological evidence suggests a date for most of these towers after the mid fourth century. Perhaps the most significant finds are from Caerwent,[150] where a coin-hoard of 351-3 was deposited in the floor make up of one of the added bastions on the city wall. This suggests that a notional date of 368-9 for all of them is too late, but a suitable historical context for their provision any earlier also proves elusive.

The provision of externally projecting towers is a move which seems calculated to provide extra protection for the walls by mounting machines to give enfilading cover against attack, or at the very least, to make the policing of the walls more effective by allowing sentries a clearer view. It is at least possible, therefore, that the bastions were related to the provision of a police force or militia within the city to act as their protectors against civil or barbarian attacks. Such a force, of foreign troops, has been postulated on evidence from the Lankhills cemetery at Winchester, where a series of distinctive graves of military type have been isolated.[151]

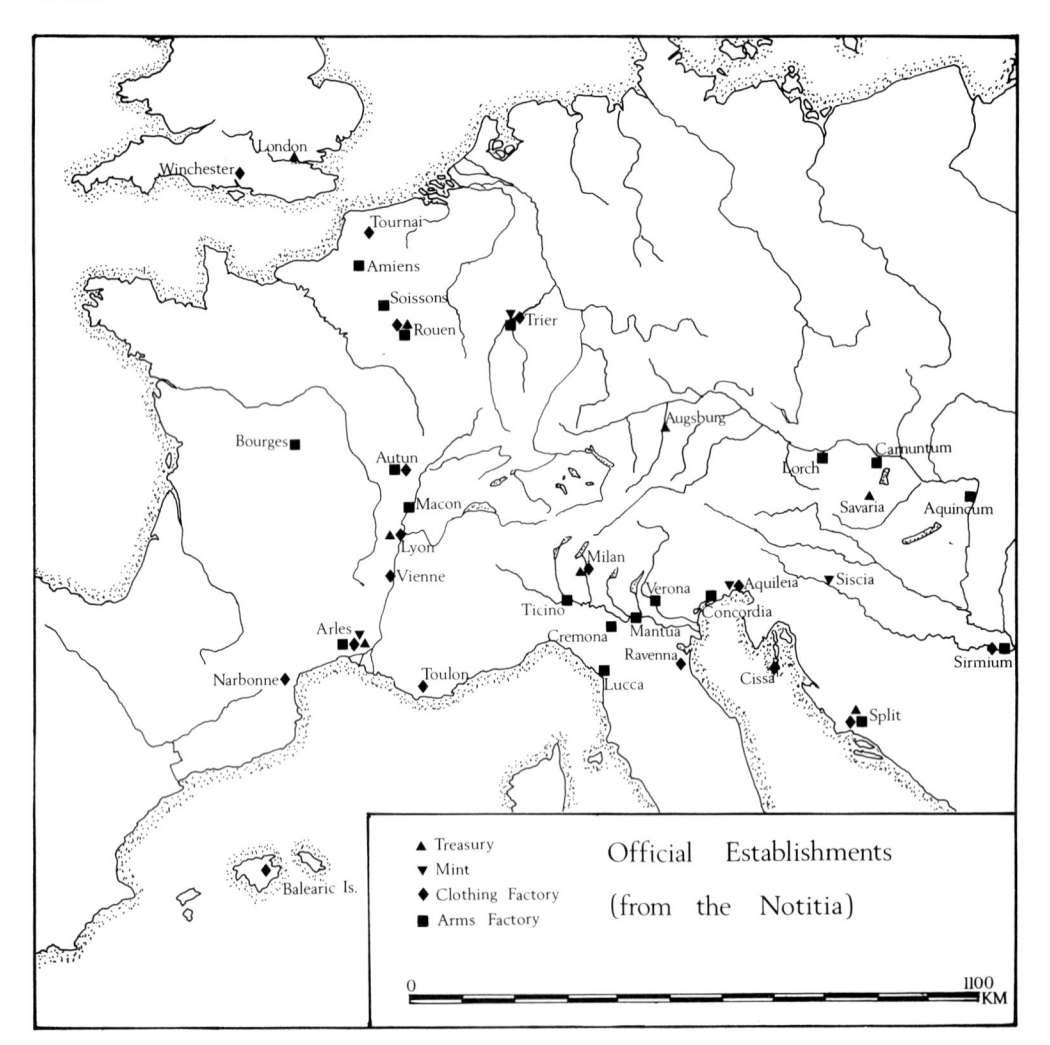

50 Map of official establishments in the western empire

This leads on to the more general question of how widely troops were deployed in late Roman cities. Military installations of various kinds—arms factories, clothing stores, or arsenals—were located within several of the towns and cities dotted round the late Roman world.[152] At all such establishments one can expect troops to have been busy, and probably on detachment to guard them.

Also within the *Notitia*, however, are a number of postings, included specifically within the military pages of the document, which show that in several areas of the empire prefects in charge of Laeti and Suebi were appointed to provincial cities.[153] The lists omit any reference to Britain, but for Gaul, Spain, and Northern Italy, these postings show a distribution supplementing the frontier guard posts by a form of defence in depth. How large such bands of tribesmen-soldiers were, and whether they actually lived within the towns themselves, rather than in more rural sites, are questions which admit at present of no answer. It may be, however, that the locations given in the *Notitia*, which normally provides only a tribal name, mean that they were posted somewhere within the tribal area, not actually in the civitas capital itself.

The overall pattern provided by the study of these urban defences in the later Roman west confirms and supplements the story told by historical sources about barbarian raiding. It shows how the need for fortification was seen in the eyes of official strategic planners often deep in the heart of the provinces. By and large, towns and cities on the frontiers and in their immediate neighbourhood had already been provided with defences. Only in the aftermath of the documented barbarian raids into Europe in the mid and late third century were the majority of mainland European towns given any defence at all. Their provision, though expensive, was carried out by an efficient and comprehensive policy involving official assistance through the medium of military labour or governmental funding, aimed at securing key installations and routes. This in itself, however, was not sufficient. The cities and towns had been updated, but many of the problems posed by barbarian raiding demanded new defensive answers on the frontiers too. Developments in both areas progressed hand-in-hand.

The Frontiers I: The Rhine

The control of the long and straggling frontier line following the Rivers Rhine and Danube across nearly 1,000 miles of mainland Europe posed great problems for the Roman army. Strategically, a river did not form a good frontier: a river in antiquity, particularly one as large and as long as the Rhine or Danube, was a main thoroughfare for shipping and trade.[1] Although it was bridged at the major cities and fortresses, such as Köln, Mainz and Strasbourg, the need for road traffic across the river was by no means as great as it is today. To raiders or invaders from Free Germany, the Rhine did not constitute an insuperable barrier. Rather it constituted a visible boundary to the Roman world to which the Romans fell back after the abandonment of all their former territory east of the Rhine, the Agri Decumates, soon after 259–60. This long stretch of frontier was separated into individual commands, based on the main towns or fortresses like Köln, in Lower Germany, and Mainz and Strasbourg in Upper Germany.[2]

The provision of defences in the Rhineland area was a progressive development, here as in other places, in response to growing pressure from barbarian raids. During the second and third centuries defences along the river had been allowed to decay. Only on the lower reaches of the river north of Bonn or Remagen, where the frontier had not appreciably altered since the Julio-Claudian period, were frontier defences still operative. Little rebuilding or refurbishment, however, is known or attested in any area in the first half of the third century. The main emergencies occurred in the late 250s and 260s. During these years some measures were taken to withstand barbarian pressure. There was some rebuilding at Trier during the period,[3] and an inscription from a *burgus* at Liesenich, the exact site of which is not known, was set up under Victorinus in 269.[4] To this period also can be assigned the beginning of construction of the defences of the civilian settlement at Mainz,[5] which lay between the legionary fortress and the Rhine. Eventually, it was enclosed by a pair of walls which enveloped a wide semicircular sweep of territory to the north and east of the legionary fort (fig. 52). A wall along the riverbank completed the circuit. The south-western sector of the walls ran through the site of the legionary fortress,

51 Map of the lower Rhine frontier in late Roman times. For numbered *burgi*, see Appendix 2, pp. 270–1

which was in occupation until about 300. The first mention of the status of Mainz as a *civitas* is found on a milestone of 293–7.

The fortification of the site at Mainz was undertaken in at least two stages: in the mid-third century, a wall enclosing all the area between the fortress and the Rhine was built. For a time, fortress and town, later to become a *civitas*, existed together. Later, the fortress was abandoned, and by the late fourth century the walls of the *civitas* were extended by the addition of a further wall to include some of the high ground on which it had stood.

Across the Rhine from Mainz, lay the fort of Castellum (Kastel).[6] A lead medallion (fig. 17, p. 43), found at Lyons, shows the twin fortified sites, separated by their bridge.[7] A fort at Kastel existed from the first century onwards, and may have been given over to civilian use relatively quickly. The small town was probably enclosed by walls early in the third century, at the same date as the other settlements on the right bank of the Rhine. Part of the late Roman circuit has been located on an alignment different from both the first century and the medieval defences.

It seems likely, too, that a series of posts along the road from Köln to Bavai was drawn up and fortified under the Gallic Empire.[8] The walls of Famars[9] and Bavai[10] were built in at least two separate stages, and the earlier of the two phases may belong to this period. Other fortifications in this area were a little less substantial. Turf and timber fortlets surrounded by ditches have been discovered along the Köln-Bavai road at Liberchies,[11] Taviers,[12] Braives (fig. 56),[13] Heerlen[14] and Villenhaus (fig. 53).[15] In several cases, the small fort was later superseded by a stone-built post—at Liberchies, for example, it was in a completely different place. The road-posts were not mere *burgi*, like the later systems of road fortification set up under Valentinian, but were also the protection for roadside villages and contained *mansiones* for staging posts along the imperial postal routes. At Heerlen, there was a substantial bath house inside the defended area, and traces of palisades and interior wooden structures at Villenhaus.

As a further protection for this route, there were also watchtowers built at Goudsburg[16] and Valkenburg.[17] Such *burgi* were not only built officially and for the protection of the road-system. A villa at Froitzheim was defended at a date before 274 with at least three watchtowers of similar type. Later these were incorporated into a continuous wall-line which may have encircled the main part of the estate.[18] Elsewhere, too, there is further evidence that the roads were among the top priority for defence at this period: apart from finds of milestones, which are relatively common for the period of the Gallic Empire, the small forts at Senon[19] and St Laurent[20] and perhaps Larga[21] have all produced finds datable to this time.

These small road-posts (fig. 54), constructed in the period 260–75, were normally of simple form, either earth-and-timber, or built of stone, but with none of the refinements normally associated with fortifications of slightly later date. At Larga, for example, the walls formed an almost perfect square, enclosing about 0.6ha, and were between 0.8m and 2m thick. At the corners were

THE FRONTIERS I: THE RHINE

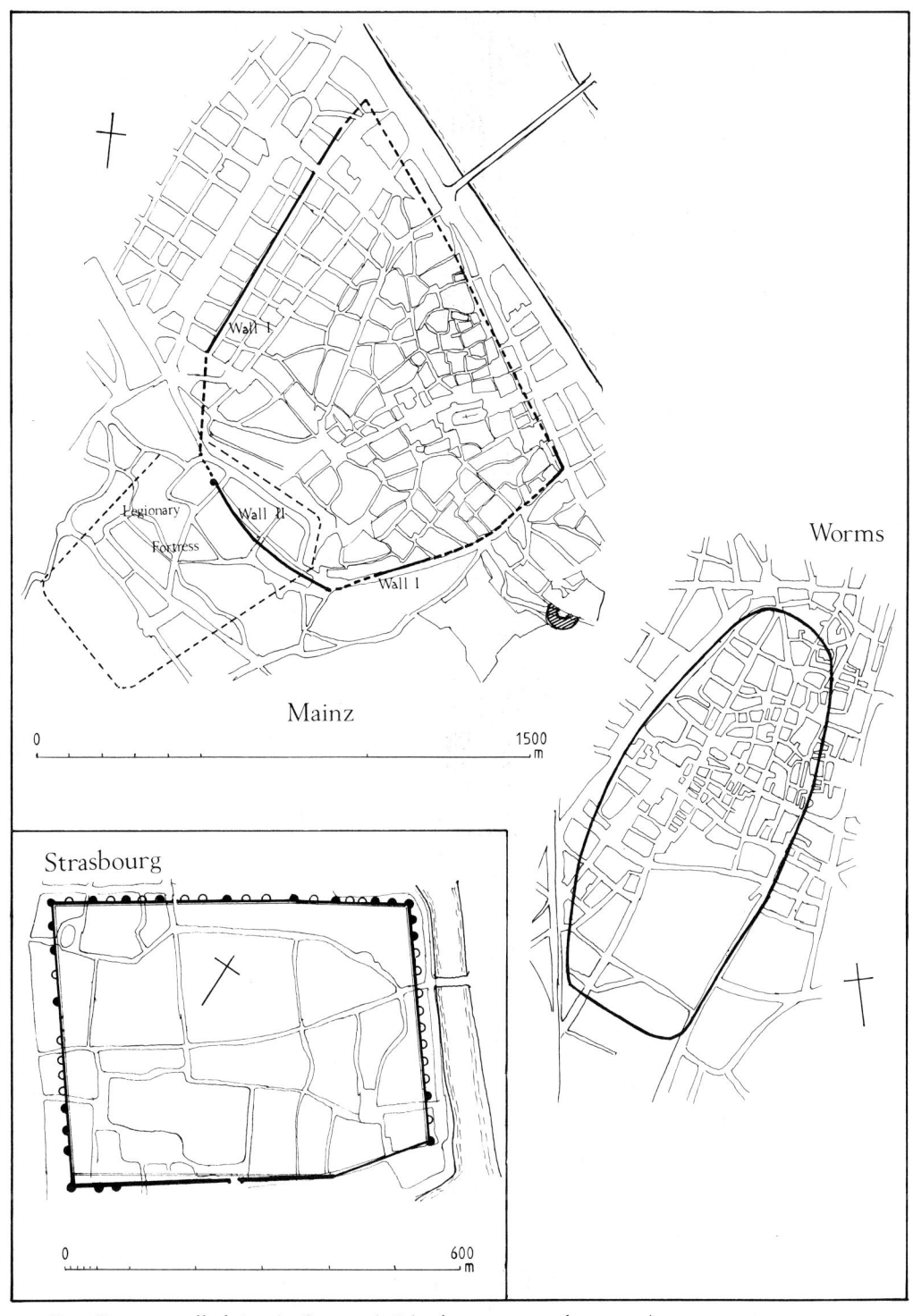

52 Late Roman walled sites in Germania I (*scale 1:20,000 and 1:10,000*)

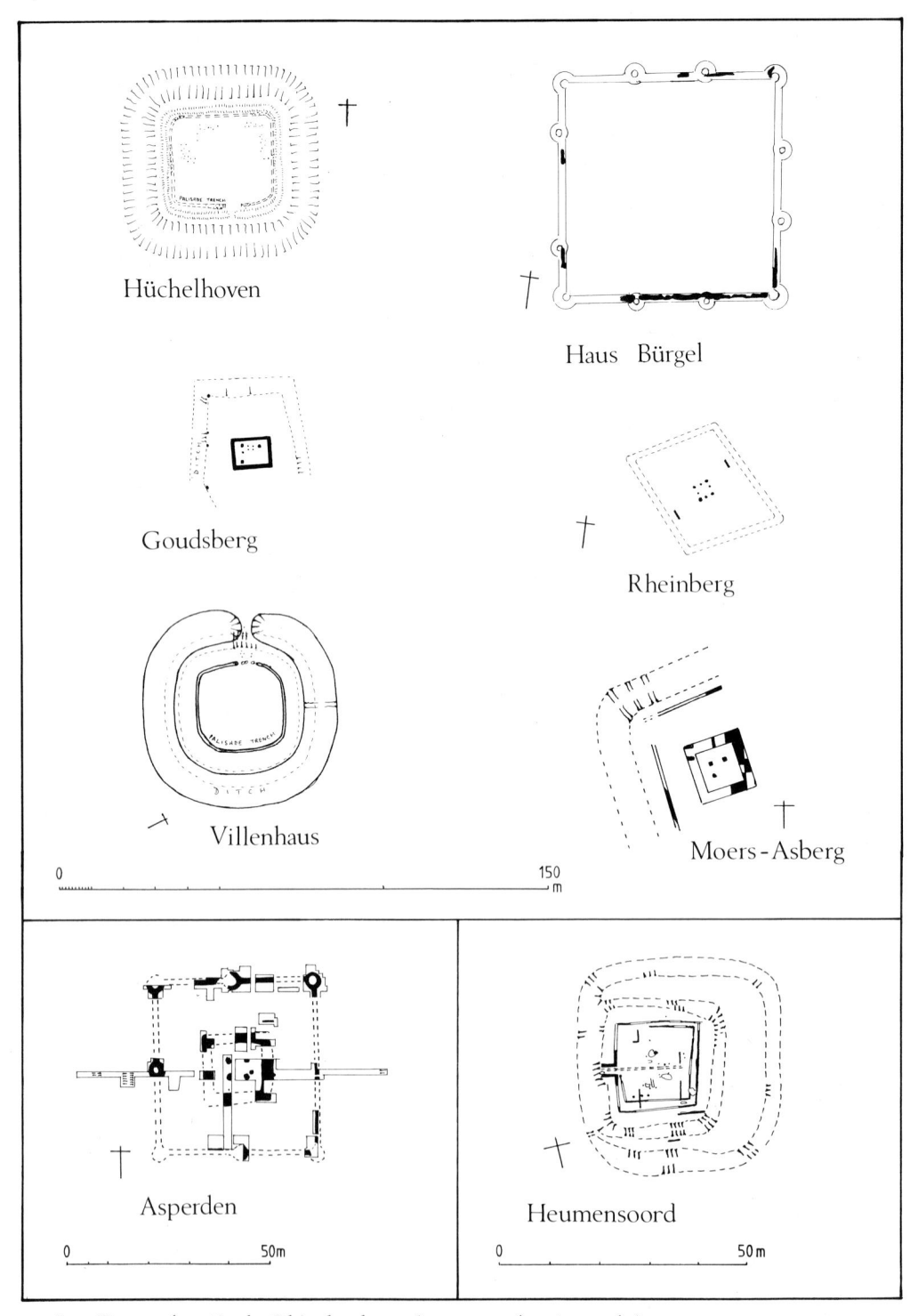

53 Late Roman *burgi* in the Rhineland area (*1:2,000 and various scales*)

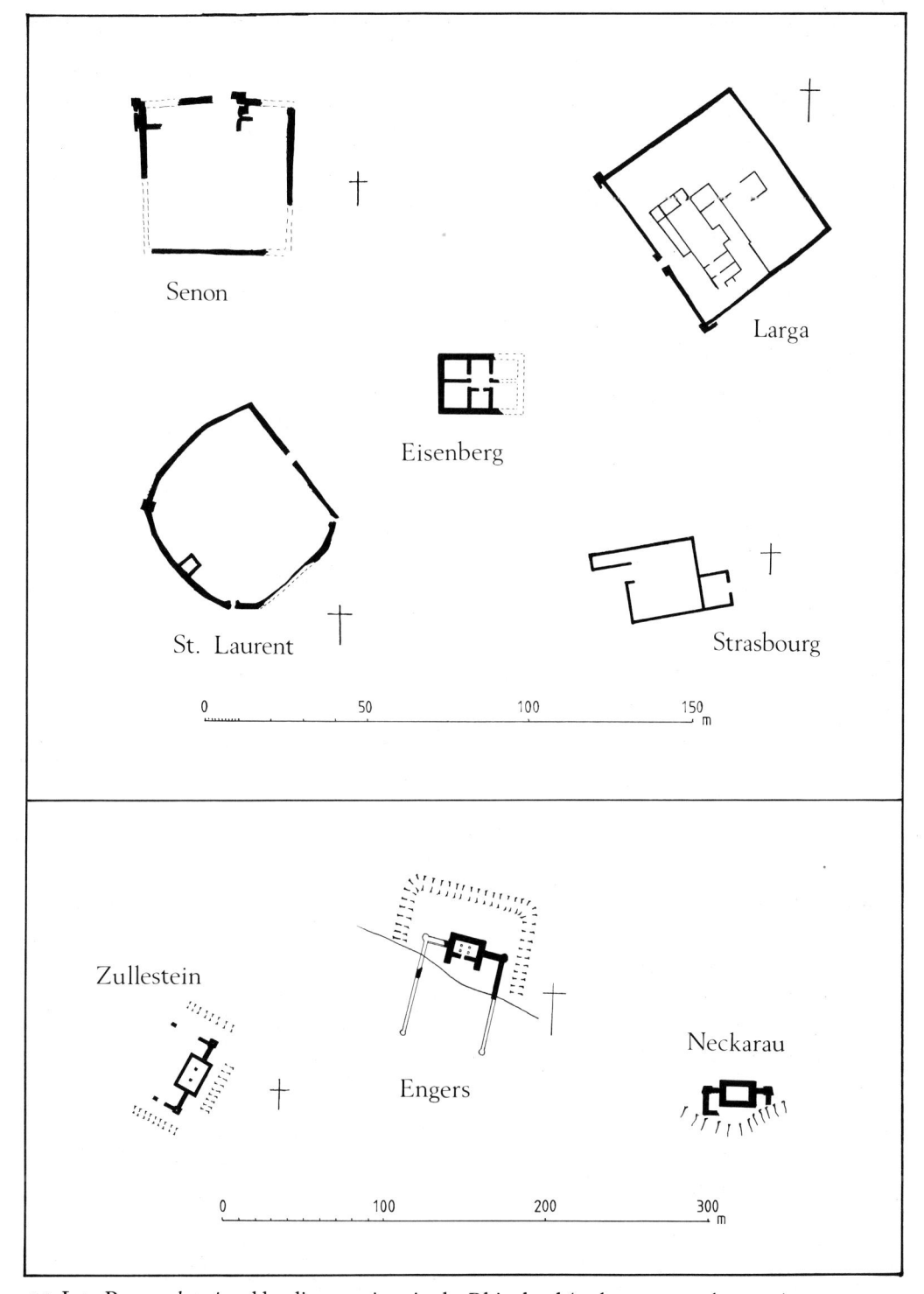

54 Late Roman *burgi* and landings stations in the Rhineland (*scale 1:4,000 and 1:2,000*)

rectangular or square projecting towers, and the main gate was flanked by a pair of projecting rectangular towers. Senon was a small *burgus* about 50m square with walls 1.2m thick built of small blockwork above a foundation of solid blocks. Re-used sculpture found in this foundation dates from before the mid third century. St Laurent is of similar dimensions to the site at Senon, roughly circular, with walls 1.55m thick containing a large amount of re-used material. Finds from the interior of the small fort were of a date no later than the third quarter of the third century.

The worst of the raiding from barbarians in the frontier zones found many of the towns and cities in the area well prepared, with defensive walls normally dating from earlier imperial days. At Köln,[22] for example, an expansive circuit of walls equipped with several large gateways and with hollow round towers astride the walls, is traditionally dated to the period of foundation of the colony under the emperor Claudius (fig. 6, p. 19), but not all such towns were given defences thus early. The walls of Xanten,[23] Tongres,[24] and Trier,[25] including the famous Porta Nigra, are all probably of second century date, and they display the concern for defence which was affecting the frontier areas virtually throughout the Roman period.

Both German provinces figure in the *Notitia Galliarum*, although the number of *civitates* they contained was few. In Germanica II, only Köln and Tongres merit the term, while in Germanica I, there were four: Mainz, Strasbourg, Speyer and Worms. At Speyer,[26] virtually nothing of the late Roman defences is known. The transformation of the legionary fortress at Mainz into a substantial town on a slightly different site, nearer the River Rhine, during the course of the third century, has already been noted.

At Strasbourg,[27] the legionary fortress of the *Legio VIII Gemina*, of second century date, itself became transformed into the later *civitas* (fig. 52). Its early defences were only 1m thick, and constructed in small blockwork and tile courses. Various phases of destruction and rebuilding inside the fort have been recognised stratigraphically, but the defences, of normal second century type, do not appear to have been drastically altered until the fourth century. At that date, possibly under Julian or Valentinian, the earlier walls were strengthened by the addition of an external wall all round, more than redoubling the thickness and bringing the total width of the defences to 3.4m. This additional wall was built of small blockwork and contained much reused material in the foundation courses, formed of two or three courses of large blocks. There were wooden piles under the walls pinning the foundations to the ground. This wall was provided with U-shaped projecting towers, spaced about 25m apart, and there were larger towers, forming three-quarters of a circle, at the corners of the rectangular circuit.

Two of the gates are known, and it seems clear that the existing gateways of the legionary fortress were adapted as the gates of the *civitas,* probably preserving the interior road layout. It is not yet clear how the interior arrangements mirrored the change from legionary fortress to *civitas*. Evidently military presence continued throughout the fourth century, since Strasbourg was the seat of the *Dux Argentoratensis*, who was in charge of this area of the *limes*, but the civilian settlement which had lain originally outside the fortress was destroyed in the later third century, and may at that date have been included within the walls. The elevation of the fortress to a new status as *civitas* under Diocletian, for example, would be a fitting moment for the absorption of the *civitas* inside the military area, but what appears to be the main reorganisation of the fort's interior has been dated archaeologically to the reign of Constantine.

Worms[28] was in the late Roman period enclosed by an elongated oval of walls, traces of which can still be seen in various parts of the city (fig. 52). Medieval walling has obliterated much of the Roman work, but the walls were built of regular small blockwork on a foundation of large irregular mortarless stones. No towers have been found, nor have the Roman gates been located, though there is an undated inscription which shows that a local dignitary, C. Lucius Victor, paid for one of the city gates.

In other provinces on or near the frontier, Belgica I and Maxima Sequanorum, the cities show a similar mix of newly walled and long-protected sites. In Belgica I, the chief city was Trier,[29] which lay uncomfortably close to the frontier, and therefore was walled by the end of the second century (fig. 6, p. 19). The second city was Metz,[30] where a late Roman circuit of walls of ambitious expanse, incorporating a large and extensive irregular area, was constructed (fig. 57). Elsewhere, there were wall circuits at Toul,[31] Grand[32] and probably at Verdun.[33] Those at Toul and Grand reflect the plan of the earlier walls of Trier in the design of their towers astride the walls.

The entry in the *Notitia Galliarum* which deals with Maxima Sequanorum, which covered an area roughly coextensive with modern-day Switzerland and the Jura, is one of the most puzzling. It lists only three *civitates*—Besançon, Nyon and Avenches, together with four further sites described as *castra*—Basel, Horbourg, Windisch and Yverdon. At all four of the *castra*, but at none of the *civitates*, some evidence for late Roman defences exists. Besançon, the capital, had been an expansive and spacious town in earlier days, but, when Julian visited it in 360, he reflected then on the former glory of the place, and the small size to which it had shrunk.[34] Finds of late Roman date have been concentrated in the area of the citadel, and this was probably the site of a small walled late Roman *civitas*, perhaps a reduction from an earlier walled circuit.[35]

No evidence at all for defences survives from Nyon,[36] on the Lac Leman, though remains of the early colony are well known. The site lay on a slight rise eminently suitable for defence. At Avenches,[37] the first- or second-century expansive circuit of walls still survives in a condition which enables it to be followed round much of the circuit (fig. 6, p. 19). To the south-east, this enclosed the modern village, nestling against the theatre at its north-western end. If a reduction in the wall-circuit was undertaken in late Roman times, this may well have enclosed the hill on which the village now stands, and which is encircled in part now by walls presumed to be of medieval date. In addition to these sites, Augst (Augusta Raurica)[38] continued life as an open settlement, only defended by walls built in the first or second century which cut off part of the southern approaches. At Lausanne,[39] too, the site of an earlier extensive settlement, traces

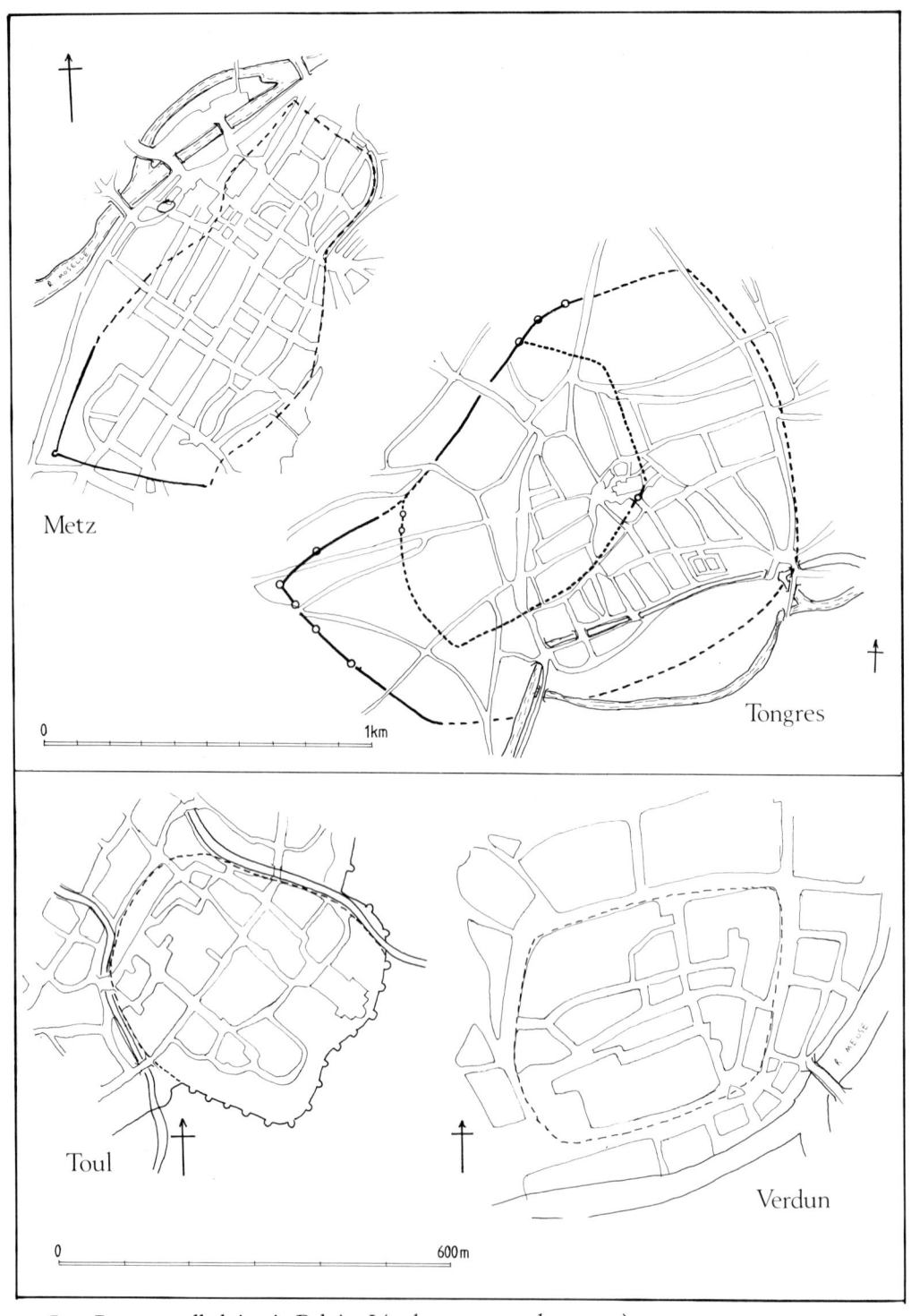

55 Late Roman walled sites in Belgica I (*scale 1:10,000 and 1:20,000*)

of a defensive wall protecting the most prominent part of the town have been found.

As far as a distinction can actually be made in the later Roman period, the specifically military fortifications on the frontiers show a comparable mix of new and pre-existing forts and installations. The main area of new building on the frontiers was the upper Rhine, where the loss of the Agri Decumates, east of Mainz and Strasbourg, left a stretch of the River Rhine, from the Bodensee to Mainz and a little northwards unprepared for use as a frontier. Surprisingly enough, however, the area north of Köln also seems to have been regarded as less important. Here, proven late Roman sites are few in number, and many of those found are small fortlets or watchtowers rather than accommodation for a sizeable garrison.

Apart from the enigmatic remains at the Brittenburg (fig. 56),[40] actually on the North sea coast, there are no known late Roman sites in the 110km between the mouth of the old Rhine and Nijmegen, despite the fact that on the late first- and second-century frontier in this area there were at least five known forts holding regular or auxiliary regiments.

Somewhere in this stretch of river or, more probably at Druten on the Waal, which now formed the frontier, lay the site of Castra Herculis.[41] At Utrecht there may have been some rebuilding of the second century fort in a later campaign,[42] and Rossum too is a potential candidate for a late Roman fort in this area.[43] None has yet been properly substantiated. Even at Nijmegen, the earlier legionary fortress, traces of late Roman occupation are slight.[44] After about 270, the former site of the second- and third-century town (Municipium Bataviorum) was abandoned, and a late Roman fortified site was built on the top of the Valkhof hill, which had held the legionary fortress until the turn of the first century. Argonne sigillata is common from the Valkhof area, and the coin finds from there date almost exclusively from the late third and fourth century. On the western slope of the hill, a defensive ditch has been found and may represent the late Roman fortification of the site. Residual pottery of the first to third centuries was discovered in its fill.

Between Nijmegen and Köln, the number of known late Roman fortifications is greater, though the problems of authenticating suspected sites of earlier periods still remains. Two of the more impressive examples of late Roman military planning lie actually on the River Maas, south of the Rhine, but in an exposed zone where the planning of defended sites on both rivers must have been conceived of as a whole. At Asperden,[45] there lay a *burgus* consisting of a single square tower with solid masonry walls (fig. 53). This was surrounded by an outer square defence, also of masonry, with circular towers at the corners and midway along each side. The outer enclosure was surrounded by at least one ditch and there were traces in excavation of a reconstruction of the site in wood later than the stone-built period.

The dating of the site rests mainly on coin evidence: of 78 coins only two were of the period of the Gallic empire in the mid-third century and 70 belonged to the period after 367, with the majority of these after 380. The site is therefore likely to be a road fort of the Valentinianic period, and is probably one of a system of such posts.

The fort at Cuijk[46] lay nearby on the river Maas itself (fig. 56). It had been the site successively of an early fort with a double defensive ditch, a settlement with Romano-Celtic temples in the second and third century, replaced in late Roman times by a fort, three sides of which have been traced. There were two main constructional phases in the late Roman period: first, possibly under Constantine I, a palisade of great stakes was set in a triple line along the inside of a double ditch. Buildings began to be placed inside this forming lean-to structures along the outer defences. In the second phase, the palisade was replaced by stone walls 1.5–1.9m thick with externally projecting semicircular towers. A stone building, perhaps a granary or magazine, with a floor supported on pillars, was built against the south wall. On the basis of coin finds, the second period of the fort's use is dated to Valentinian.

On the Rhine itself, the nearest fort east of Nijmegen lay on the Qualburg hill.[47] Here a series of late Roman ditches (but no palisade or wall) in connection with pottery and finds datable to several periods of the later third and fourth centuries have been found. The ditch, 16m wide, had filled up gradually in Roman times with consecutive layers dating from different periods. The earliest of these contained pottery datable to 270–300, and following layers contained material of 300 or just after, and of 330–40, including some coins of that period. Later layers contained pottery of the late fourth and early fifth centuries. The site was probably first fortified in the period 260–70, and the limited examination to which it has been subjected suggests it had a long period of use, though the form of the fortification itself is not yet known.

South of the Qualburg, following the course of the Rhine, late Roman occupation, perhaps a *burgus*, is known from Altkaltar,[48] the site of an earlier *ala*-fort. Traces of a small fortlet, a wooden tower enclosed by a rectangular ditch, have been revealed at Rheinberg, while on the site of the *limes*-fort of Moes-Asberg (Asciburgium) there was also a small *burgus* in the fourth century.[49] A tower, 18m square, with central pillar supports for an upper storey, was surrounded by a wall and ditch (fig. 53). Although examination has been limited to a small portion of the site, coins of Valentinian I and late fourth-century pottery confirmed the dating.

Occupation of a more substantial nature in the late Roman period is attested or to be expected at other sites on this stretch of the frontier between Nijmegen and Köln. At Xanten, the site of the late Roman fort of Tricesima has been located within the area of the Colonia Ulpia Traiana (fig. 61, p. 157).[50] The fortification, 357m square, clasped the central nine insulae of the Colonia within thick walls, protected by projecting bastions. Included within the area were the baths, *forum* and *basilica* and in part, their outer walls may have been used to buttress the defences. Not all of the circuit has been studied, but the indications are that there were four main single-portalled gates on the line of the cardo and decumanus. A single large ditch, normally following the line of the former street, lay outside the wall. The construction date of this fortification was difficult to determine, but

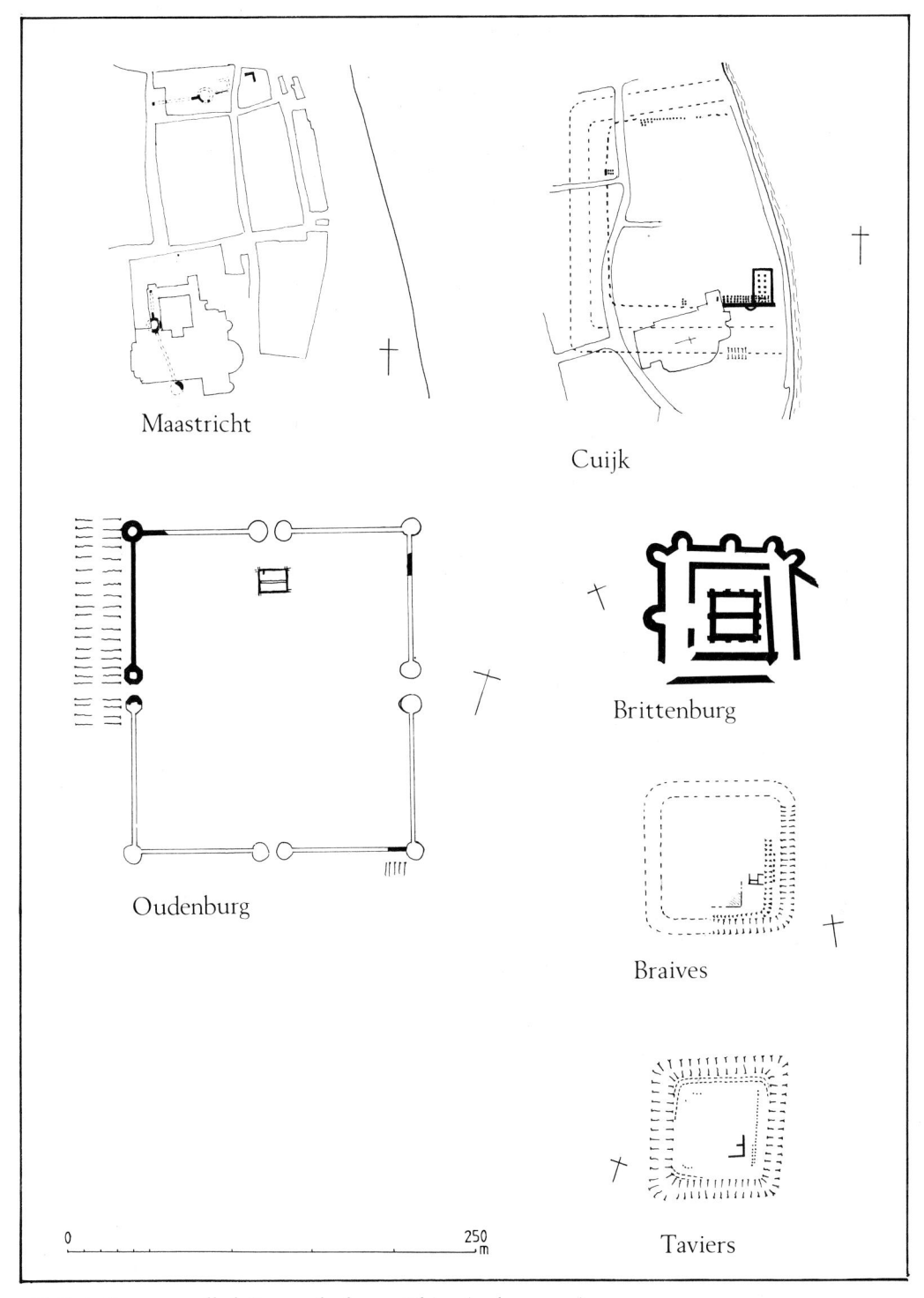

56 Late Roman walled sites on the lower Rhine (*scale 1:4,000*)

coins dating from after 268 were found in markedly greater numbers from the whole of the inner, defended area, by contrast with the rest of the site. This may suggest a late third-century date for the concentration of settlement in the central area, and thus a date for the building of the walls. Coin-finds from the site ceased abruptly with issues of Magnentius (350–3).

At Gellep (Gelduba),[51] the next site southwards, no trace of the late Roman fort has survived at all, through an extensive cemetery, heavily used in particular from the fourth to the seventh centuries, has been the subject of long and notable excavations. The relationship between late Roman fort and cemetery is unknown, but men buried in the cemetery must have formed part of the fort garrison.

More enigmatic is the site of late Roman Neuss (Novaesium).[52] The legionary fortress of early imperial date was abandoned by the second century, and remains of a small rectangular but undated defensive enclosure, surrounding little more than a *principia* of the fortress, have been found. It has been suggested that this was of second century date, and abandoned by the 260s. If this is correct, late Roman Neuss may already have been moved to the site which became the medieval town, an area which has produced a considerable number of late Roman coins. By contrast, late Roman finds from the legionary fortress site are unknown.

Between Neuss and Köln, at present on the Rhine's eastern bank, but originally, before changes in course of the river, on its 'Roman' bank, lay the small fortlet of Haus Bürgel.[53] It was of rectangular shape, almost 64m square, and its walls, a concrete core faced with small blockwork and double tile courses, are now incorporated within farm buildings (fig. 53). Excavation has been very limited, and has succeeded only in establishing the outline plan of the walls, which were originally protected by circular towers, set astride the walls. No interior buildings are known, but a clue as to the construction date is provided by the discovery of one stamped tile of the Treveran tile makers 'CAPONIACI', dated elsewhere in the region to the Constantinian period.

Opposite Köln itself lay the bridgehead fort of Deutz.[54] This lay astride the main road leading eastwards across the Rhine bridge (fig. 57). It was square, with thick walls of rubble concrete faced with small blockwork and tile courses. Round towers were built astride the walls, and like those of Haus Bürgel, they project further beyond the walls externally than internally. The gates, to east and west, were single portalled, and protected by a pair of heavily-built U-shaped towers. Traces of internal buildings suggest that a row of rectangular stone barracks lay on each side of the main central street. The construction date of the site, according to a lost inscription from it which survives as a hand-written copy[55] is the period 312–315, a date otherwise confirmed by tile-stamps discovered within the fort walls. Some of the stamped tiles found from the site are of a date somewhat later than this, and suggest later rebuilding took place.

Late Roman defensive arrangements south of Köln along the river bank continue with forts of more or less regular pattern and outlook, and roughly equally spaced. Within Germania II lay two further sites, Bonn and Remagen. At Bonn[56] in the early imperial period, there had lain a legionary fortress. The site

Boppard

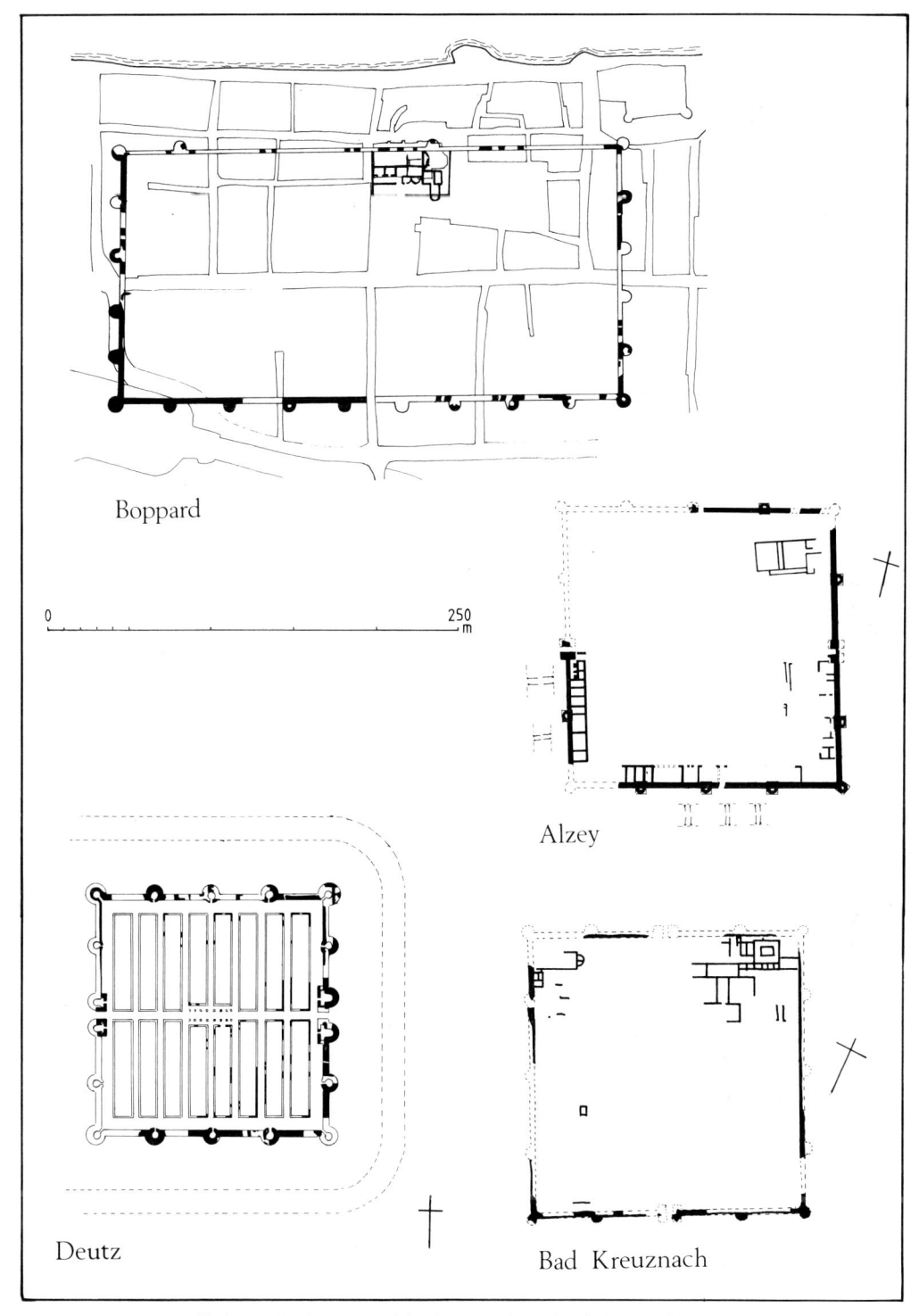

Alzey

Deutz

Bad Kreuznach

57 Late Roman walled sites in the area of the lower Rhine (*scale 1:4,000*)

was still occupied by a legion in the late Roman period, when a fortress of playing-card shape, protected with double ditches, was in use. The walls were relatively thin, and had no earthen rampart backing. There were internal rectangular turrets at intervals, and a number of small masonry projections on the interior face of the walls to support a rampart-walk. Only one internal building belonging to this phase of the fortress has been found: it lay in the north-east corner, and was long and narrow, subdivided into a range of small rooms and one large hall. It has been interpreted as a *fabrica*. The construction-date of this fortress cannot be precisely determined, but it was probably in use in the Constantinian period. Raids by Alemanni in the 350s clearly affected the site, for a buried hoard of gold rings and coins was dated to this period. Coin-finds from the site ceased with issues of Honorius and Arcadius, thus suggesting continued use into the early fifth century at least.

Remagen[57] was originally the site of an earth-and-timber fort until AD 70 when it was replaced by a stone-built fort which lasted at least until the late third century. Only three sides of this fort have been located, with the Rhine side missing. At some later date, probably in the late third century, these walls were more than doubled in width (from 1.28m to 3m) by the addition of a new exterior face (fig. 58). At the western angle, there appears to have been a large circular tower. Some of the facing now visible near the western angle of the fort is in small blockwork with triple tile-courses, and this is likely to belong to the late Roman phase. Fragments of interior buildings have been discovered, but none can be positively assigned to the late Roman period. Only few of the finds belong to the latest phases of occupation.

In Germania I, a similar pattern is maintained. A military commander called the *Dux Germaniae Primae* is mentioned in the *Notitia Dignitatum* but geographically he must have had the same territory as the *Dux Mogontiacensis* who has a surviving chapter in that document.[58] Thus many of the sites along this stretch of the Rhine, both those known from archaeological remains, and those of which no trace now survives, are known to have held a military garrison in the late fourth century at least. The majority of sites named in the *Notitia* lay actually on the Rhine itself, but roads further inland around Mainz were also protected by fortifications of martial aspect. One of these, Alzey, is recorded as holding a garrison.

On the Rhine, south of Remagen, the first major defended site was Andernach.[59] In general, the Roman walls followed the line now taken by the medieval defences, and enclosed an irregular D-shaped enclosure, its straight side along the Rhine (fig. 58). In Roman times, the area to the north west of the defended site, an old arm of the Rhine, probably formed a good harbour. This area was marshy, which helps to explain the irregular development of the fort towards the south and east. Most of the Roman walling is buried under medieval re-facing. Circular hollow towers sat astride the walls at intervals of 30–38m, and two of the towers were provided with small posterns through them. The main gates are only imperfectly known, but they consisted of a single entrance passage between a pair of rectangular towers projecting both inwards and outwards—the

Altrip

Andernach

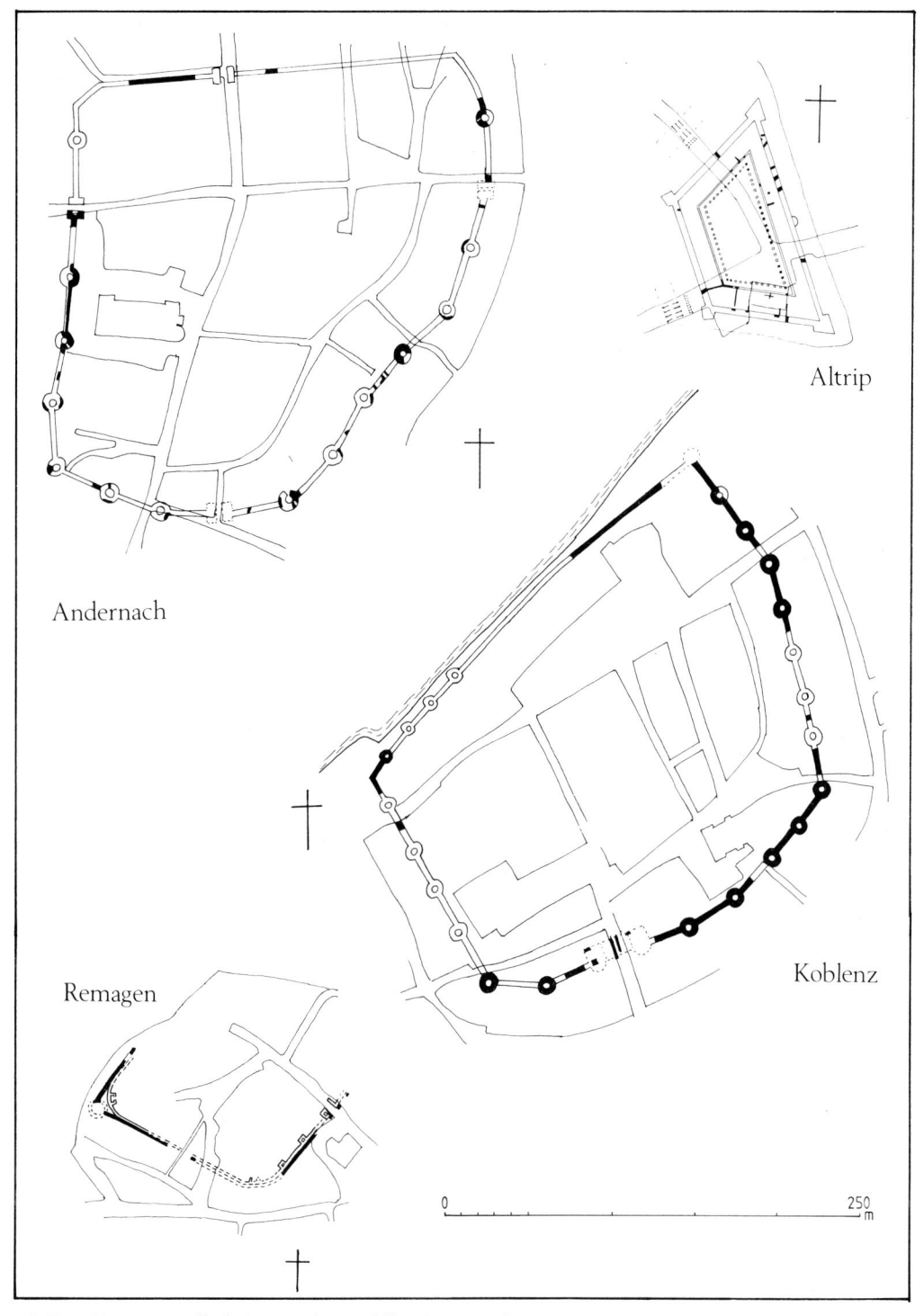

Koblenz

Remagen

58 Late Roman walled sites on the middle Rhine (*scale 1:4,000*)

so-called 'Andernach type'. The wall along the Rhine front appears to be of slighter build than the rest of the defences, and as yet no towers have been found on it.

On the south bank of the river Mosel, the fort of Koblenz[60] lay at the heart of the modern town, near the point where the Mosel joins the Rhine (fig. 58). The chosen site was a raised hillock which was surrounded by a circuit of walls. roughly D-shaped, with its straight side parallel to the Mosel. The walls were built of coursed small blockwork and had circular towers set astride them at intervals of 25–34m. Little survives above foundation level and research has been hampered by the present height of water in the Mosel which is above the bottommost Roman layers. Most of the remains are preserved in cellars, although in the construction of the thirteenth-century town walls, portions of the Roman walls were rebuilt and refaced. No archaeological evidence has been found which helps to give a firm date for the construction of the fort.

The fort at Boppard[61] lies on a small area of flat ground next to the Rhine (fig. 57). The site is uniformly rectangular, and parts of the walls are well preserved and have been meticulously surveyed and published. The foundations were composed of a thin layer of natural sand, capped by a packing of river boulders set in clay. Above this were two or three courses of squared blocks, some re-used from previous buildings. The wall, faced with small blockwork, was then built above this foundation, with slight offsets both internally and externally. The towers were semicircular and regularly spaced. One still preserves traces of Roman window arches: they were large openings 1.25m wide and approximately 1m apart at rampart walk level.

Little trace has been found of interior buildings within the modern village, although excavation under the church has revealed the remains of a late Roman church and font. This must date from the fifth century, and probably signifies a period when the military were no longer in occupation of the fort. The construction date is uncertain, but tile-stamps of the Constantinian period have been found in the walls. The comparative absence of a distinctive style of Rhineland pottery, typical of the Constantinian period, may suggest, despite this, that the site was not in use until after this period.

South of Boppard, the next fort site lay at Bingen,[62] at the crossing of the river Naue. Although imperfectly known, the late Roman site appears to have occupied the same site as the earlier imperial fort. Ausonius, journeying along the Mosel in the 360s, refers to Bingen's 'new walls', but no trace of these can now be seen. Between Bingen and Mainz, no forts are known on the river itself, but a series of watchtowers and an enigmatic fortification at Wiesbaden, possibly intended as a town wall, extended the cover of late Roman defences into the territory east of the Rhine round Mainz.[63]

Deeper within the province, however, two sites west of Mainz merit mention as members of the frontier defensive system. Both were new foundations of fourth century date, and occupied part of the site of previous open settlements. At Bad Kreuznach,[64] the fort was almost square (168 × 170m) with U-shaped towers astride the wall at intervals of approximately 40m (fig. 57). The gates, set centrally in the north and south sides, were each protected by a pair of rectangular towers of 'Andernach' type. Part of the walls, now called the 'Heidenmauer', still stands on the north-east. Excavations have shown that the fort was defended by double ditches.

A complex of interior buildings has been discovered in the northern corner of the fort, among which are recognisable parts of a bath house, and the remainder of which consists of an *atrium* with rooms of small size leading off a central courtyard. It is possible that this is a *mansio*. Its relationship with the rest of the fort, and in particular with the defences, is not well established. Although it fits neatly enough into the corner, it could be an earlier building, part of which was included inside the defended area by the builders of the fort.

Excavations on the fort defences have been aimed at providing a firm construction-date. Because of underlying earlier material, it has proved difficult to be certain which of the layers belong to the occupancy of the fort. The construction trenches, however, cut through layers deposited in the Constantinian period, and may not even have been laid down before the 350s, although the site's stratigraphy was not precisely clear about this. The coin series from the site shows a comparative lack of issues of coins round the 350s, and it may be suggested, therefore, on the basis of such evidence as has so far been published, that the fort was built in the late Constantinian period (after 337), and rebuilt or reoccupied in the late 350s or 360s.

Like Bad Kreuznach, Alzey[65] was also the site of a *vicus* before the construction of the late Roman fort (fig. 57). It is also of similar proportions, an almost perfect square of 165m. Projecting semicircular towers were spaced round the curtain wall at intervals of about 40m. These stand on bases 6m square which project slightly inside the fort. The known gates, to east and west, are of 'Andernach' type, with rectangular projecting towers flanking a single passageway. Of the interior buildings only the casemate buildings lining the inside face of the walls and one or two others apparently aligned with the fort walls can be assumed to belong to the fort.

On excavation of the fort, it was originally thought to be clearly datable to a late Constantinian period. Two periods of building were distinguished in the construction, and the barrack-blocks were adjudged to have been secondary to the main construction of the walls. The number of Constantinian coins found on the site suggested that the first period of military occupation took place early in this reign, beginning *c*310–315. This was followed by destruction at some time in the middle of the fourth century, and by rebuilding and the addition of stone-built barrack blocks possibly under Valentinian.

The publication of a section through the fort defences cut in 1925 completely revised the dating of the site, and it is from this section (fig. 59) that many later conclusions about the dating of the fort have been reached. It has been interpreted as showing that the fort was built after the mid fourth century on the grounds that the walls cut through a layer of burnt material and debris containing Constantinian and slightly later pottery, interpreted as the levelling of *vicus* buildings for the construction of the fort walls.

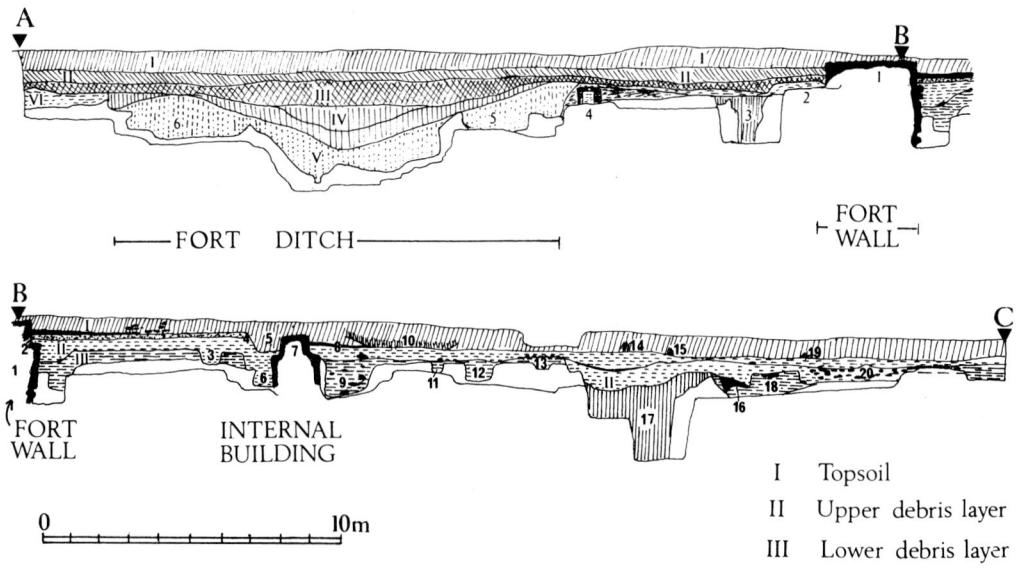

59 Section excavated through the fort wall and ditch at Alzey

I   Topsoil
II Upper debris layer
III Lower debris layer

Even in this section, which is so vital for the dating not only of Alzey but of the rest of the late Roman forts in the Rhineland, the evidence is not as clear as one would wish. The layer of debris II which the wall foundation was supposed to cut through is shown as overlying the offset of the wall foundation, and on normal stratigraphical arguments, should therefore post date the construction of the wall trench. In addition, if the layer of debris is interpreted as being the original levelling of the site, it should be equally visible both inside and outside the walls: oddly enough, however, it seems to stop short inside the fort. Despite the almost universal acceptance of the validity of the evidence of this section for a Julianic or Valentinianic date, the original interpretation of the fort's dating on the basis of this published section is still possible.

Between Mainz and Strasbourg, despite a number of names in the *Notitia*, which show which sites were in use, only one fort site has been excavated. At Gemersheim and Rheinzabern,[66] the production centre for late Roman tiles, no trace of the late Roman defensive arrangements have been found, and at Seltz,[67] nearer Strasbourg, the town on the hill top was enclosed with thick walls of late Roman style. Only fragmentary portions have been found. These three sites, however, complemented the towns of Worms and Speyer, which also lay right in the frontier zone. One small additional site lies near the Rhine at Altrip.[68] Limited excavation here has established that the fort was of trapezoid shape, enclosing an area of only 0.7ha (fig. 58). The corners were possibly defended by circular towers, but the evidence for this is very slight. The southern and eastern sides were protected by the Rhine.

Little of the actual wall remained: the foundations were pinned to the ground by upright stakes, and in the portions where the rare fragments of the wall facing remained, it could be seen that the wall was built of carefully faced blockwork.

On the inside of the walls were barrack-blocks 9m deep, with pillar bases 3m in front of them all round forming an internal arcade. Many tile stamps of the later Roman period were found on the site, as well as pottery and coins of the mid to late fourth century. Argonne sigillata was prominent among the finds, which suggest a construction date for the fort in the reign of Valentinian.

In addition to these larger forts, a number of smaller posts lay along the river frontier. Four of these, at Rheinbrohl, Engers, Niederlahnstein and Zullestein,[69] differ in detail, but their basic plan is the same (fig. 54, p. 141). A rectangular tower, very close to the river bank, stood in front of a walled enclosure, its corners normally defended by projecting towers, with an open side along the Rhine. These are normally interpreted as fortified harbours or landing stations. Little dating evidence from any of them has been found, but finds of Mayen ware from Engers have suggested a Valentinianic date, at least for a second period of occupation. At Zullestein, there was evidence that the tower had been two storied, with storage areas on the ground floor and living quarters above. At none of the sites has evidence been found for buildings within the small courtyard. Antiquarian records suggest that the wooden posts and jetties of a landing stage were once seen in Engers.

A distinctive feature of the frontier area in the later Roman period was the large number of small walled posts dotted along the major roads. The majority of these sites had been no more than *vici* in the earlier Roman period, and the provision of walls enabled them to command and control the key points in the road system at a date when communication and control of movement, particularly of potential invaders, was all important. It is difficult to distinguish these road posts from regular forts: in fact Alzey and Bad Kreuznach lie in a position a little behind the frontier line of the Rhine, and could be thought of not as frontier forts but as road posts. Others, of varying sizes and shapes, do not appear to have occasioned such regular planning. The majority describe a rough oval shape, like Bitburg,[70] Jünkerath[71] or Neumagen,[72] and the sizes vary, from the small, possibly unfinished fortification at Saarbrücken,[73] to a site like Saverne,[74] or Arlon,[75] both of which could be ranked as towns (figs. 60, 61). In almost all cases, nothing is known of the internal arrangements or of the dating of these sites: at Jünkerath, however, excavations early this century revealed a series of long narrow rectangular buildings flanking a central main street within the walls. The stone buildings had replaced earlier timber structures, probably in the late third century, but the defences had overlain the sites of some of the buildings. It is not certain therefore that all (or indeed any) of them were in use when the defences were constructed.

The only site of this type which can be dated with any accuracy is Neumagen, and the archaeological remains provide little help (fig. 61). In Ausonius' poem, the *Mosella*, however, the site is described as the famous camp of the divine Constantine.[76] This may mean only that Constantine was associated with the site because he stayed there, but it is usually taken to suggest that Neumagen was built in the reign of Constantine. By extension, but with very little solid evidence, the other fortifications of similar type[77] might be suggested to date from the same

156 THE FRONTIERS I: THE RHINE

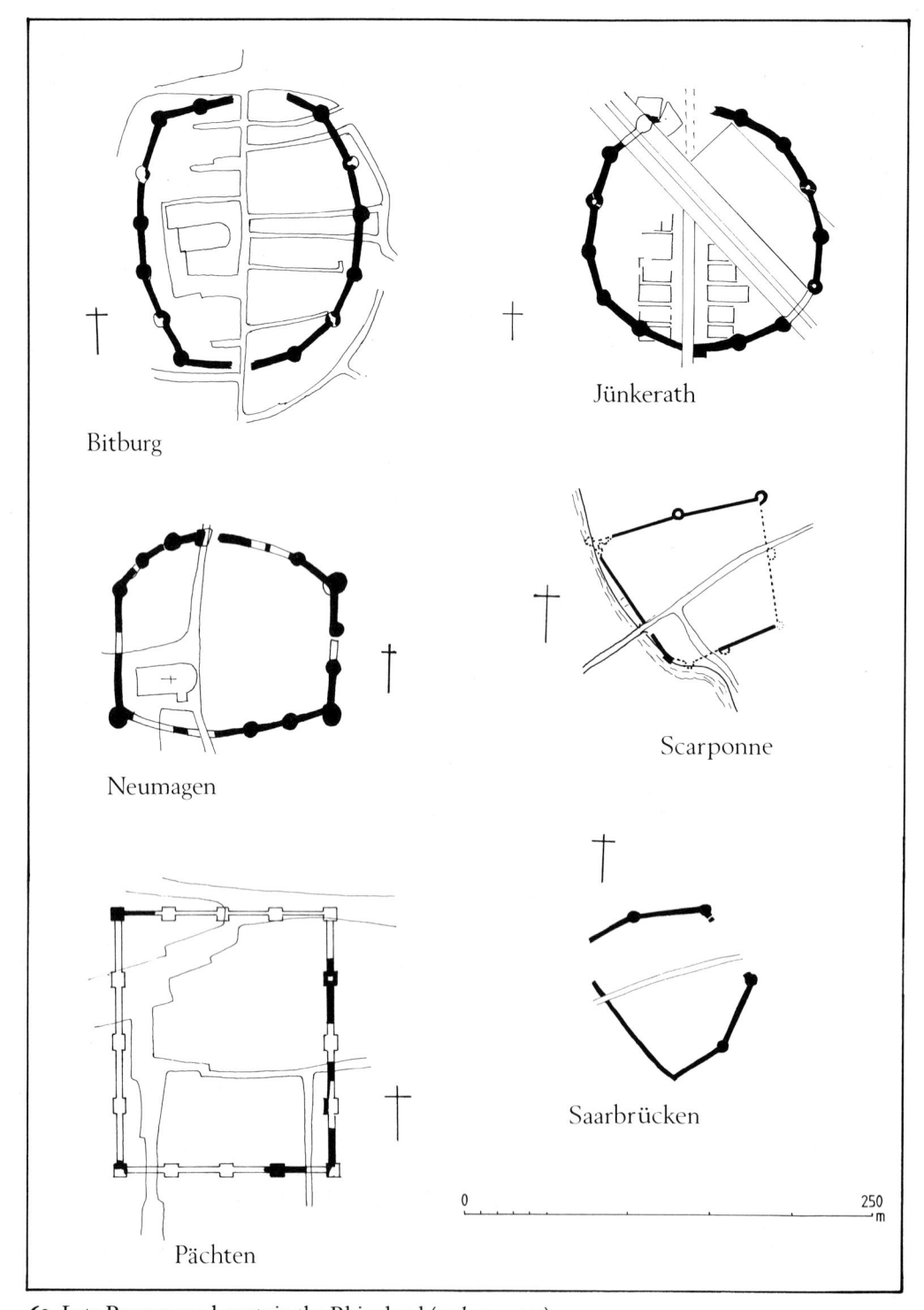

60 Late Roman road-posts in the Rhineland (*scale 1:4,000*)

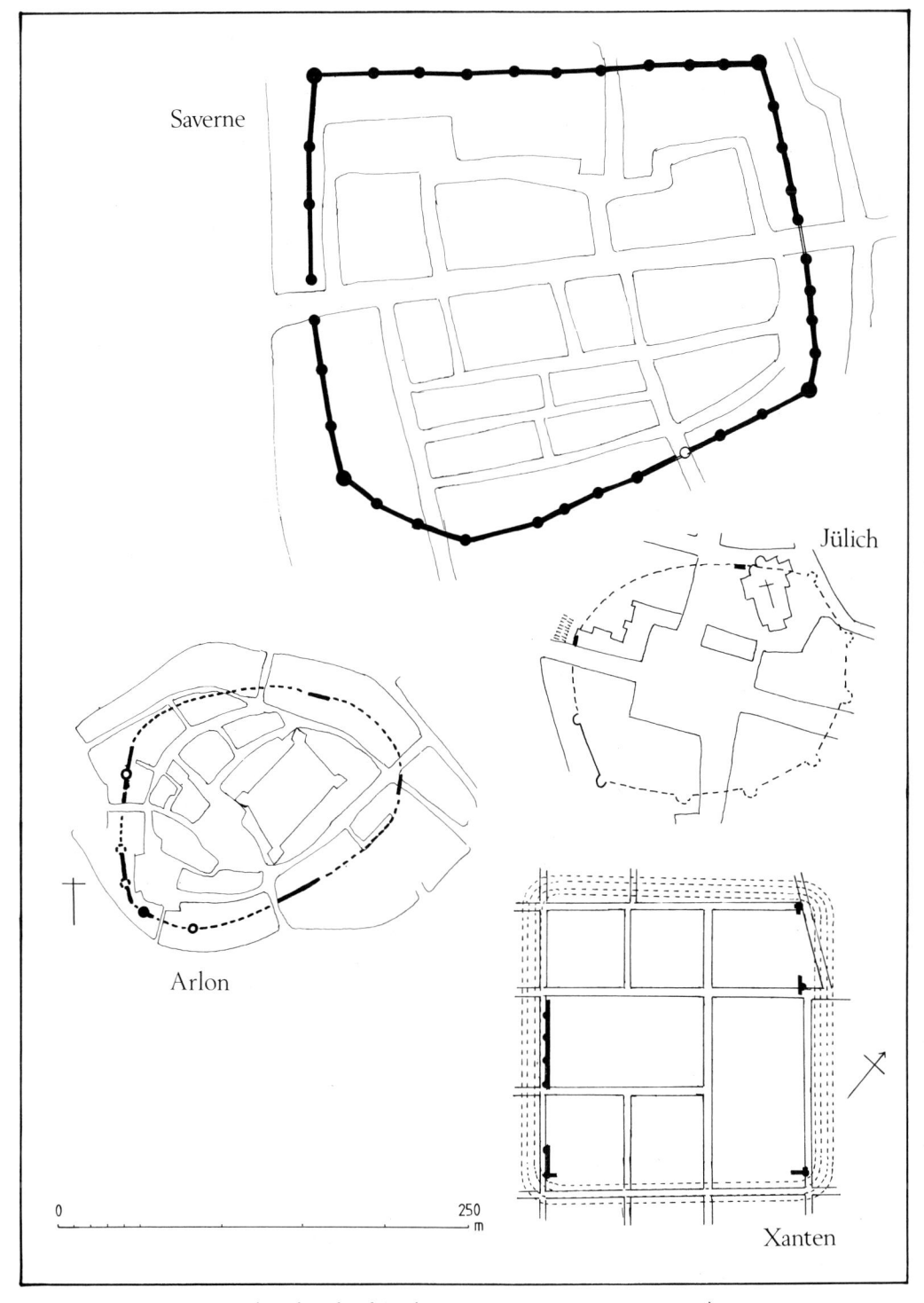

61 Late Roman posts in the Rhineland (*scale 1:4,000, except Xanten 1:10,000*)

period as part of a Constantinian programme to provide more security for inland areas, and defence in depth rather than on the frontiers alone.

This overall pattern in the German frontier areas is repeated further up the Rhine in Maxima Sequanorum. Here, a basic series of posts along the Rhine bank was established and supplemented by road forts, often of smaller dimensions, further inland. South of Strasbourg, the first fort encountered is Horbourg.[78] This fort is imperfectly known, but was almost square, similar to Alzey or Bad Kreuznach, and built like them on the site of a former *vicus* (fig. 63). The strong defensive walls were protected with U-shaped projecting towers. At a later stage, the gates were walled up, and smaller posterns opened nearby.

Two further late Roman installations lay at Sponeck and at Breisach, close to the Rhine, and now lying on the eastern bank. At Sponeck,[79] under the medieval castle, and just south of it, there are the remains of a rectangular tower with a wall branching off it incorporating a round tower. At this tower, the wall turns and runs to a second round tower lying at the very edge of the Rhine. Finds from the site, comprising coins and pottery of Valentinianic date, suggest the context of its construction and use. The whole layout is strikingly reminiscent of part of one of the small 'landing stations', such as Zullestein or Engers.

The Münsterhügel at Breisach[80] was defended in late Roman times by a wall which completely encircled it, slighter in build round the southern, irregular portion of the hill, but with a substantially thick straight wall backed by an earthen rampart cutting off the northern approaches (fig. 63). Roughly centrally in this wall was a gate formed by a single passageway flanked by a pair of small rectangular projecting towers. Double ditches outside the walls completed the defences on this side. Excavations within the walls, near the Münster, have revealed an extensive series of buildings of unknown purpose. The site was personally visited by the emperor Valentinian in 369, and archaeological finds— pottery sealed within the earth bank on the northern side—also suggest a Valentinianic date. This may be no more, however, than a date for refurbishment or renewal of the defences: the coin-series includes a list of issues of earlier fourth-century date.

Basel,[81] Civitas Basiliensium of the *Notitia Galliarum*, has revealed little trace of a late Roman settlement which might qualify as a *civitas* (fig. 63). On the Münsterhügel, however, there lay a relatively small late Roman walled post of irregular shape. Its walls, normally only 1.8m thick, were built on a foundation of re-used material, and little above this survives. No external or internal towers survive, although the plan incorporates some as yet unexplained changes of angle which may be caused by the presence of towers or gates. Only one building—a granary with internal floor-supports—has been located within the fort. Dating evidence is provided only by coin-lists from the site, which suggest that the site may have been in use from the early fourth century onwards. This might be used to argue for a Diocletianic or Constantinian date of foundation.

Almost opposite this fort, north of the Rhine, the traces of a late Roman fortlet—basically a tower 13m square with three-quarters round corner bastions—has been found.[82] Only the lowest course and the foundations have

62 Map of late Roman sites in the Upper Rhine and Danube areas. For numbered *burgi* see list of sites in Appendix 2, pp 271–3

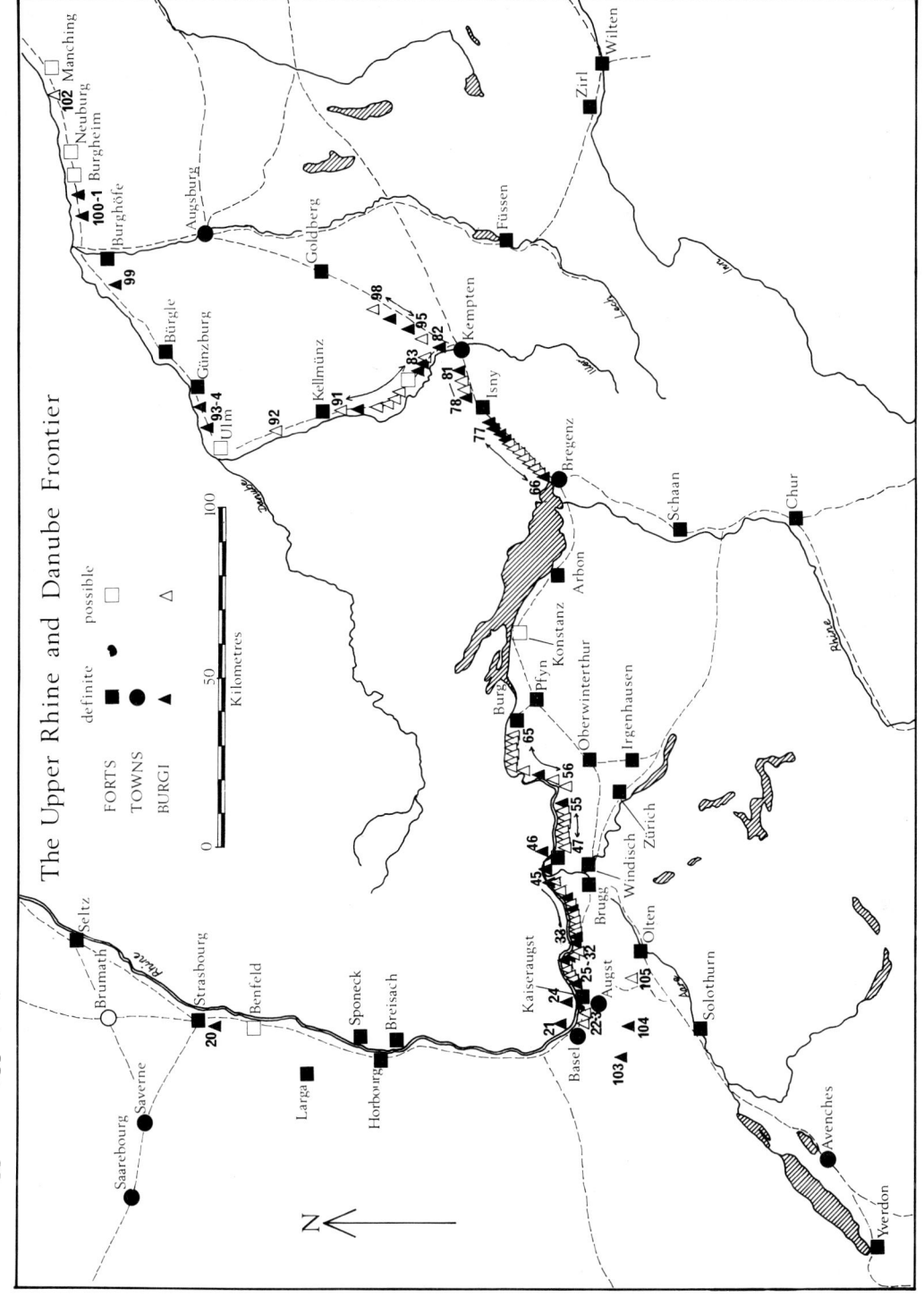

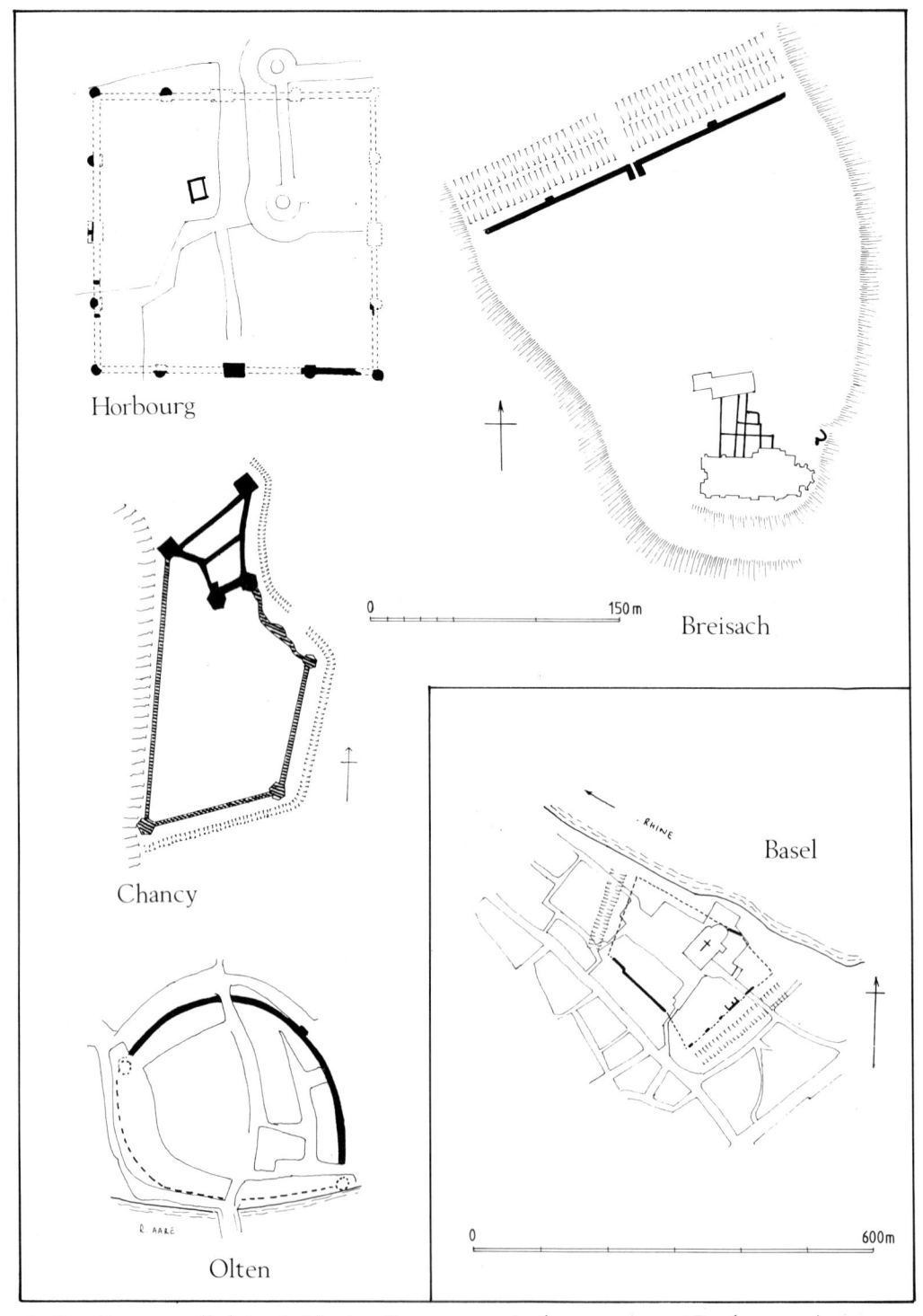

63 Late Roman walled sites in Maxima Sequanorum (*scale 1:4,000 except Basel 1:10,000*)

been discovered, but on the clay of the foundations were clear impressions of timber beams forming a lacing course. It is highly likely that this site, protecting the bridgehead from Basel across the Rhine, was the Valentinianic fort of 'Robur', whose construction is recorded in 374 by Ammianus Marcellinus.

On the Rhine's south bank near Augst lies the late Roman fort-site of Kaiseraugst,[83] whose main road also extended northwards to bridge the river (fig. 64). Opposite, the small fortlet of Whylen provided bridgehead defence, though most of it has been swept away by the Rhine, leaving fragments of only three towers in place.[84] The fort at Kaiseraugst was trapezium-shaped, with thick, freestanding walls protected by polygonal fronted towers on square bases which project externally from the walls. Externally, little of the facing survives, but internally the facing was of small blockwork with occasional tile courses. The gates were complex; to east and west they were probably single portals flanked by rectangular towers, while the south gate had two phases. At an earlier date, the gate was recessed within the fort, with a small courtyard in front of it, while later on this was replaced with a larger rectangular tower, astride the line of the fort wall.

Inside the defended area, several buildings have been found, including a bath building and other large structures which have been interpreted as magazines, official halls, and houses. There were, however, earlier buildings on the site, and the second and third century settlement had been more extensive than the later fort walls. The irregular layout of buildings within the fort may be explained by the re-use of earlier buildings whose foundations or lower portions were adaptable to the uses of the fort. One building of decidedly late date, however, was the church, found near the modern successor near the north wall of the fort alongside the Rhine. It probably dates to the fifth century, and contained a baptistery in one of the rooms next to the fort wall. One of the most noteworthy finds from the fort is a hoard of late Roman silver ware, found near the inner face of one of the towers on the south wall. This had been deposited in about 350.

Many finds of re-used material come from the walls, but none has produced clear dating evidence. Finds of tile stamps of the *Legio I Martia* had been thought to be indicative of a date in the reign of Diocletian for the fort's construction, since that legion is commonly supposed to have been raised under him, and the fort is approximately the correct size for a legion of the small, Diocletianic, type. Recent studies have shown, however, that this tile stamp occurs throughout the fourth century, and need not date the fort's constructional phases. Recent analysis suggests that the fort was constructed early in the reign of Constantine, but rebuilt later in the fourth century.

Between Kaiseraugst and Lake Constance, today the Swiss portion of the Rhine, there were two further small forts. Both defended river crossings, and both have accompanying traces of bridgehead fortifications on the north bank of the river. At Zurzach,[85] the road leading northwards and crossing the Rhine led between a pair of small bastioned forts (fig. 64). The eastern one was small and square and little is known of its interior arrangements, but the western fort was larger and more irregular. Inside it, excavations have revealed several buildings,

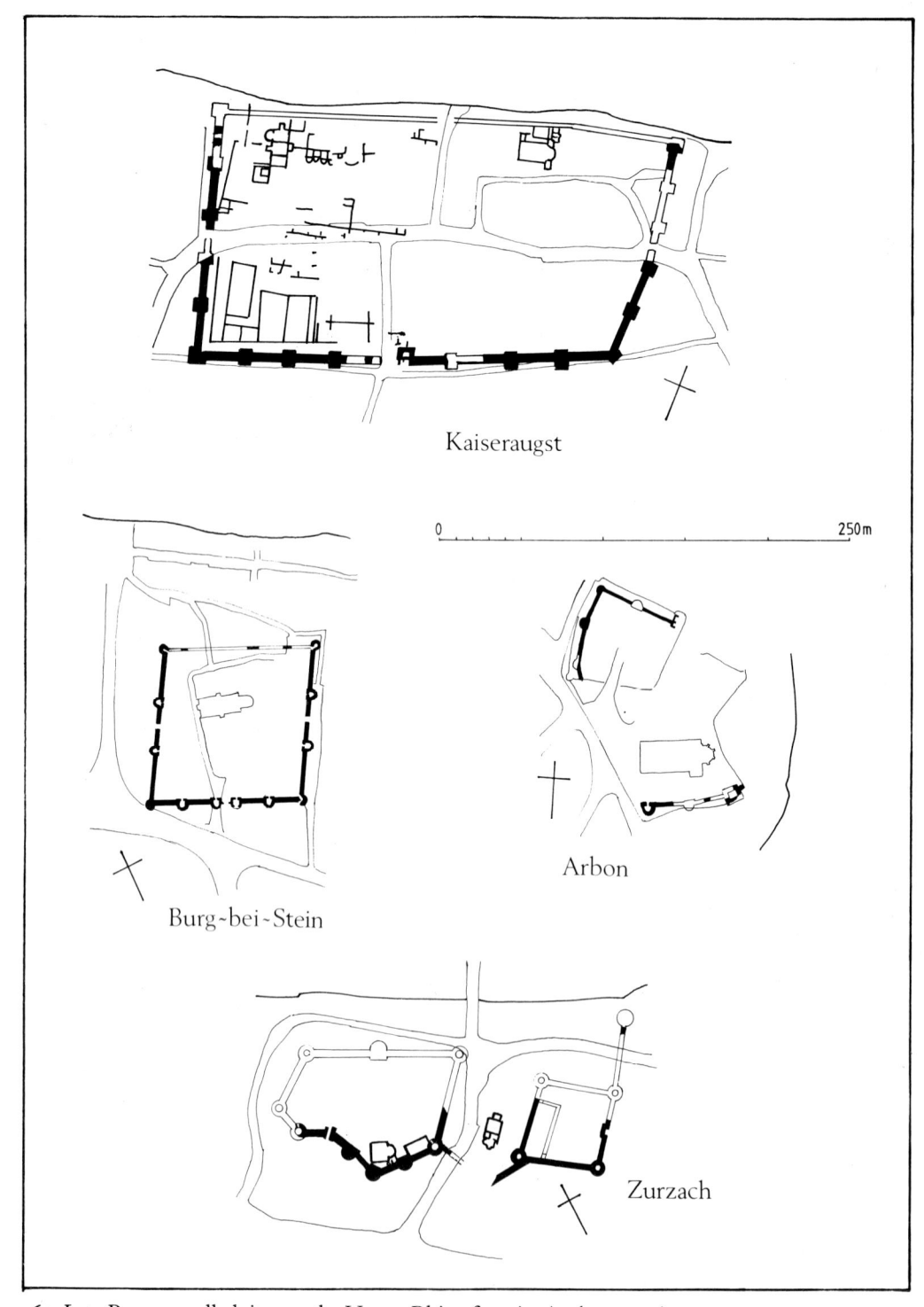

64 Late Roman walled sites on the Upper Rhine frontier (*scale 1:4,000*)

including a church with a baptistery in one of a series of small rooms between it and the fort wall. Earlier buildings around the church site had included barrack blocks which had been demolished and rebuilt at least twice. A date in the late third or early fourth century has been suggested for the fort's construction. On the opposite river bank, at Rheinheim, traces of a small square fort, protecting the northern end of the bridge, have been located.

Burg-bei-Stein am Rhein[86] is the site of another small fort, rhomboid and strong-walled, with projecting polygonal towers (fig. 64). Its facing was of medium blockwork, carefully coursed, and though they enclosed only 0.8ha, the walls were 3.3m thick. The south gate was a single portal between a pair of semi-octagonal towers. Recent excavations have concentrated on the fort cemetery. Outside a defensive ditch flanking the south wall, a cemetery of more than twenty graves, haphazardly oriented, was found to contain many rich burials, which included fine examples of fourth century glassware among the grave goods. The main feature of the fort, however, is the fragmentary building inscription[87] which, though incomplete, is closely paralleled by one from Oberwinterthur.[88] Both date to the reign of Diocletian and show that this fort was built between 285–306.

Lake Constance is formed by the Rhine as it progresses northwards from the Alps before turning westwards to run through Switzerland. The late Roman frontier ran along the lakeside and there were further small posts at Konstanz,[89] of which little is known, and at Arbon,[90] which lay on a small promontory jutting into the lake (fig. 64). Here there was a small and irregular bastioned fortification under the medieval castle. The remains of walls and towers, though still visible, have not been accurately dated.

A particular feature of the Rhine frontier from Basel eastwards is the number of small watchtowers and related posts lying close to the line of the military road and river.[91] The line is continued east of the lake of Konstanz in Raetia along the course of the road from Bregenz to Augsburg, and is an indication of the defensive importance of this awkward corner of the late Roman frontier.[92] The majority of the sites of the small towers, which varied in size from 7 to nearly 18m square, have produced finds of fourth-century pottery often diagnosed as of Valentinianic date. The tower at Kleiner Laufen has produced a building inscription which reveals that it was built in 371 (in the reign of Valentinian)[93] and that the site's name was Summa Rapida. A similar building inscription has also been discovered at Rote Waag, though the burgus from which it came has not been located.[94] At their closest, these towers, 42 of which are known in the stretch of Rhine between Basel and Burg-bei-Stein am Rhein, are a matter of 1–2km apart. Two of the sites in particular were curious rectangular blockhouses with a pair of apsidal ends. These were surrounded by a ditch and one had a separate bathhouse (fig. 65).

As in the German provinces, the network of roads in the area immediately next to the frontier was protected by a number of small posts and forts. Of these, only one, at Oberwinterthur,[95] is securely dated by its inscription of Diocletianic date. Curiously enough, this fort is of completely different shape and layout from the

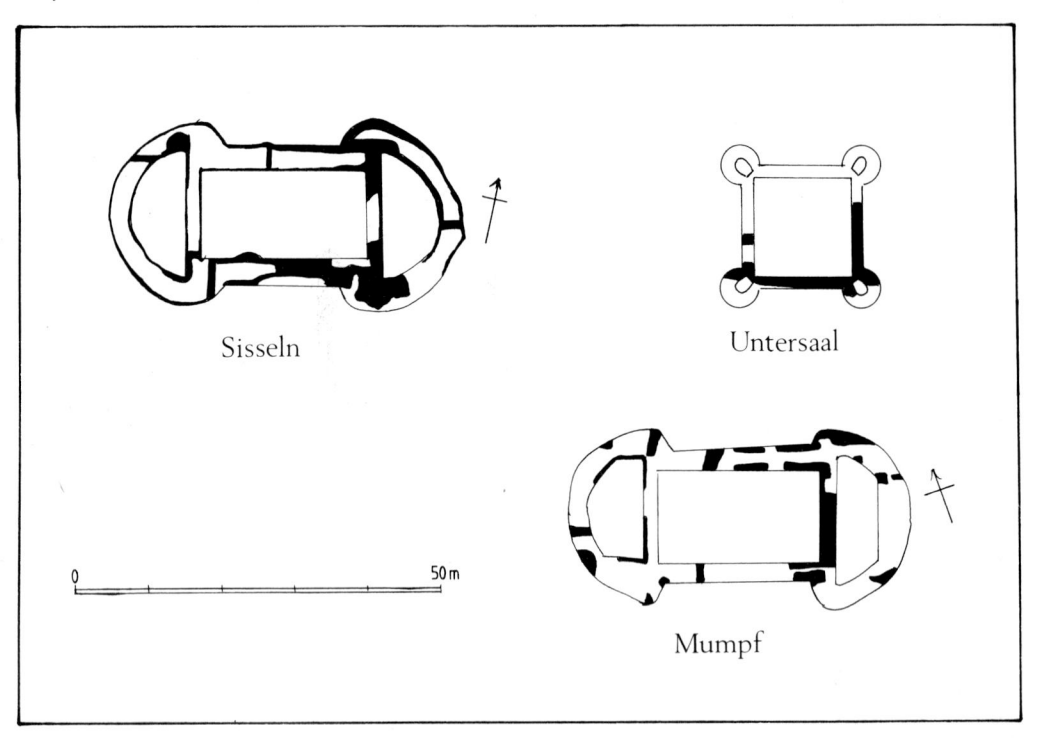

65 Late Roman *burgi* in the Upper Rhine area (*scale 1:1,000*)

rhomboid frontier fort at Burg-bei-Stein which was its contemporary. Its walls clasped an irregular low hill in a roughly oval shaped circuit. Elsewhere in the hinterland, the road forts display a variety of shapes and forms: a group which line the valley of the River Aare were D-shaped, with their flat side following the river's course. None of the three sites, Solothurn,[96] Olten[97] or Altenburg[98] near Brugg, have been dated archaeologically (fig. 66). At Yverdon,[99] a late Roman fleet base which also appears in the *Notitia Galliarum* as one of the *castra*, formed a rhomboid enclosure on the site of an earlier *vicus*. This, too, is not firmly dated.

Apart from a site at Pfyn,[100] on the borders of Raetia and Maxima Sequanorum, there are three further sites which merit mention. At Zürich,[101] the site of the late Roman fort lay on the Lindenhof, which has been comprehensively excavated. The defences completely encircled the whole of the small hill, forming an irregular rectangle adapted to the terrain. The walls were strengthened by rectangular towers, three of which housed gate-passageways, and U-shaped towers astride the walls. No interior buildings were discovered. The finds from the site were so similar to those of late date from Alzey that the excavator had no hesitation in embracing a Valentinianic date for the construction of the walls. More recently, however, this dating has been questioned: arguments based on the sparseness of the coin finds and on finds of Argonne ware point to a foundation in the late Constantinian period.

The fort at Irgenhausen[102] was built on the site of a *vicus* which had been occupied since the beginning of the first century AD, but which was destroyed in

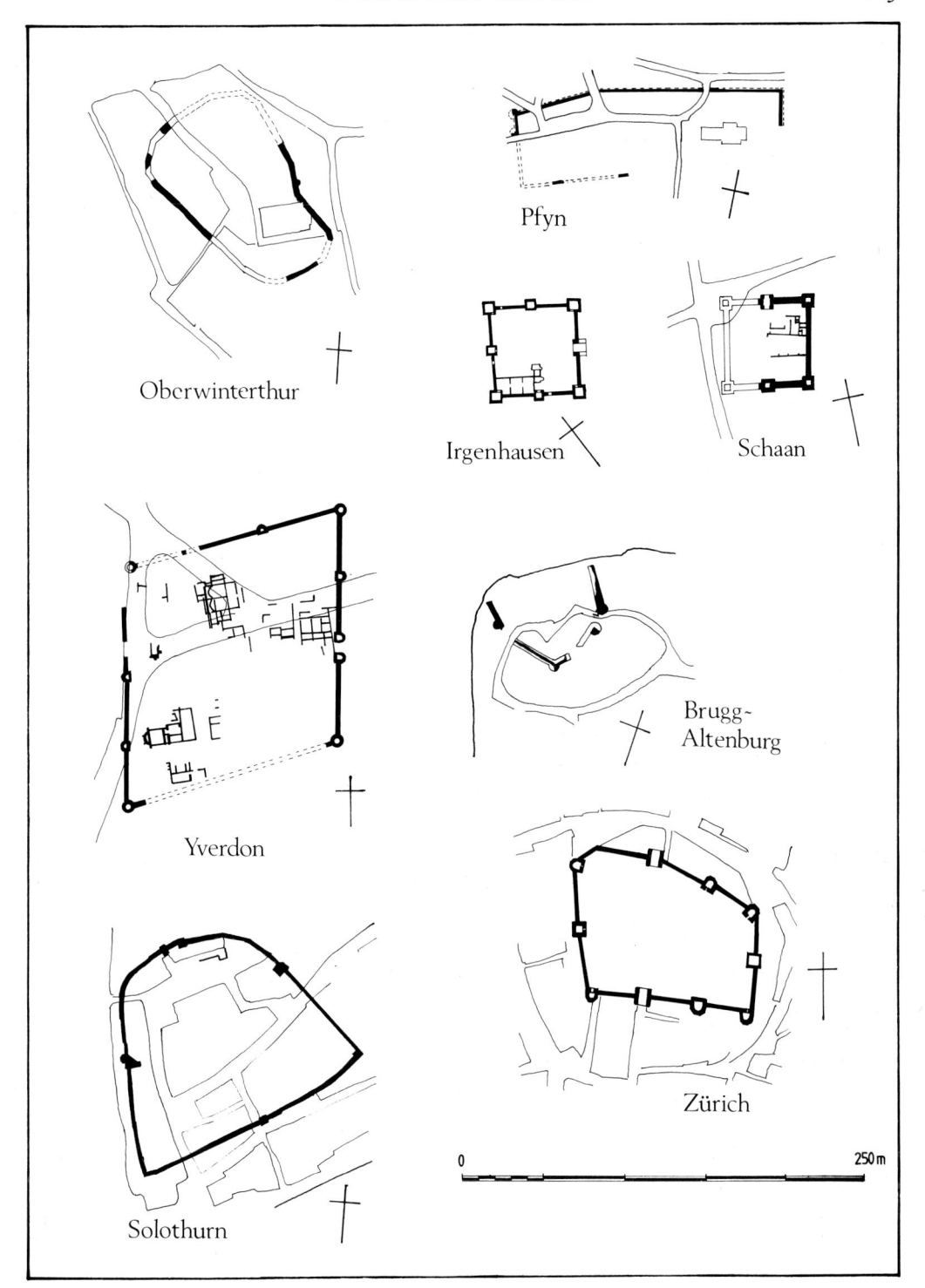

66 Late Roman walled sites, the Upper Rhine area (*scale 1:4,000*)

the invasions of 260–76. It occupied the top of a slight hill with a clear view in all directions, guarding one of the Roman roads leading from the frontier areas towards Italy. The fort was small and square, with square towers at the corners. It is similar to the type known as the *quadriburgium* commoner on the African and Eastern frontier areas than in the western Roman empire. Another fort of closely similar type, however, lay further south in the Alpine passes, at Schaan.

Schaan,[103] enclosing only 0.3ha, was almost square, with square towers at the corners and at the mid points of the sides, housing gates. Inside, a series of interior buildings has been recognised, some of stone including one large hall with pillar bases along it, and other buildings of timber, which appear to line the outside walls. A bath-building rested against the east wall. Study of the small finds suggests that the fort was built fairly late in the fourth century, at the earliest in 350.

Archaeologically speaking, this mass of data from sites on the German frontier does not admit of easy conclusions. There is little substantial evidence for any concentration on frontier defence in the middle of the third century, though a scatter of smaller sites, including the Qualburg, the posts along the road from Köln to Bavai, and the defences at Mainz, can be dated to this period. The picture is much the same for the reigns of Aurelian and Probus, despite the well-documented barbarian raiding of the period.

Firm dating is encountered only with the Diocletianic inscriptions from the Swiss sites of Burg-bei-Stein and Oberwinterthur, but there are surprisingly few signs of any other Diocletianic strengthening of the Rhine frontier. On a general scatter of finds of late third century date, one might suggest that other Swiss sites, Arbon and Basel on the frontier, and Yverdon, Brugg and Pfyn inland, were contemporary. It is not possible, however, to claim that Irgenhausen and Schaan, both four-square forts typical of attested Diocletianic sites on other frontiers, belonged to this period. Of the two, Schaan certainly, on the basis of excavated material, cannot be so dated, and the use of typological argument in the case of Irgenhausen may conspire to produce a wrong date.

Despite this lack of sound evidence from the areas bordering the Rhine, Roman tradition as a whole had it that Diocletian was a prolific builder of forts.[104] It is thus the more curious that the weight of archaeological evidence from the area should assign to his successor Constantine the major role of construction. The bridgehead fort of Deutz, opposite Köln, is the one fort attested by a lost inscription to the reign of Constantine, but Haus Bürgel and Cuijk may also have been near contemporaries. Constantine is largely credited with the provision of many of the small walled posts in the area. Neumagen was described by Ausonius as 'Constantine's camp', and other sites like Bitburg, Jünkerath, Jülich, Arlon, Saarbrücken and Pachten, can also be seen as part of the same regional grouping, partly from their typological similarity and partly from the evidence of excavated finds. On the frontiers themselves, however, archaeology as yet shows no sign of a concerted campaign of construction at a similar date. Except for Zurzach and Kaiseraugst, no sites on the upper Rhine can be claimed from the evidence of finds to belong to the reign of Constantine.

Julian's campaigns in Gaul in the years following 356 are fully documented by

Ammianus Marcellinus, and the parlous state of fortifications on the lower Rhine at the time is highlighted by his comment that, on Julian's arrival in Gaul, there was no city or fort in the area round Köln, except Remagen, and a single tower near the Colonia (Köln) itself.[105] In a four-year campaign, Julian wrested control of the area from the Alemanni, repairing forts round Mainz in 357, rebuilding forts along the Meuse in 358, and reoccupying mainly frontier posts on the lower Rhine in his last campaign in 359.[106]

Signs of this activity are visible in the archaeological record. At Alzey and Bad Kreuznach, there is evidence for a phase of building immediately after severe destruction in the mid fourth century. Saverne and Andernach have both provided some evidence for a date in the mid-fourth century, and both are mentioned in Ammianus as forts which were now rebuilt or recaptured. Saverne, reportedly built by Julian, was of distinctive type: the fort was irregular, with projecting semicircular towers at intervals, and with larger towers placed at the corners. Andernach was of the same general type, and several others of the Rhine forts seem to have been similar: Koblenz, Boppard (also dated from the pottery finds to this period), and perhaps Maastricht seem to belong to the same group, and may therefore have been part of the same Julianic design. Incidentally, Neumagen is also of this type, with larger corner towers, and it may be that it too had to be rebuilt. This set of fortifications has a distinctive style. It is clear from Ammianus that fortifications in this part of the Roman world had decayed by the 350s, and this series is probably the product of a single building campaign carried out by Julian's troops and set in motion by the Caesar before he left the area in 359.

Of the sites further north in Germania II, the archaeological record is unfortunately scant. At no period is there abundant evidence of late Roman control of the area, yet the list of sites which Julian regained from barbarian hands lists four sites north of Köln, one of which, Castra Herculis, has not been positively identified.[107] In addition, there were three forts on the Meuse which Julian fortified,[108] which probably lay in this part of Germany, and granaries which he rebuilt at the mouth of the Rhine, or at least on its lower reaches.[109]

Both Julian and Valentinian were active in the defence of the frontier in Gaul and Germany. Since the two emperors' reigns were separated by only a short space of time, it is difficult to be certain, in some instances, of the precise context of forts built about this time. Stylistically, there appears to have been little change in the third quarter of the fourth century, though there is no plan of a fort securely dated to Valentinian's reign from the area. Ammianus records that he was active in building forts along the frontier and some way inside barbarian territory.

Forts usually attributed to the energies of Valentinian are grouped mainly in the Upper Rhine area. Altrip, with its unusual shape and its interior ranges of buildings against the fort walls, has been held to be of typically Valentinianic pattern, but in few other places is there such an arrangement known. Breisach is also quoted as a Valentinianic fort, merely on the strength of the fact that Valentinian issued edicts there in 369.[110] One particular group of defences in Maxima Sequanorum along the Aare valley, at Olten, Solothurn, and possibly Brugg-Altenburg, may date from this time.

Judging by the dated fort at Schaan, which is of *quadriburgium* shape, this series of small rectangular forts with square corner and interval towers may also belong to the Valentinianic period. Dating by typological means is dangerous, however, and the other sites, close parallels to Schaan, might belong to any phase of the late Roman period. Further forts, possibly of Valentinianic date, are at Alzey and Bad Kreuznach; they were certainly rebuilt in the period after 365.

Inscriptions and pottery finds from small watchtowers along the Swiss section of the Rhine prove that one of the concerns of Valentinian and his officers was the provision of this series of guard posts. There is some evidence that some of these posts were already in existence as early as the second century, and the Valentinianic provision here may have been no more than the renewal or strengthening of an already functioning network. Often claimed as Valentinianic also are the small fortified landing bases on the lower reaches of the Rhine: finds from these sites, by and large, do not support their attribution to any particular period. A series of closely comparable sites on the Danube has been claimed to be of Diocletianic date.[111]

Archaeological traces of new fortifications dating from the reigns of succeeding emperors are still to seek. Normally the excavation record does not greatly assist in the definition of occupation periods of very late fourth century or early fifth: there are few finds distinctively different from those of a date slightly earlier in the fourth century, and the coin series tend to cease around the end of the century. Apart from the study of cemeteries such as the rich and interesting one at Krefeld-Gellep,[112] questions as to the date of abandonment of the forts on the Rhine frontier and their use, if any, in the fifth century remain obscure.

7

The Frontiers II: The Danube

Late Roman defences on the Danube split neatly enough into three separate sections. The first is the portion of frontier which lies between the Rhine at the Bodensee (Lake Constance) and the Danube; this is known as the Danube-Iller-Rhine frontier, since it links the two major rivers by following the Iller, one of the smaller Danube tributaries. The second portion of Roman frontier covers the long stretch of the Danube in the Roman provinces of Raetia and Noricum, from Ulm to Vienna, a linear distance of some 500km: The third portion is the Pannonian frontier, covering the area thickly studded with late Roman forts and small posts at the Danube bend and in the region of Aquincum (Budapest).

Perhaps the weakest link in the whole chain of Roman frontiers in the west, or at least the area where there was no visible boundary, such as a river or a wall, between Roman and barbarian was between the Rhine and the Danube.[1] From the foot of the Lake of Constance, at Bregenz, where traces of a late Roman fortification have been located,[2] a road led north-eastwards to Kempten and Augsburg. Between Bregenz and Kempten, this road was the frontier (fig. 62, p. 159), and, like the military road running parallel to the River Rhine in Switzerland, it was heavily guarded by a series of watchtowers at frequent intervals. Although many of these have been discovered, few have been dated accurately, and it is not clear whether they were the product of a single late third century campaign of construction and layout, or whether they were built and replaced at various times during the late third and fourth centuries as part of a developing strategy on this portion of frontier.

At least three sites along this road were defended strongpoints. The most important, and least well known, was Kempten.[3] This was a town rather than a military post, and although the early Roman town was expansively laid out on level ground next to the River Iller, in late Roman times the site of the prehistoric hill fort became the new nucleus of a small town with a circuit of defensive walls of which small portions have been found, notably one straight stretch of wall containing a semicircular tower.

Roughly midway between Bregenz and Kempten lay the site of Isny.[4] It lay on

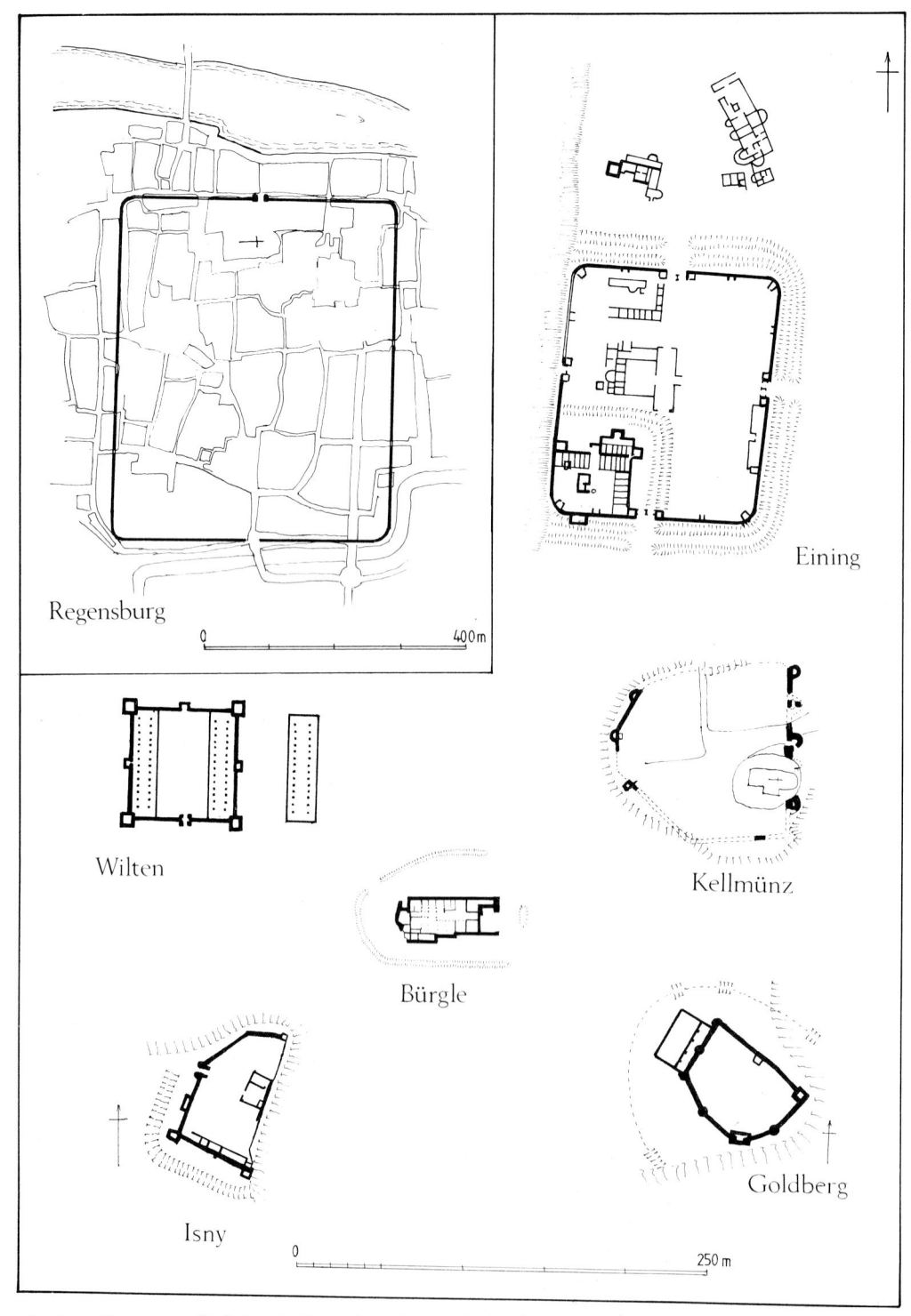

67 Late Roman walled sites in Bavaria and Austria (*scale 1:4,000*)

a small pentagonal plateau, 60–80m long and 40–45m wide (fig. 67). This was completely surrounded by walls, with a ditch to the south and west only. The main gate, on the west, lay between a pair of semi-circular towers which had been added to the fort wall in a second period of construction. Other towers of various shapes were found elsewhere, though they were absent on the north-west and east walls, where the slope is steepest.

Inside the fort, few stone buildings were found, but a large rectangular structure against the east wall, and a series of long lean-to type buildings against the south wall were perhaps the commandant's quarters and barracks (or stables?) respectively. Central within the fort was a well, and the interior also bore traces of a series of post-holes, suggesting that there had been several different periods of timber construction inside (see p. 53–4).

Finds from the fort included those apparently of a smithy. From the relatively copious coin-finds, some assessment can be made of the periods of the site's use. It was first occupied at the beginning of the third century, and perhaps received its first defences around 260. These took the form of ditches, the exact course of which is uncertain, but traces have been found underlying the stone walls. It is probable that the first stone building was constructed under Probus, since there was a particular concentration of coins of the period 260–82 from the site. At some time in the period 283–8, the fort suffered some destruction, perhaps in an Alemannic invasion, but it was rebuilt by Diocletian. Troops were possibly withdrawn again by Maximian in 296, and again the fort was destroyed, or fell into disrepair: a coin-hoard of 302–3 suggests the insecurity of the times. Throughout the fourth century, there was an unbroken sequence of coins from the site, with particular concentrations between 330 and 340, and 363 and 390. The latest coins belonged to the year 406, the approximate date of the end of Roman military influence in Raetia.

The third site on the road from Bregenz to Augsburg lay north-east of Kempten, at Goldberg near Turkheim.[5] This small fort site lay on an elevated plateau dominating the road, and the whole building complex included structures inside and outside the fort walls (fig. 67). From coin-finds on the site, and particularly those found in a group in the ditch dating from 283, the first military organisation of the site can be assigned to Probus, and since the walls and towers are of similar dimensions to those of Isny, it may be that they too date from a similar period. Construction of the stone defensive walls of the fort however—a D-shaped enclosure with projecting semi-circular towers and rectangular gate-towers—is dated to Constantinian times. Extra buildings were also now added outside the main defended area, then at some later stage, a similar building, with a floor mounted on pillars, was added against the exterior of the north west curtain wall. These exterior buildings were later destroyed, and possibly rebuilt under Valentinian.

This road, and the posts and towns along it, formed one part of the frontier defences in this area: at Kempten, however, another road turned northwards and followed the line of the River Iller. A post near Memmingen has been commonly supposed to be the site of Cassiliacum,[6] mentioned in the *Notitia Dignitatum* as the limit of command of the *Praefectus* of the *Legio III Italica* who was based at Kempten. As the roads round Kempten were heavily guarded by watchtowers, the site named as Cassiliacum may refer only to one of these. Further north, midway between Kempten and Ulm, on the Danube, lay Kellmünz.[7] It occupied high ground above the Iller (fig. 67): the shape of the small, heavily walled post, was basically rectangular and enclosed about 1ha. The walls were of exceptional thickness, and included projecting semicircular towers on the eastern face. The main gate was of unusual plan, for although the entrance lay between a pair of externally projecting semicircular towers it also combined a right-angled passageway to ensure maximum protection. Coin-finds attest occupation on the site from the first century onwards, but the main concentration of finds belonged to the middle years of the fourth century from Constantine to Valentinian. It has been suggested, however, that the site dates from the period of the Tetrarchy.[8]

The watchtowers themselves in this area continued the pattern established on the Swiss frontier.[9] They occupied strategically useful ground, often dominating the road, and were normally small rectangular towers surrounded by a ditch. Only few have been excavated, but the presence of a larger, rectangular *burgus* with corner towers, at Untersaal (fig. 65, p. 164), suggests that the organisation of such chains of small towers may have been linked with a smaller number of forts (as at Isny) or larger, better defended *burgi* which could have acted as stores or magazines. A similar function for the *burgi* at Mumpf and Sisseln in Switzerland, or for Oberranna in Austria, might be suggested: in Raetia, as well as Untersaal and Isny, the fort at Turkheim had a large store-building added in the Valentinianic period which could have been used for such a support role. There are considerable problems, of course, in establishing that chains of such *burgi* were in contemporary use, and there are no inscriptions from the frontier in this area to assist such a hypothesis. Although most scholars suggest that these *burgi* were Valentinianic, the process of setting up the frontier in the region between Rhine and Danube would suggest that here, at perhaps the weakest point in the late Roman frontier, such a chain of *burgi* may date from an earlier period, possibly contemporary with the foundation of forts like Isny, Kellmünz and Goldberg.

From Ulm in Southern Germany eastwards, the late Roman frontier followed the River Danube to the point where it met the Black Sea, some 625km away. The Roman provinces along the river thus all had a military road and frontier zone flanking the river, with only rare bridgehead fortifications on the 'barbarian' bank. In the fourth century, five provinces of the western Roman empire lay along this stretch; progressing from west to east, these were Raetia, Noricum, Pannonia I, Valeria, and Pannonia II. The greatest concentration of late Roman posts lay in the Pannonian provinces at the Danube bend. For convenience, however these arrangements will be dealt with progressing from west to east.

In Raetia, although the *Notitia Dignitatum* lists many garrisons at forts under the *Dux Raetiae*, the number of those which can be identified with the few archaeological remains known in the province is slight.[10] In fact, of 12 postulated late Roman military sites along the 156km of Danube within the province, only four have any definite archaeological evidence to substantiate their late Roman use (fig. 62, p. 159). Suggested sites are Ulm, where the River Iller meets the Danube, and ought to be a candidate for a late Roman base, possibly Febiana of the *Notitia*. Günzburg, the next site east of Ulm, has been suggested as the site of late Roman Guntia. Burghöfe, actually 6km south of the Danube, on the main road from Augsburg northwards, may also have been the late Roman base of Summuntorium. Further sites, equally tentative, have been postulated at Burgheim, Neuburg and Manching. Of these, Neuburg alone has produced late Roman evidence of substance, in the form of a cemetery which may have belonged to a fort.[11] East of Regensburg lay the two earlier Roman fort sites of Straubing and Künzing, both of which have been suggested to have late Roman successors. Neither has been traced with certainty, and any late Roman establishment did not lie on the site of the earlier forts. At Künzing, for example, the Life of St Severin records that the late Roman site of Quintanis was already partly under water in the fifth century.[12] It may well, therefore, have been totally swept away by the Danube.

Surer ground is reached by the excavated sites at Bürgle near Gundremmingen, Eining, Regensburg, and Passau. Apart from Regensburg, all these sites were small, and the smallest of them was the Bürgle.[13] It lay near the Danube road, on a slight hill which was in Roman times further defended by ditches (fig. 67). Excavations in the 1920s located a roughly rectangular fort, somewhat similar in type to a *burgus*. Its dimensions were 62 by a maximum of 29m. The stone from all its walls had been entirely robbed, but it was possible to ascertain that the interior of this irregular fort was divided into a series of barrack rooms, each with its own hearth. Inside was also a well, and more spacious accommodation including a hypocaust system provided for the commandant. The space inside the fort is calculated to have been capable of holding a garrison of 150 men. This was most likely a *centuria*, for which the size is admirably suited. Finds of *ballista* bolts and the suggestion from excavation that there were towers at at least one end of the fort show that even such a small site as this was strongly defended. Evidence for the occupation of the site comes from coin-finds, most of which date to the years 335–83, at which date the site was probably finally abandoned. Despite this, the excavator suggested that it may have had its beginnings in Diocletianic times, when the Raetian frontier as a whole was re-organised.

The late Roman fort at Eining[14] was little more than a small rectangular *burgus*, tucked into the south-western corner of a Trajanic stone-built auxiliary fort (fig. 67). The site was of strategic importance, since it guarded one of the major Danube crossings. The late Roman fort was a mere 48 by 37m in size and though it had a rounded corner to the south-west it was basically rectangular, with square towers projecting at the corners where space allowed. On the south side, the western tower of the former *porta Principalis Dextra* was used as a corner tower. In the earliest period of the late fort's use was a simple opening in the north wall, opening into the area of the former fort.

This small *burgus* was later extended by the additon of a large projection to the north, between the tower at the north-east corner and the gate, which also was protected at the same period in part of the re-arrangement. Another rectangular tower was added on the south side, involving the partial filling of one of the ditches. Now the whole site was surrounded by a new ditch: barrack blocks of timber were built against the inside of the walls, and from the slight evidence of burning inside the fort, it seems likely that it met its end from a fire in the fourth century.

At various points elsewhere inside the area of the early fort evidence of late Roman occupation has been found, and some of its buildings seem to have again been pressed into use. The state of the former defences cannot have been totally ruinous, and it may be that some of the troops lived in this area of the fort, treating the smaller better defended area as a citadel.

The side of the late Roman fort at Passau lay on the right bank of the River Inn.[15] It was a small, heavily defended, but irregularly shaped fort (fig. 69, p. 177), with projecting fan-shaped towers at the corners and what appears to be a range of interior buildings against the eastern and north-western walls. The published plans suggest that this fort covered in total no more than 0.25ha. On the Altstadthügel of Passau, however, traces of late Roman walls some 2m thick have also been found.[16] They cut off the western approach to the promontory formed by the Inn and the Danube, the only side where access was easy. Only a small section of these walls has been located, and thus it is difficult to tell whether this formed an encircling wall round the whole promontory, or a defence along this exposed side only. Opposite this defended area lay the site of Passau-Innstadt,[17] which lay in Noricum. Both Batavis (Passau) and Boiodurum (Passau-Innstadt) are named sites which occur in the *Notitia*, and also in Eugippius's *Vita Sancti Severini*. There it is clear that by the later fifth century, while Batavis still retained a garrison of some kind, the Norican site was now no longer occupied by the military, but had a monastery on it.

Regensburg[18] was the site of a late second-century fortress for the *Legio I Italica* built in the wake of the Marcomannic Wars (fig. 67). It lay immediately south of the Danube, and its main north gate, the *Porta Praetoria*, afforded access to a river bridge. Little is known in detail of the interior buildings of the fortress, as a consequence of the medieval and modern town having occupied precisely the same spot. Portions of the defences however, including the *Porta Praetoria* and at least one other monumental gate, have been discovered by excavation or are still visible. The gates are probably contemporary with the construction of the fortress, but there was probably some late Roman rebuilding both of gates and walls, in the wake of substantial destruction, attested both by contemporary sources and by archaeological finds, in the period 260–88. In the late third and fourth centuries, the settlement which had lain south of the fortress was virtually abandoned, and legionary garrison and civilian town both found space within the refurbished walls. This may have resulted in an unusual layout, for it has been suggested that the late Roman fortress *principia*, incorporating accommodation for the Prefect, lay in the north-east quadrant.

Inland in Raetia, surviving traces of late Roman defences are equally hard to find. South of the frontier, the rolling Bavarian hills soon give way to the steeper Alpine foothills. The few Roman routes through the Alps to Italy were protected

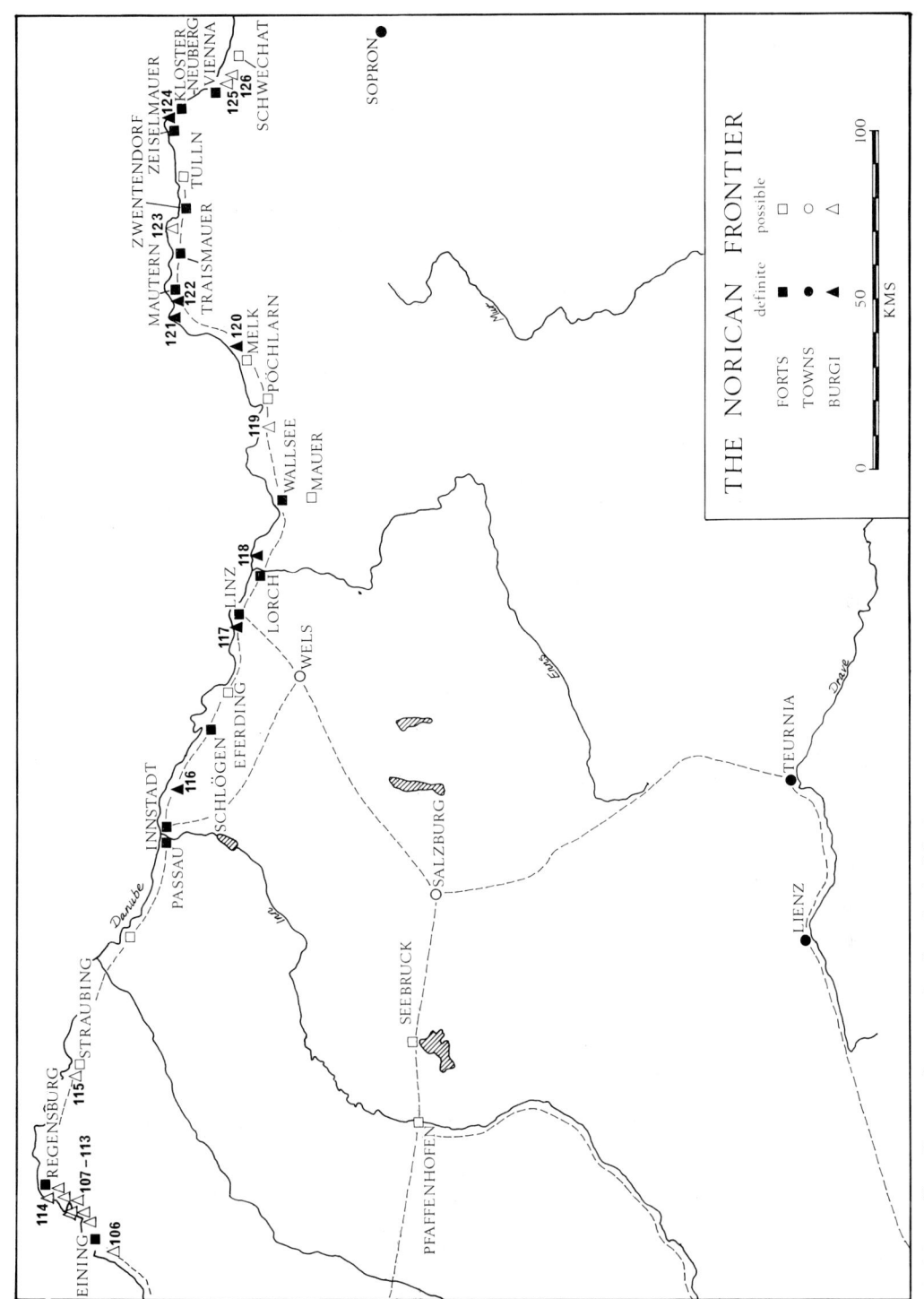

68 Map of late Roman forts on the frontier in Noricum

by forts and hill-top fortifications of rather less regular form. Two of these, the Moosberg and the Lorenzberg, have been comprehensively excavated.[19] The major Alpine pass was that of the Inn. Here two sites, Zirl and Wilten, both near Innsbruck, were defended in the late Roman period. At Zirl,[20] the site of a prehistoric fort on the Martinsbühel was re-used in the late third century. A mortared wall 2m thick ran all round the hilltop, but the Roman remains on the hill have been heavily overlain by later use.

At Wilten,[21] there lay a fort of square shape with square corner towers (fig. 67), in plan very similar to Irgenhausen and Schaan, two forts in the hinterland of the Rhine in Switzerland. Within the walls lay a pair of long rectangular aisled buildings. A third similar building lay outside the walls, and though there was no stratigraphy which linked all parts of the site, it was thought to be of later date than the others because it is slightly smaller in width, and appears to have been less carefully laid out. Coin-finds from the fort, none of which came from securely dated deposits, ran from Constantine to Theodosius, but from areas around the fort they were generally earlier. The aisled buildings may have been granaries or armouries, forming a supply base for troops on campaign in Raetia and on the frontiers.

East of Raetia, a similar pattern of late Roman forts continued into Noricum (fig. 68). The most westerly site was Passau, at the mouth of the River Inn.[22] Although the early Roman fort is known, its late Roman successor, unless it was the tiny irregular fort with heavy walls and fan-shaped angle towers, has not been located, and does not seem to have lain on the same site. From this point onwards, the *limes* road was protected by forts and watchtowers, as far as present knowledge goes, relatively widely spaced.[23]

There was a particular concentration of fortifications at the approaches to Pannonia, but a number of the named garrison-sites in the *Notitia* refer to places in the western part of Noricum which have as yet revealed little or no evidence for their late Roman use. Such sites are Linz (Lentia), and Mauer an der Url (?Locus Felicis). Watchtowers, at Wilhering, Au, Bacharnsdorf, and Rossatz of plain rectangular shape and one at least of circular shape at Spielberg, are recorded along the whole of this stretch. A fortlet of rather more sophisticated form lay at Oberranna. Overall it was 12.5 by 17m, and had three-quarters round corner towers, somewhat similar in form to Untersaal in Raetia or to the bridgehead fortification opposite Basel. Only one *burgus* site in this area is securely dated, on an inscription of 370 discovered at Ybbs which records its construction under Equitius, *magister militum*, by the *milites Auxiliares Lauriacenses*.[24] The site of the *burgus* to which this inscription refers has not been located.

Four fort sites in the western part of the province merit fuller description. At Schlögen (fig. 69), excavation has revealed the remains of a small fort occupying a small plateau at the Danube edge.[25] Of two main periods of construction, the first, probably Hadrianic, has only been revealed in fragmentary traces underneath the second, which was an almost exact rebuild. A layer of burnt material marked the end of the first period of the fort, and this produced coins of late third and early fourth-century date. The second phase of construction must

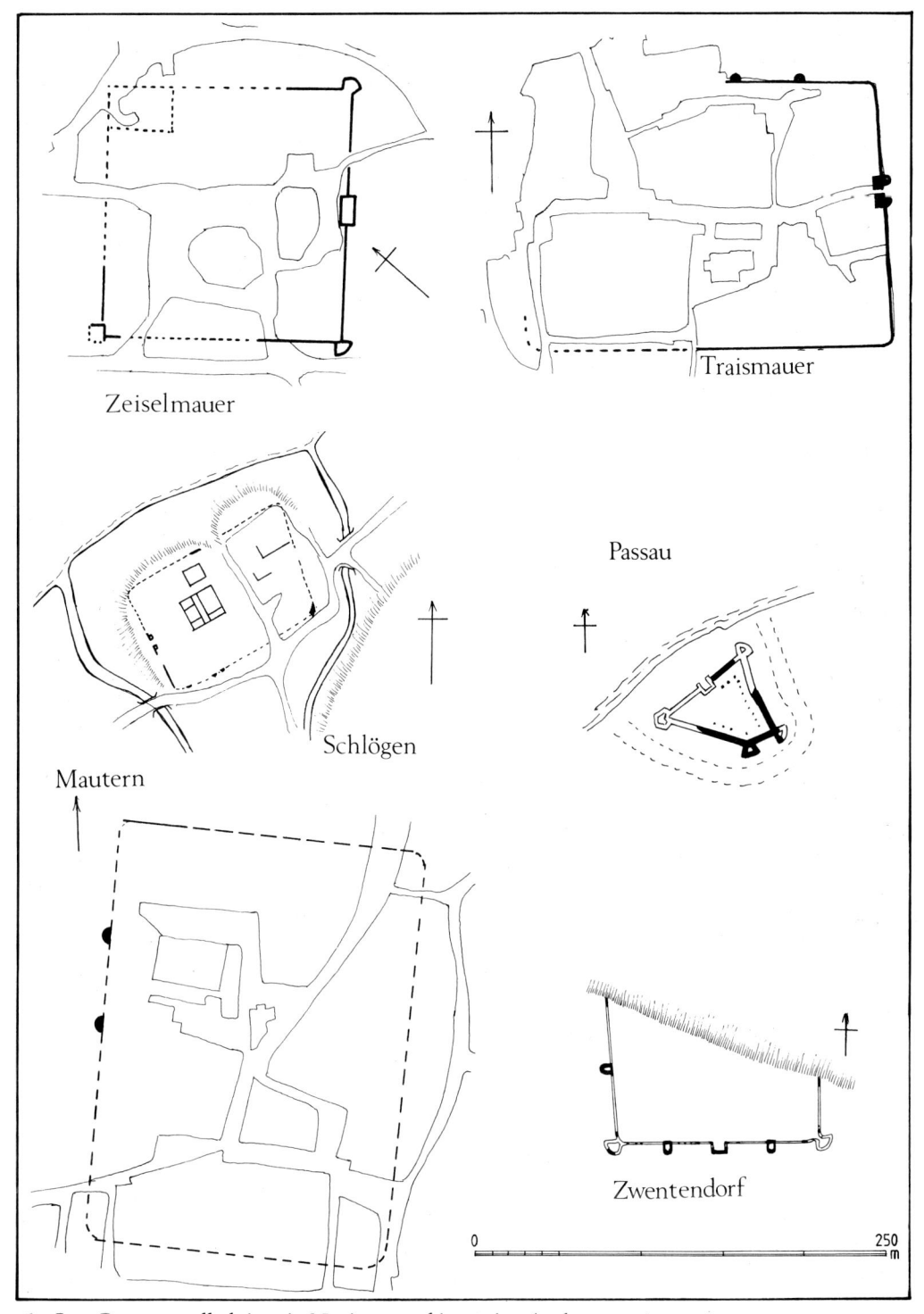

69 Late Roman walled sites, in Noricum and its region (*scale 1:4,000*)

have been of fourth century date, and possibly as late as the Valentinianic period. Reconstruction of the defences seems to have been a priority, but even so, there was no attempt to bring them up to date, since the old interior rectangular flanking towers were rebuilt, and no projecting towers were added. At this period there was also a large interior building but there was no immediately recognisable headquarters.

According to the *Notitia*, at Enns-Lorch (Lauriacum), there was a fleet-base as well as a legionary garrison.[26] The legionary fortress remained in use from the time of its construction, after the Marcomannic Wars, until the fourth century,[27] but the late Roman period within the fortress has not been systematically studied. The fortress was of parallelogram shape, and though several buildings do not appear to belong to the street grid encompassed within these walls, they cannot with certainty be assigned a late Roman date. A civilian settlement, which had grown by the early third century to a sizeable *municipium*, lay next to the fortress. This was probably never walled, but was in continuous occupation until the mid-fifth century, despite suffering destruction in or about the years 230, 270 and 350. After that date, three emperors, Constantius II, Valentinian and Gratian visited the site and stayed there whilst on campaign.

Wallsee, the site of a Flavian and Hadrianic auxiliary fort, was transformed in the late Roman period to a small *burgus*, nestling in the south-eastern corner of the former defended area.[28] It has been identified as the Ad Iuvense of the *Notitia*. At Pöchlarn,[29] probably Arelape, the Roman fort, built on the site of an earlier installation, has been partly swept away by the Danube, but portions of the defences have been recognised at various points along the south, east and west walls. Two apparently Roman towers, round and standing astride the walls, have been incorporated into the medieval line. Coin-finds from the site ran from the first to the fourth century.

There is a concentration of late Roman sites on the Danube as it approaches Vienna. The most westerly of these, Mautern, has long been thought to be the site called Favianis in both the *Notitia Dignitatum* and the *Vita Sancti Severini* of Eugippius.[30] The town contains considerable portions of a circuit of walls, partly of Roman date although refaced in medieval and later times (fig. 69). Sections across the defences in the central areas have located Roman walling on a different alignment from the medieval town defences, and excavation has revealed a fan-shaped tower added to these walls. Late Roman occupation of the site is further confirmed by finds of late Roman stamped tiles. The fort's north wall showed signs of two periods of building, the second possibly Valentinianic. At this period, the ditches may have been recut, and semicircular towers added to the walls.

Traismauer[31] was originally a Vespasianic fort, built in stone in the Trajanic period (fig. 69). The overall shape of this fort survived into the medieval period. The east gate is a single portal flanked by a pair of large U-shaped towers, and probably took the form of its late Roman predecessor. At other points round the fort walls a semicircular projecting towers typical of late Roman fortification.

At Zwentendorf,[32] two main phases of use were distinguished in excavation, the first a stone-built fort of early second-century date (fig. 69). The late Roman period was marked by developments parallelled at many other sites in the region. There was a thickening of the walls, and projecting towers were added between the corners and the south gate. At the corners, fan-shaped towers were added, and the south gate was converted into a large rectangular tower or courtyard. A narrow foot-passage still formed the entrance, though the main entrance probably lay elsewhere. Narrow rectangular stone buildings were encountered in the fort's interior, and in the south-western corner there was a building of three wings arranged around a central courtyard. No precise date was afforded by the excavators for these late Roman developments, but the construction of fan-shaped corner towers has been assigned to the reign of Constantine.

No evidence for late Roman Comagenis has been recovered from Tulln, and further evidence comes from Zeiselmauer.[33] Here, an early imperial fort lay under the modern town, but this has been the subject of only limited investigation. Published plans suggest that in the late Roman period its gates were converted into large rectangular towers, and that there were projecting U-shaped and fan-shaped towers round the walls and at the angles. At the north-western corner of this circuit, however, closer inspection has revealed the remains of a substantial block house or *burgus* of Roman date, parts of which still stand up to 9m high clasped within later buildings. The external dimensions of this *burgus* were 20 by 21m, and it had an interior butressed central 'tower', which possibly acted as a light well for first floor rooms. As well as a doorway at ground level, traces of at least 5 slit windows with large interior responds were found at first floor level. Joist holes on the interior suggest that there was at least one further storey above this. These remains have not been accurately dated and their relationship with the fort is not yet clear.

The exact boundary-line between Noricum and Pannonia is not known but Vindobona, Vienna, was certainly in Pannonia. It is probable that Klosterneuburg,[34] between Zeiselmauer and Vienna, also lay in Pannonia. Here there was an early Roman auxiliary fort of standard pattern. In the late Roman period its defences were updated by the addition of projecting towers.

In the late Roman period, as in earlier times, Vienna held a garrison of *Legio X*.[35] The fortress occupied the centre of the modern city, but because of the difficulties of examining the site under modern buildings, relatively little is known in detail about the late Roman phases of occupation. The known plan of the Roman layout at Vienna follows the normal shape of a legionary fortress, except that along the northern edge, a diagonal wall drastically reduced the amount of space within the walls. The date or the reasons for this curiosity of plan, however, are not known, but if it signifies a reduction made to an earlier fuller wall-circuit, this might be a feature of the late period, when the reduction of legionary size would leave unused space within the fortress.

Two further early Roman forts are known on the frontier road between the legionary fortress of Vienna and Bad Deutsch-Altenburg (Carnuntum). The site at Schwechat has been located, but apart from a scatter of late Roman finds, has revealed nothing of its late Roman use. The other, thought to lie at Fischamend, has not been found. The same is also true of the auxiliary fort Gerulata, which lay east of Carnuntum. Although excavations have taken place in the cemetery, no work has been undertaken on the fort site itself.

Carnuntum[36] (Bad Deutsch-Altenburg) lay on one of the most important Roman trading routes between the Baltic and Italy. It was the site of a fortress for *Legio XV Apollinaris* from the later years of the first century onwards, but *Legio XIV Martia Victrix* soon replaced it, and, according to the *Notitia*, was still in occupation towards the end of the fourth century. Excavations at the site have produced a reasonably full, if rather simplified, plan of the interior buildings, which more detailed work is now beginning to refine. Buildings of the late Roman period have been recognised within the fortress, but until the very latest phases, the interior arrangements remained fairly orthodox and standardised. Only in the late, possibly post-Valentinianic occupation were new buildings laid out which failed to follow the alignment or pattern of earlier barrack blocks. Here, too, it has been claimed that a small portion of the fortress was taken up with a smaller *burgus* type of fortification (see p. 52).

Along the stretch of the Danube within modern Hungary as far east as Brigetio (modern Oszöny), the remains of the Roman frontier works, though by and large little excavated in recent years, have been well listed and recorded.[37] Because of the unfamiliarity of the Hungarian tongue to English readers, sites are best identified whenever possible by their Latin names, even though this may sometimes presuppose a degree of certainty of identification with named sites in the *Notitia Dignitatum* or in other documentary sources which may not always be completely justified. In addition to the named forts, there are many fortlets or *burgi* known in this stretch, and further details of them can be found by comparing the numbered *burgus* sites on figs. 70 and 72 with the corresponding Appendix of sites at pp. 275–9.

Progressing eastwards into Hungary, the first fort-site was at Ad Flexum, of which little is known, though its use in the fourth century is likely. At Quadrata,[38] a Trajanic fort was retained in occupation at all Roman periods. Most of the buildings found within the fort so far have proved to be late Roman. At the north west corner, first a round, and later a fan-shaped tower was added. Tentative dates of the early third and the second quarter of the fourth century respectively for these developments have been suggested.

A main Danube crossing point lay at Arrabona (Györ).[39] This was a legionary fortress whose remains lie deep under the modern city, and which in consequence have been rarely examined. Parts of its vicus and cemetery, both of which prove occupation reaching into the fourth century, have been excavated. About 13km east of Arrabona lay Ad Statuas,[40] a late Trajanic cohort fort which was occupied and rebuilt at various phases of the Roman occupation. At a late date, possibly the reign of Constantine, the corners were strengthened with fan-shaped (or more accurately light-bulb shaped) corner towers, and the east and west gates were walled up. A new defensive ditch, wider and deeper than the old one, was dug. In one of the fan-shaped projecting towers, a coin-hoard of mid-fourth-century date was found buried. Of the next fort eastwards, Ad Mures, virtually nothing apart from the site's location itself is known.

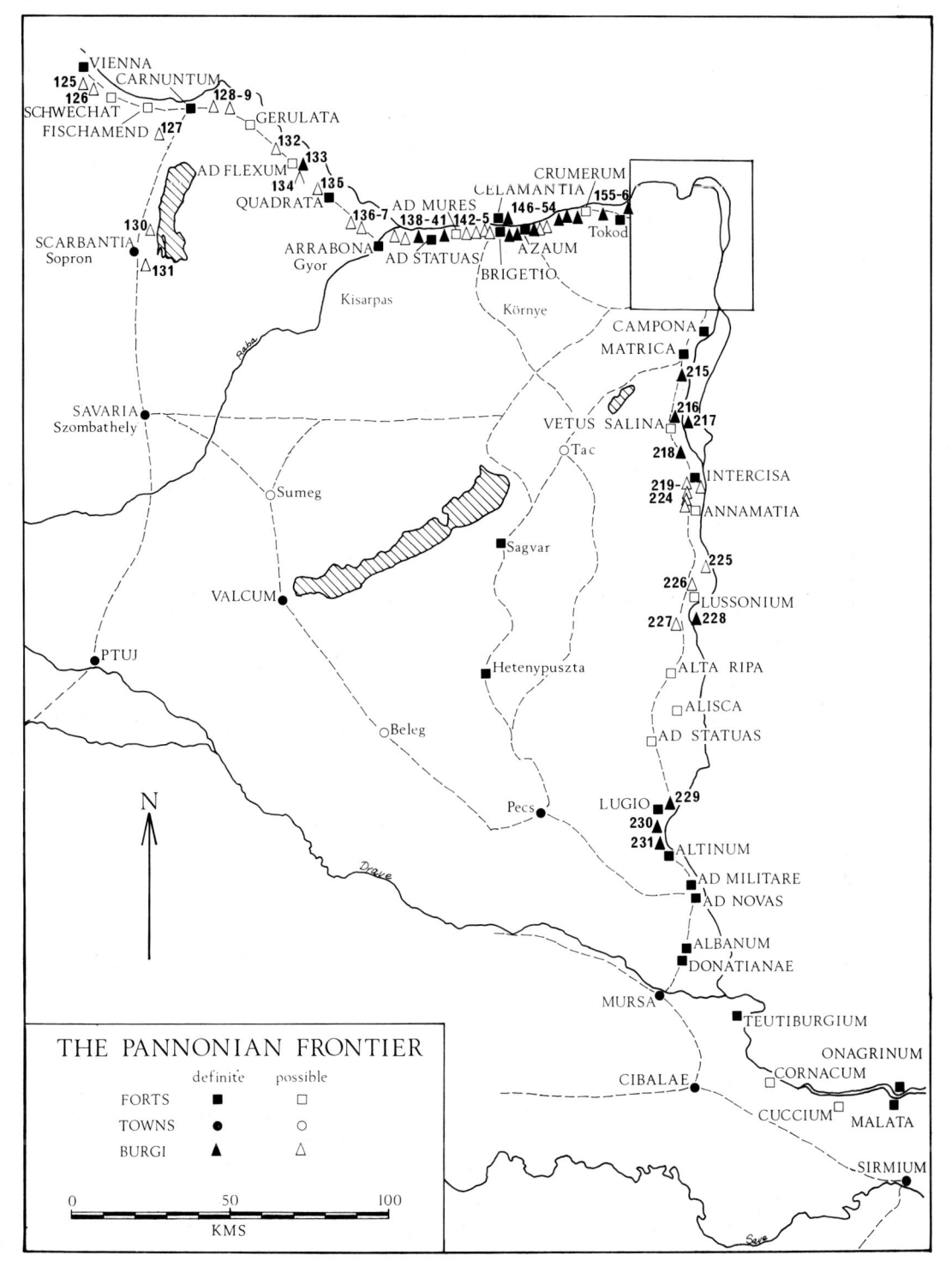

70 Map of the late Roman frontier in Pannonia and Valeria (see, e.g. fig 72 for inset area)

Brigetio[41] was established as a legionary fortress in the first century. It was probably built in stone under Trajan, and has so far revealed none of the characteristics which signify later rebuilding at other Danube forts. Research has concentrated mainly on the defences, but such interior buildings as have been discovered seem to be of irregular construction, though usually aligned with the fort's orientation. A bronze plate bearing an edict of 311 was found in one of the buildings and there is other evidence from the cemetery and from tile stamps of increased occupation at the fortress around the turn of the fourth century (270–320).

Celamantia, the bridgehead fort for Brigetio, lay at Leányvár[42] immediately opposite and was built and garrisoned by a legionary detachment (fig. 71). The earliest period of fort occupation probably dates from the early second century. It suffered under the Marcomannic wars when parts of the walls were probably demolished, and it was afterwards rebuilt. Subsequently, the north-east corner tower was reconstructed, and an exterior fan-shaped corner tower added, partially overlying a filled-in ditch belonging to the earlier fort. Inside, the regular building of the second century were partially reconstructed, but those thought to belong to a late Roman phase were on a different alignment from the rest of the fort.

Azaum,[43] some 7km east of Brigetio, was excavated between 1971 and 1973, when it was discovered to have followed the normal development of a fort on the Pannonian frontier. An earth-and-timber fort, probably of late Trajanic date, was replaced under Marcus Aurelius by one in stone. This had the usual internal interval towers, and four main gates with dual carriageways. At the beginning of the fourth century, three of the gateways were walled up, leaving only the main northern one open. Large semicircular towers were built out in front of the gates, making use of irregular stone work and much re-used stone. At the same time, fan-shaped towers were added at the corners. After the Valentinianic period, a small fort was built in the NW corner. Its walls, 2.2m thick, enclosed an area of only 31.8 by 32.5m, and these too incorporated resued stone, including some inscriptions.

Little excavation has been done at Crumerum,[44] but on the hill now occupied by an eighteenth-century fort, aerial photography and earthworks have combined to suggest the plan and layout of the Roman fort. The plan incorporates traces of the main gate towers and apparently 'fan-shaped' corner towers, but is despite this most irregular. Elements of it may have been confused by traces of buildings belonging to the eighteenth century fort.

The fort of Tokod[45] lies on gently rising ground near the junction of the frontier road and the main road from Brigetio to Aquincum (fig. 71). The walls, 1.65m wide, enclosed a near rectangle of 136 by 121m. The main gate lay to the north-west, and its threshold was formed by gravestones of the second and third century. In the middle of the north-east and south-west sides were further gates, later walled up. The rectangular gate towers were built of coursed blocks, interspersed with tile courses, but the walls and the U- or horseshoe-shaped towers which sit astride the walls and projected forwards from them were of

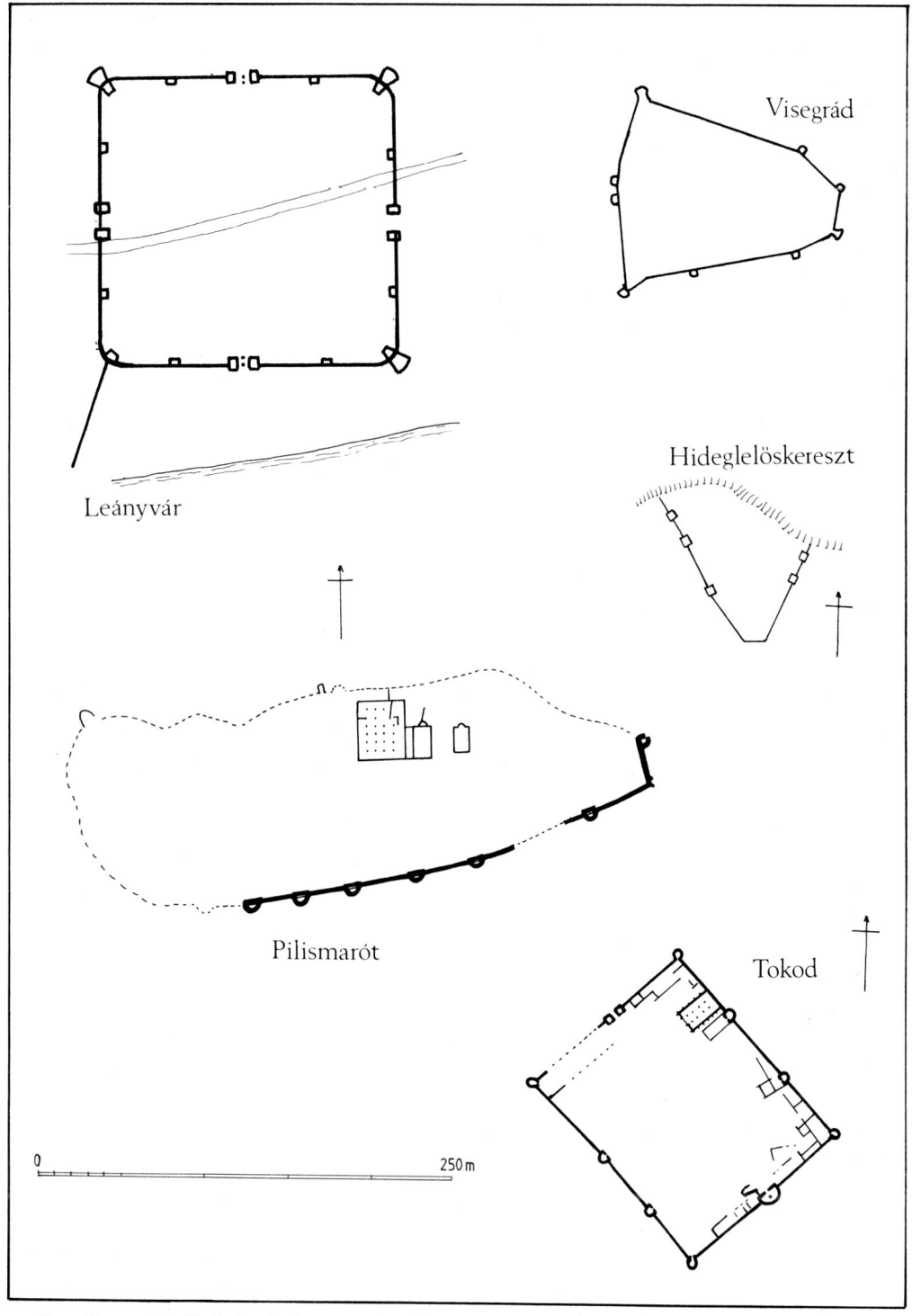

71 Late Roman walled sites in Pannonia (scale 1:4,000)

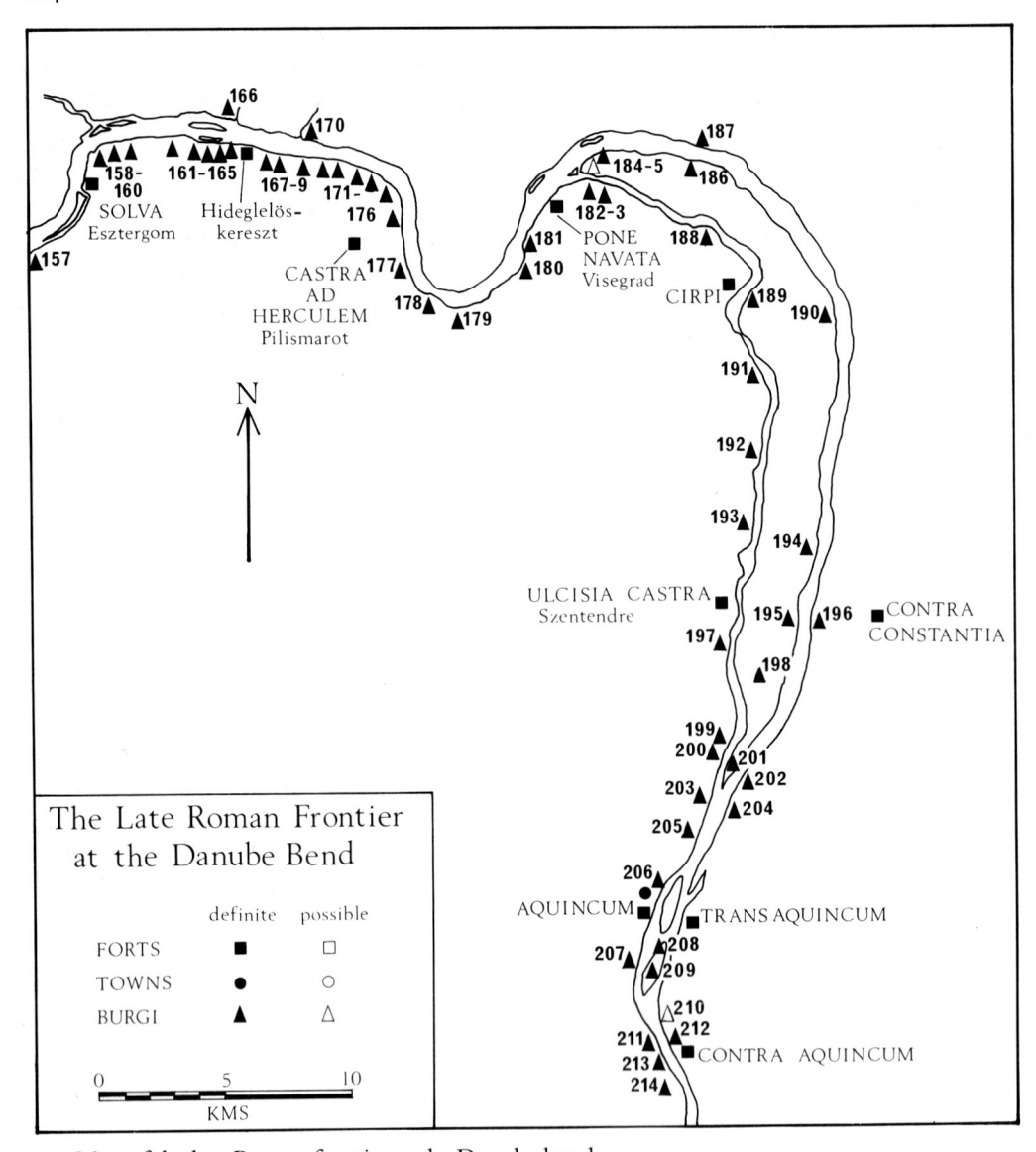

72 Map of the late Roman frontier at the Danube bend

limestone only. The interior buildings, with the exception of a stone building against the east wall, were built in dry stone without foundations, and may have formed part of a fifth-century settlement. The construction of the fort can be assigned to the Valentinianic period.

Only a tiny fragment of the interior buildings of the fort at Solva,[46] within modern Esztergom, has so far been found, and there is scant evidence for the fort's layout. It lay under the medieval royal palace on the Burg hill. Four occupation periods have been distinguished including a last rebuilding under Valentinian.

The northern part of the small fort at Hidegleloskereszt[47] has been swept away by cliff falls into the Danube (fig. 71). The site is possibly the original findspot of an inscription recording the construction of the walls of a fort in 367. The fort lies on a substantial hill, and has square towers set astride the wall, with a good viewpoint over the river and several nearby watchtowers. Only a small triangular area of the fort's interior now survives, with the only known buildings lying next to the walls.

Castra Ad Herculem[48] was a large irregular fort lying on a hill site above Pilismarot (fig. 71). Along the east and south walls were found traces of U-shaped towers, but no main gate. On the evidence of coin finds, which run from Constantine to Valentinian, a construction date in the first quarter of the fourth century has been suggested. There was also some refurbishing of the defences at a later period. The fort cemetery contained graves from the period 370–90. Inside the fort, the remains of three large buildings of unknown purpose have been found. One was a large hall with four rows of equally spaced columns along its long axis. Next to it was a pair of rectangular buildings with apsidal ends.

The fort of Visegrád[49] lay on a high hill and was roughly triangular in shape (fig. 71). At each corner there was a fan-shaped tower, and all the known interval towers were U-shaped in plan. A main gate lay in the centre of the shortest, western side. It had been placed on the site of an earlier interval tower and had a pair of U-shaped towers flanking a double entrance passage. Later still a large rectangular tower was built where the gate had stood. Interior buildings of casemate type were encountered against the south-eastern curtain wall. Coin-finds from the site indicate that the fort was built in the years immediately following 322, the year of the Sarmatian invasions.

Cirpi[50] was built as a regular fort and garrisoned in the second and early third centuries by an auxiliary cohort, though it is possible that the Danube fleet also had a base here. Archaeological examination has been confined to the fort's southeast corner. The fort was built according to the normal pattern of a Trajanic or Hadrianic foundation, and probably had an added fan-shaped tower at the corner. Later, a small *burgus*, only 17 by 16.5m, was inserted. It occupied an irregular quadrilateral shape and may date from the post-Valentinianic period.

Twelve kilometres north of Aquincum lies the fort of Ulcisia Castra (Szentendre),[51] a trapeze-shaped quadrilateral with rounded corners (fig. 73). In the late Roman period, projecting U-shaped towers were added to its walls, and fan-shaped towers were added at the corners. Both types of tower have yielded reused inscriptions, of late second and mid third century date. At the same period the main gates, apart from the east, were converted into large projecting towers. Interior buildings, including a principia building, are known, but imprecisely dated. During the period of its use the name of the fort changed from Ulcisia Castra to Castra Constantia, and this may have happened either under Constans or under Constantius II.

Aquincum[52] was the site of a stone built fortress of early second-century date, built for *Legio II Adiutrix*. The site was used well into the late Roman period, at which time no alterations to the fortress defences themselves are known, but the area between the fortress and the river Danube was enclosed by a new thick wall, festooned with a series of U-shaped towers, and linked to each other by a series of

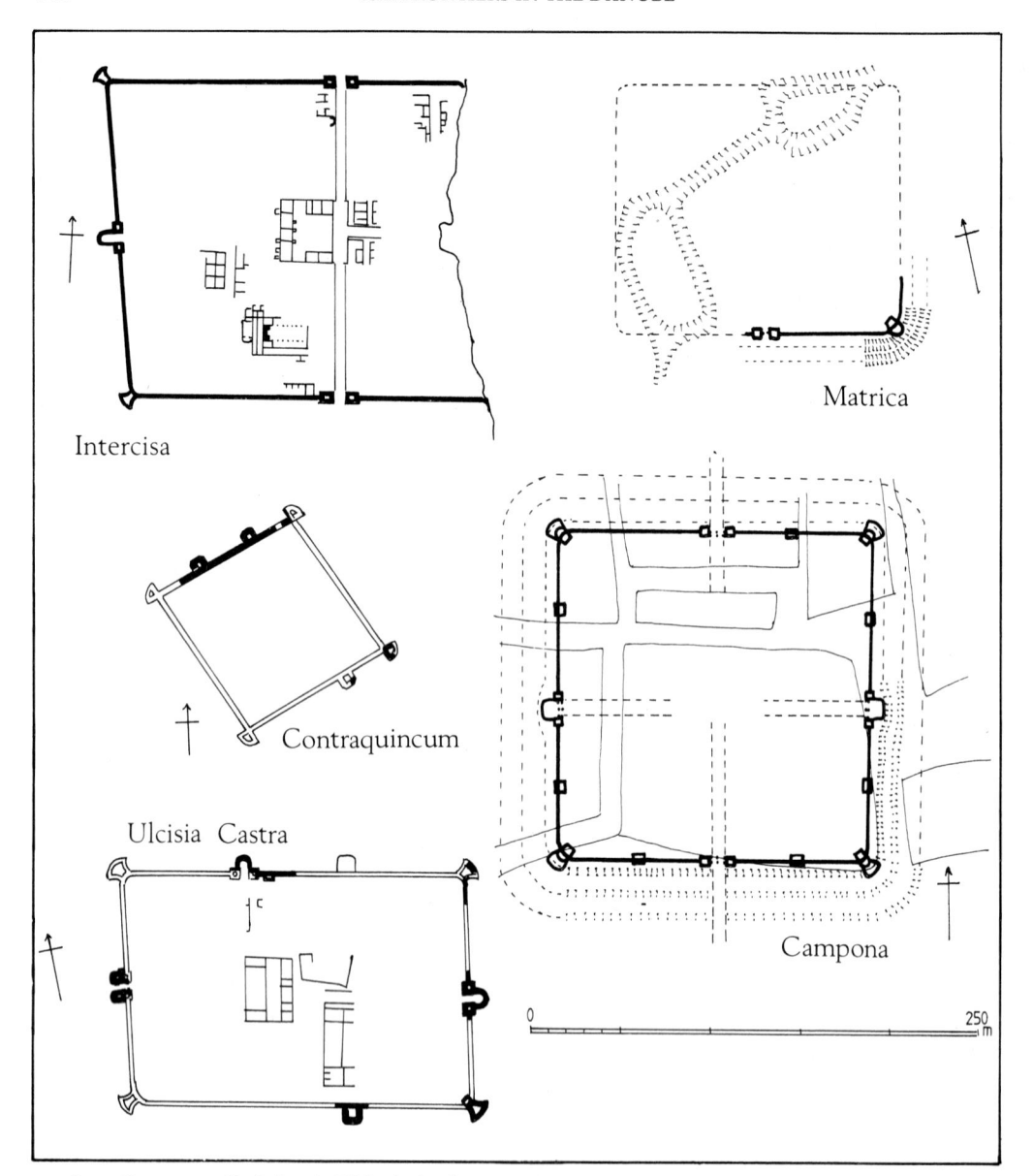

73 Late Roman walled sites in Valeria (*scale 1:4,000*)

looped stretches of curtain (fig. 45, p. 122). The use of this area near the fortress probably signalled the abandonment of the civilian town of Aquincum, which lay a couple of kilometres to the north.

On the opposite bank of the Danube from Aquincum, linked to the fortress by a timber bridge, lay a small fortification thought to be the Roman Transaquincum.[53] It was recorded in the nineteenth century, and three sides have been located. It was a rectangular fort of small size, with strong exterior walls (fig. 74). On the interior a pillared colonnade lay just inside the fort walls. No gates or external towers are known.

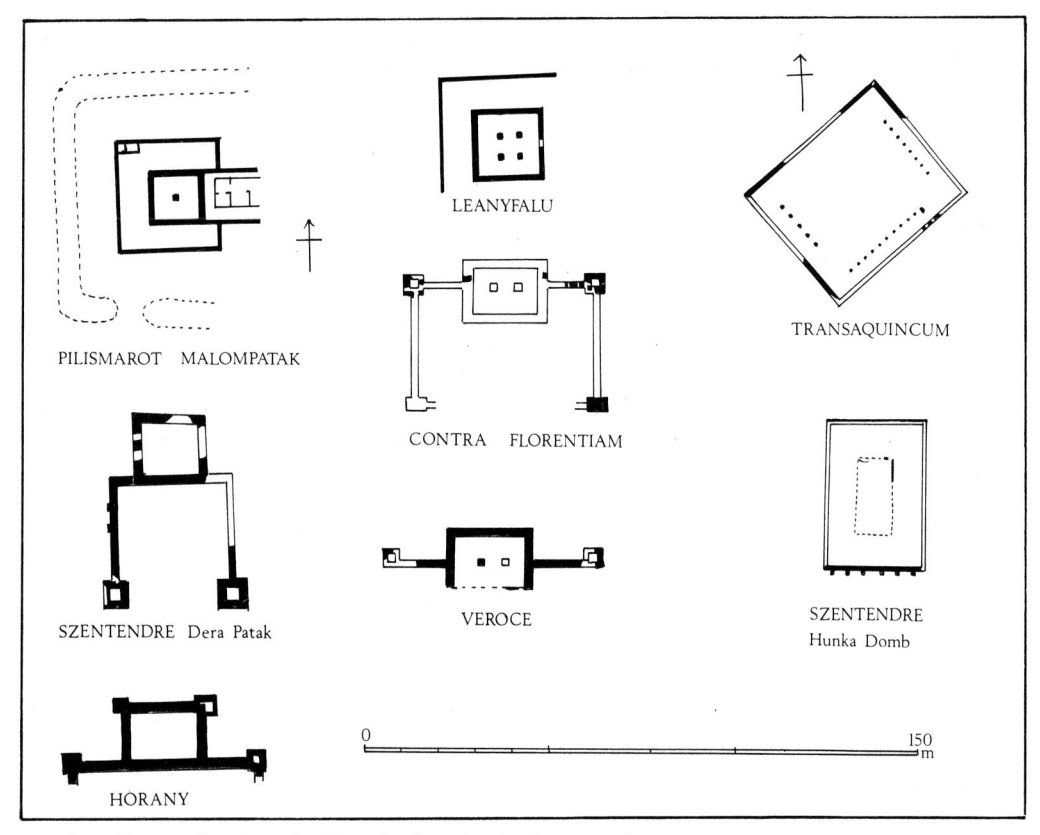

74 Late Roman *burgi* on the Danube frontier (*scale 1:2,000*)

A further small fort lay on the east bank of the Danube within modern Budapest, partly under the Church of St George.[54] This may have lain on the site of an earlier and larger fort, traces of which were found in excavation in the 1940s. The late Roman fort was of parallelogram shape, with projecting U-shaped towers at intervals along its sides, and with fan-shaped towers as the corners (fig. 73). These were built contemporaneously with the fort walls.

Campona,[55] in the late Roman period the next fort southwards, was a regular rectangular structure with rounded corners at the Danube edge about 12km from Aquincum (fig. 73). In the early third century, projecting semicircular towers were added to the fort walls at the corners. Later still, the inner ditch was filled, and a larger, fan-shaped tower was added at the corners in place of the semicircular one. At the same times a larger, outer ditch was added and the gates on two sides were also blocked off by the addition of a large semicircular wall from one guard tower to the other, forming a large U-shaped bastion. Historical sources suggest that Campona was overrun by the Sarmatae in 322, and it has been thought that these defensive developments followed as a conseqeunce. Excavation in recent years has concentrated on clarifying the sequence of the defences, and little has been done inside the fort, much of which lies under modern buildings.

Although both named in the *Notitia Dignitatum*, neither of the next two forts south of Campona has produced any late Roman finds of note, nor have typical late Roman developments been recognised in their construction. At Matrica,[56] where all but about a quarter of the fort has been destroyed, the normal development of earth-and-timber fort, followed by Trajanic or Hadrianic building stone, was followed (fig. 73). At the single surviving corner, a semicircular projecting tower was added in the late second or early third century, but no later developments are recorded, and few later Roman finds from the site have come to light. At Vetus Salina,[57] four periods of earth-and-timber fort have been distinguished, all pre-Hadrianic in date, followed by a single stone phase. Only a small corner of the fort has escaped being swept away by the Danube. No finds of note later than the early third century have been made.

Intercisa[58] is perhaps the most thoroughly excavated fort on the Hungarian stretch of the Danube frontier and as a consequence much is known of its history. It began, like many others, as an earth-and-timber fort, which was converted surprisingly late, not until the 170s or 180s, to stone. This fort formed the basis for the later use of the site (fig. 73). The eastern side has been swept away by the Danube, but the majority of the interior remains. On the south side a sequence of ditches has been examined and this has been related to the development of the defences on the south wall of the fort. Fan-shaped towers were added to the defences at the corners, and their construction overlay a filled in ditch. At a point on the south wall, virtually mid way between the fan-shaped towers, coins with terminal date of 345 were found in a layer of earth filling this ditch. In the late Roman fort the main west gate was blocked off by a larger U-shaped tower, but the main north and south gates were retained in use.

Several interior buildings have been examined, including the headquarters building at its centre, and an aisled building and associated rooms to its south-west, which also lay within the fort's central range and was of late Roman date. Against the south wall, a range of rooms was built at a date after the mid fourth century, and kept in use until the end of the Roman period.

South and east of Intercisa on the frontier road up to the border of Pannonia and Moesia at Belgrade, Roman forts were spaced at frequent intervals, but although their Roman names and some of their locations are known, little excavation or research has taken place on them.[59] Even those sites which have been located have revealed little evidence of the late Roman period. The most important site on this stretch of the river was the legionary fortress of Malata/Bononia,[60] at Banostor, where *Legio V Jovia*, a unit raised by Diocletian, was stationed. Some scant remains of this fort have been found within the town. Opposite Bononia there lay a bridgehead fort, possibly the Roman Onagrinum, though no exploration of this site has taken place. Otherwise only two sites have been more than superficially examined: at Cerevic[61] a small fort has been located and at Cortanovci[62] a small, rectangular fortlet 70 by 100m in size was found to have a single large round tower at one corner. Although both belong to the late Roman period, neither site has been precisely dated.

After suffering heavily in barbarian raiding in the late 250s and 260s, and possibly again a decade or so later, the Danube area seems to have been strengthened with new frontier defences only gradually, and, from the current state of archaeological knowledge, patchily. The most problematical area for the Roman military authorities was the flatter stretch of terrain between Lake Constance and the upper reaches of the Danube. Here a new frontier road with watchtowers and forts was established, probably by Probus, and reinforced later by Diocletian with forts at Kellmünz and possibly Goldberg.

Many of the documented sites occupied by troop units listed in the *Notitia* for this area have so far remained unlocated. Even at the sites whose names are recognisable, and particularly in those areas of the upper Danube where the frontier had remained unaltered from Trajanic and Hadrianic times onwards, firm archaeological evidence for a late Roman military presence is scant, and cannot be tied down with any confidence to dated periods.

Frontier defences were supplemented from the late third century onwards by occasional hill-top defences, particularly in the Alpine river valleys and passes.[63] It was here, in Raetia, Maxima Sequanorum and Noricum that many such sites were established not only as local refuges, though all may have had this as their origin, but also as look-out posts and even as chains of defended strongpoints supplementing the frontier. A series of hilltops near the Rhine in Maxima Sequanorum—Mandacher Egg, Renggen, Portiflüh and Stürmenkopf—appear to have been no more than towers which might carry signals included within an enclosed area on the hill-top itself. Other sites also contained similar towers, but with no recognisable signalling system linked to them. A number of sites in Raetia were similar to the Moosberg, linked more closely with the road system. In the Alpine passes themselves and in the foothills, there were numerous hills suitable for defence by a populace seeking refuge from Germanic invaders. Although once secure enough in their isolation in these valleys, the local populace now had the misfortune to be on the few routes through the Alps into the richer areas of Italy. From north and east, these passes were barred effectively enough by a series of forts and hilltop sites in the upper Rhine valley (Schaan, Füssen, and the neighbouring group of hill-top fortifications), in the Vorarlberg (Zirl, Innsbruck-Wilten, and Kuchl) and in the south Tirol and Kärnten (a large series of hill-top sites). It has been suggested that the hill-top sites in Kärnten are closely linked with the system of barrier walls of the Limes Italiae, the *Claustra Alpium Juliarum*.[64] Whether deliberately organised as such or not, the local strongpoints seem to have provided at least subsidiary defensive arrangements linking them with the Italian frontier.

Few of such sites have been dated with any precision, but many must have had their origins at a period of greatest pressure, in 260–70. A great many were intermittently occupied from the end of the third century until the fifth, when, with small churches often inside their enclosing walls, they formed the nucleus of villages linked to the larger episcopal centres. In some of the mountain valleys, life changed little in the late Roman period: once re-instated in their safer hill-top forts, the local people were reluctant to come down again to the flat and vulnerable valley floor.

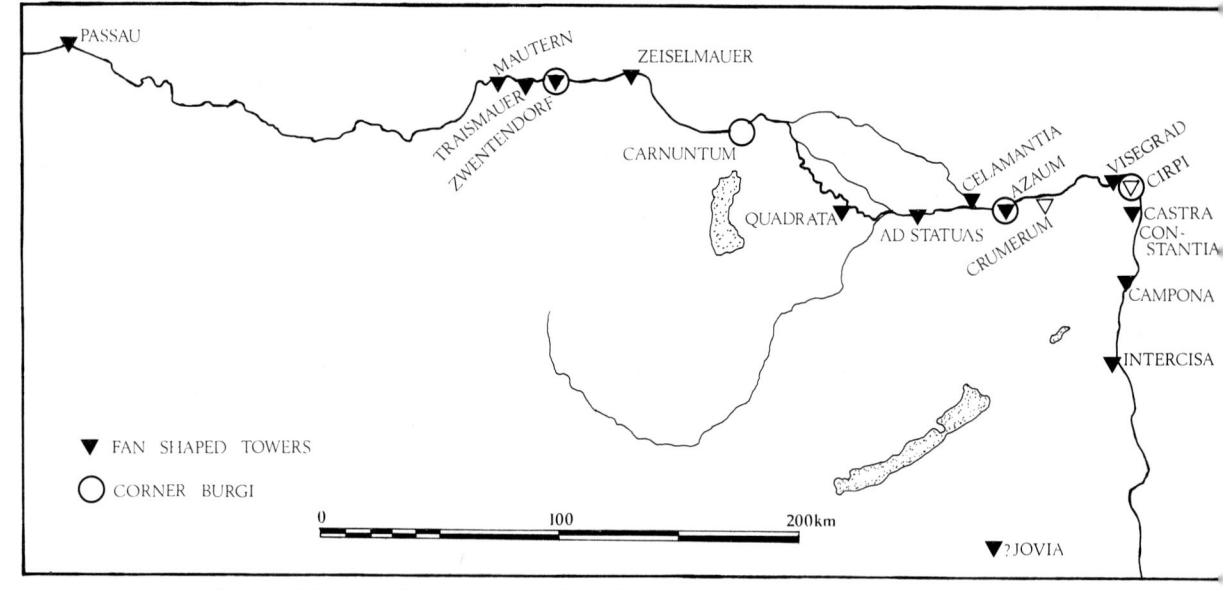

75 Burgi and fanshaped towers on the Danube frontier

After so much energy expended by Probus on the Danube and Iller frontier in Raetia, Diocletian merely had to keep up the impetus. Possibly the only new fort which he added was at Arbon, on Lake Constance. Goldberg near Turkheim, a small fort between Kempten and Augsburg, was also either built under Probus or under Diocletian. Diocletian's main achievement however, was the continuation of the schemes put in hand by Probus, and the manning of many of the newly built frontier forts. The garrisons of several Raetian forts bore names which suggest that were originally formed by Diocletian or his co-emperor Maximian. The name *Valeria* which appears in the garrisons at Isny, *Piniana*, Neuburg and Manching, is Diocletian's own; those of Kellmünz and Burgheim bear the cognomen *Herculia* which was adopted by Maximian.

The major bulk of archaeological evidence comes from further down the Danube, in Noricum and Pannonia. There, many of the existing frontier forts, normally of early second-century date, were used and refurbished in the late Roman period. As far as the fort defences were concerned, the major significant alteration which is evident to the archaeologists is the addition at the corners of fan-shaped towers, and the construction of large projecting towers from some of the gates. It is usually assumed that because they all occur on forts and can therefore be seen as part of a military scheme, the fan-shaped towers are all contemporary additions and belong to a comprehensive building policy by one particular emperor. This is perhaps likely, but not necessarily so. Fan-shaped towers are well known not only throughout Pannonia, but there are instances of them in Noricum (fig. 75) and many more further east along the frontiers of Moesia. It cannot therefore be automatically assumed that they were a sudden creation to meet a particular crisis in Pannonia.

The only way in which to bring a fort of 'playing-card' shape up to date was to add corner towers which could enfilade both adjoining walls as well as provide a platform for the emplacement of ballistae. To gain sufficient distance from the walls to provide flanking cover, a large tower was necessary, and the most economical method of providing this was a tower which was a quarter segment of a circle. In Pannonian forts, not all towers are of exactly similar shape. This might suggest that there was no 'standard plan' produced by the Roman military planning office, but that commanders of each fort built the kind of tower they thought fit for the proper defence of their base.

The forts on the Danube frontier which have produced towers of this type now form a long list, but few have provided any clear dating evidence. The consensus of scholars who date the towers to a Constantinian period do so by reference either to similar shaped towers at other sites at the mouth of the Danube which bear Constantinian building inscriptions or because the Pannonian sites themselves have yielded little dating evidence. Capidava, for example, at the mouth of the Danube, is commonly cited as a fort of the Constantinian period. It had several different types of towers; rectangular, U-shaped and fan-shaped, but not all belonged to the initial stages of fourth-century rebuilding, and the discovery of an inscription of 235–8 and the finding of Constantinian coins from one of the added rectangular towers does not necessarily date all the rest.[65] This assumed date from the site has been used, however, to date many other similar forts in the area, even though Adamklissi, the only walled site which has produced a building inscription of the period of Constantine, has no strictly fan-shaped towers. Drobeta, a bridgehead fort across the Danube in Dacia, was rebuilt several times; one such period of construction included the addition of fan-shaped corner towers, but the date of this, too, is disputed.[66]

As far as Pannonia, Valeria and Noricum are concerned, the dating evidence for fan-shaped towers is not wholly consistent. Two forts so far located on the Danubian frontier were newly built with fan-shaped towers in the late Roman period. One, at Passau, was considered to be of late third century date, while the other, Visegrad, had both fan-shaped and U-shaped towers, and was dated to the reign of Constantine from coin and pottery finds made on excavation. Archaeological evidence from regular forts with added fan-shaped towers comes from further sites. At Ad Statuas, a coin-hoard of 354 was found buried on the floor of one of the corner towers, thus proving that the tower was built before that date. At Ulcisia Castra, the fan-shaped and U-shaped towers are supposed to have been added at the time of the fort's renaming as Castra Constantia. The date for this has been suggested to be within Constantine's reign, but the name would be more appropriate under Constans or under Constantius II. Campona's fan-shaped towers may have been built as a consequence of the destruction of the fort in the Sarmatian raids of 322, but the indications from Intercisa suggest a later date for similar developments there. An inner ditch, which would have had to be filled in, at least in part, in order to construct fan-shaped or projecting U-shaped towers added to the fort walls, contained coins of the mid-fourth century. Reports on the excavation do not reveal whether the whole of this ditch was deliberately filled in to allow projecting towers to be added to the fort walls. The evidence for the date of filling the ditch came from an area well away from any known projecting towers.

It is still just possible to argue that the introduction of fan-shaped towers was a single building campaign dating to the reign of Constantine. It is equally possible, however, that fan-shaped and other projecting towers were added at various stages as and when the frontier commanders decided they were necessary. The fact that at Matrica normal developments were not followed fully, suggests that the addition of fan-shaped towers was not automatic; if it was not automatic, it may not have been a development of a single date at all sites.

Although it is a piece of evidence as equivocal as much of the others, a late Roman document records the construction of two forts in the *barbaricum* in 294, opposite Aquincum and Bononia. The fact was clearly so unusual an achievement that it warranted inclusion in the recorded annals. By the time of the compilation of the *Notitia*, nearly a century later, these two sites were still regarded as lying in the barbarian territories.[67] There is some force in the argument that the insistence of these documents on the siting of the forts 'on barbarian land' should mean that they were placed deep in enemy territory and not merely on the river bank, but there are equally powerful contrary arguments. Forts deep in enemy land would not normally be called Contra Aquincum, or Contra Bononiam. The word *contra*, according to normal latin usage, should mean 'opposite'. It was rare at this date for Rome to seek to expand her control outside the empire's bounds, and to establish a permanent garrison on barbarian soil was an achievement which could be used to good propaganda affect. Contraquincum is the name traditionally associated with the fort south of Aquincum on the east bank of the Danube. It was built in a single campaign and has fan-shaped towers. It is a fair assumption also that a fort opposite Bononia was called Onagrinum because of the provision of externally projecting towers for stone-throwing. If these two sites are correctly identified, these fan-shaped towers were built under Diocletian.

The provision of these forts on Sarmatian soil was probably part of a more comprehensive Diocletianic programme to control the area east of the Danube.[68] A system of ditches, known as the Devil's Dyke, may be part of this attempt at Roman control.[69] They are useless for defence, but could be a border defining an area of Sarmatian settlement in the area immediately east of the Danube. The ditch systems strike out eastwards from the Danube bend, enclosing a broad sweep of territory of a roughly rectangular shape, bounded on the east by the mountains cutting Dacia off from the Hungarian plain. Excavations have failed precisely to date this line of banks and ditches, but it overlay Sarmatian graves of the period 220–300. Other buildings were also built on barbarian territory. A small watchtower has been excavated at Hatvan, many miles into Sarmatian territory east of Aquincum. It has yielded Valentinianic tile stamps, but it may have been built rather earlier.[70] This, like other similar watchtowers, may have been placed to control the trade routes. Other posts, possibly of a military nature, beyond the Danube are known at Oberleiserberg,[71] Stillfried,[72] Milanovice,[73] and Cifer-Pac.[74]

More important militarily were the defended harbours which the Roman fleet maintained on the east bank of the Danube. Several such sites of similar plan to those at Engers and Niederlahnstein on the Rhine are recorded on that part of the Danube within the ditches which seem to delineate Sarmatian settlement. From discoveries of tile-stamps from one of them, Nogradveroce, it has been suggested that this harbour at least belonged to the campaigns of Diocletian rather than those of Valentinian.[75] Most of them have also produced tile stamps of Valentinianic date, but may only date interior buildings or reconstruction at a later period.

The reign of Constantine heralded a period of comparative peace and prosperity for Raetia, broken only by domestic difficulties over the succession of Emperors and Caesars. On the frontier, the Bürgle bei Gundremmingen was now built, though whether the diminutive size of fortifications such as this was the norm is not certain. Inland, the fort at Goldberg near Türkheim was rebuilt. In Pannonia this period was influenced by continuing campaigns against the Sarmatae. Sirmium was the site of Constantine's imperial palace and therefore his visits to the area were frequent. The most serious raid by Sarmatians during his reign was led by king Rausimodus in 322, and this affected an area of frontier round the fort of Campona.[76] This temerity was handsomely repaid by Constantine, who gave chase across the Danube and brought a successful campaign to its conclusion by striking a treaty with the Sarmatians and issuing coins which proclaim the defeat of Sarmatia.

The struggle for power between Constantius II and Magnentius in 351 had equally disastrous consequences for the Danube frontier, as it had done for the Rhine. Magnentius' defeat at Mursa resulted in the death of a reported 50,000 Roman soldiers, a considerable weakening of Roman defensive capabilities. This strengthened the hand of the barbarians: the Rhine and north-eastern Gaul fell target to Franks and Alemanni in 352. This also affected parts of Raetia lying nearest to Maxima Sequanorum. The hill-top settlement at Schaan was destroyed, as also was Chur, capital of Raetia Prima. Constantius II set his campaigns of restoration on a sound footing by appointing Julian as Caesar to deal with the Gallic situation. Campaigns were launched on several fronts against the Alemanni: Julian dealt with the situation in Gaul, while Constantius spearheaded an advance against the Suebi, Quadi and Sarmatians.

Constantius II immediately turned his attention to Pannonia, which in 356 had been overrun by a dangerous invasion of Quadi.[77] After a show of force against the Sarmatae, Constantius arrived to make peace with the Quadi at Brigetio. The Sarmatae were reduced to the status of a client kingdom. Constantius stayed in Pannonia for a further period, strengthening the frontier of Valeria against a further tribe the Limigantes, who were threatening to strike through Moesia. By 360, the Danube frontier was again settled.

The final major period of reconstruction and development on the Danube was in the reign of Valentinian, from 364–75. Relatively rich inscriptional evidence for construction at this period has suggested that a large series of frontier works were set in hand. The fort at Hidegleloskeresz is probably the findspot of an inscription dated to 365–7.[78] Tokod was also built at this time. Two further inscriptions, as well as finds of Valentinianic tile stamps at many sites, particularly watchtowers, attest an active period of construction, or reconstruction. A number of inscriptions of Severan and earlier times record that some watchtowers were of considerably earlier origin, but perhaps at no other time was there such a comprehensive amplification of the system. The Pannonian frontier was now at its strongest.

In Raetia too, a series of watchtowers supplemented the frontier forts. Finds from several of the towers on the line between Bregenz and Kempten and on the Danube itself show the renewed presence of soldiers in the Valentinianic period though some, like those in Pannonia, may have had earlier origins. No inscription from this part of the limes exists to confirm the impression given by pottery finds and coins, but at Ybbs in Noricum an inscription of Valentinianic date attests the construction of a *burgus* in 370.[79] Another nearby watchtower or fortlet, at Oberranna, may be its contemporary. Valentinian was concerned to a great degree with the frontiers themselves, and personally inspected the defences on Rhine and Danube, adding new forts where he considered it necessary.[80] In Raetia, archaeological evidence suggests that he added Schaan and Wilten, both probably established as fortified granaries.[81]

Despite this spate of activity by Valentinian, there is some evidence that defences inside the provinces were in a rather worse state. The important amber route, from Carnuntum to Aquileia, had been reinforced by a series of towns with strong defences; Sopron, with town walls of typical late Roman style, enclosing a reduced area and ringed with external towers, was still relatively secure. Savaria, by contrast, had a wall-circuit which has not been located. Portions at least were in a poor state in antiquity. Valentinian stayed there in 375 and was unable to leave the town by the same gate as he had come in, since it had fallen down in the interval.[82] In the same campaign, Valentinian passed through Carnuntum, which is recorded as 'a dirty and deserted town'.[83]

Valentinian made several inroads into barbarian held territory beyond the Danube, mainly into territory of the Quadi. The major grievance felt by these tribes was his construction of Roman forts on their territory itself.[84] These seem to have been watchtowers rather than major forts, possibly some of those actually lining the Danube bank. This imposition caused a great furore among the Quadi, who invaded Pannonia with a thoroughness which had not been seen for many years. Valentinian, who arrived in 375 and attempted to restore the situation, was killed near Brigetio in a barbarian attack. From then onwards, there is little recorded or attested building on the frontier or in the Pannonian provinces before their final abandonment in or around 405. The history of the provinces between the death of Valentinian and this date is chequered by accounts of ever deeper incursions by barbarians, Quadi, Sarmatae and Huns who gradually moved in and took over the flatter lands to the north and east of Italy.[85]

The end of Roman control in Raetia, too, presents a similar picture: invasions and attempts at restoration of the frontier continued throughout the later decades of the century, until a point early in the fifth century when it was not possible to withstand such pressure any more. German settlers in the area became more and more frequent. The last epigraphic evidence from the area is a gravestone of the late fourth century recording the burial of soldiers of regular Roman troop units in Augsburg. The pattern of Roman coin-finds suggests that the province was cut off from monetary connections with Italy by the end of the fourth century, and cannot have maintained a financial or political independence for long after 410.[86]

Of particular interest is the evidence for a form of continuation of Roman life in fifth-century Noricum. A local bishop, Severinus, organised a form of spiritual and physical defence for scattered Roman communities, some of whom lived within former forts of the Roman frontier.[87] The events related in the almost contemporary saint's biography show how local people still continued to hold on to some form of Roman existence for as long as they could. Such small communities were possibly organised under Severinus' control as an official episcopal function.

There are also traces of barbarian settlement, however, on some Roman fort-sites. At Carnuntum, there were buildings in a late Roman phase which contained so-called Alemannic pottery, and in Vienna, buildings of a similar type have been located. At Lauriacum, the fort took on a new use as that of refuge for the local townspeople. This, too, was one of the sites where Severinus brought moral support to the inhabitants. Many hill-top refuge sites in Tirol and Kärnten, too, continued in flourishing use. The construction at an early date of churches inside many of them, as well as inside some of the Danube forts may have been due to the increasing connection, illustrated in the career of Severinus, between the late Roman episcopate and survival in the face of barbarian threats. Some forts acted in part as *castra* for the bishops themselves to use in times of danger.[88]

The story of Severinus shows that the last vestiges of a Roman style of life in heavily defended settlements were in the process of being whittled away, as more and more communities were swallowed up in the barbarian advance through Noricum. Rural economy was not yet ruined, and peace was for the most part maintained by payments to the tribes beyond the Danube. The survival of groups of settlers—soldiers and farmers, all descendants of the men who were once there as limitanei—in the Roman frontier forts is a last glimpse of the sort of tenacity which once was able to keep the barbarians at bay.

8

The Frontiers III: The North Sea

One of the areas of the Roman west which came increasingly under barbarian pressure in the third and fourth centuries was the south-eastern tip of Britain and northern Gaul. These seem to have fallen victim to the raids of seaborne pirates from at least the early third century onwards, though historical sources do not specifically mention attacks at such an early date. Control of the seas between Britain and mainland Europe had always been of considerable importance, and the Roman fleet, *Classis Britannica*, based on Dover and Boulogne, was the main vehicle for policing the channel straits in the second and early third centuries. Knowledge about the Roman British fleet, its method of working, and the scale of its presence, is scant, but an inscription attests its survival at least into the mid-third century.[1]

By the early third century, there are signs that early fleet-bases at Dover and Boulogne, supplemented possibly by other sites such as Brough-on-Humber and London, were proving inadequate for the necessary amount of policing.[2] A series of towns and other settlements lining the eastern coastlines of Britain was put on the defensive, at about the same time as the majority of other British towns received their first earthwork defences. More specifically, however, forts were built at Brancaster,[3] guarding the approaches to the Wash, and at Reculver[4] at the approaches to the Thames (fig. 77). An inscription found at Reculver serves only to suggest a construction date no more precise than some point in the third century and more detailed archaeological argument for its date has yet to be presented.[5] Typologically speaking, the two forts are similar: they were of almost equal dimensions, almost square, with rounded corners, no projecting bastions, and a rampart of earth behind a thick enclosing wall. This design has much in common with forts in Britain of the Hadrianic, Antonine, and Severan periods, and contrasts strongly with more irregular bastioned fortifications of the later Roman period. It is probable, therefore, that these forts belong to the earlier decades of the third century. The pressures on the eastern coastline to which they were a response cannot be quantified, nor is it clear how the fort garrisons were organised to combat any seaborne threat.

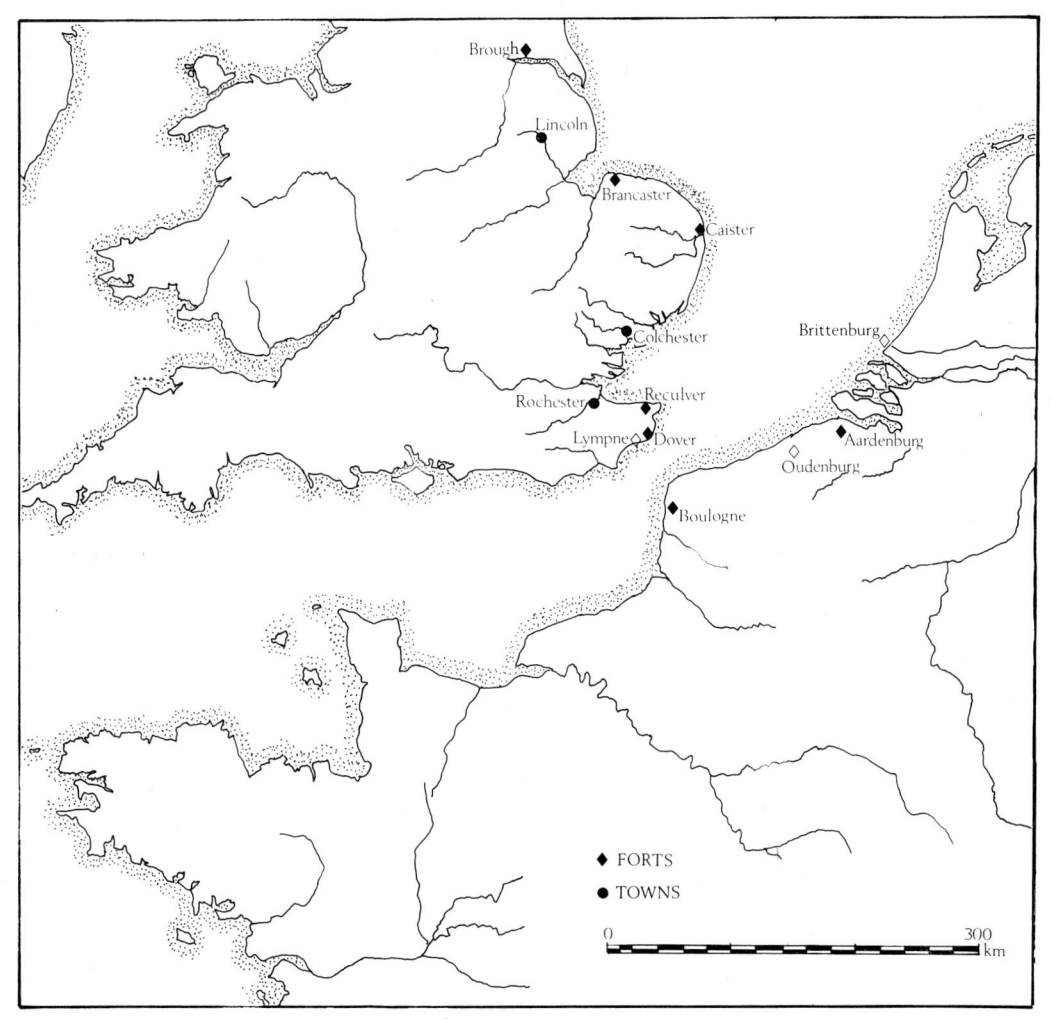

76 Map of the coastal areas of Britain and Gaul, early third century

In parallel with these measures on the British coast, there were defensive developments on the coastlines of the Continent at about the same time.[6] At Oudenburg, where the site of a later bastioned fort and its cemetery has been examined, limited work has established that there were also earlier phases of use of the site, including two earlier earth-and-timber forts.[7] These were not closely dated, but the earliest of the series may have belonged to the early third century, perhaps as much as a century earlier than the bastioned fort. At Aardenburg, a busy trading port not far from the Rhine mouth, excavations have revealed one side of an apparently rectangular fortification with large projecting circular bastions at the corners (fig. 77), and, at its centre, an irregular building, probably the headquarters.[8] This suggests that the installation was intended as a military or naval post. The form of the fortification does not find a ready contemporary parallel elsewhere in the Roman world. According to the abundant coin-finds from the site, it was constructed in the early or mid third century, and must have

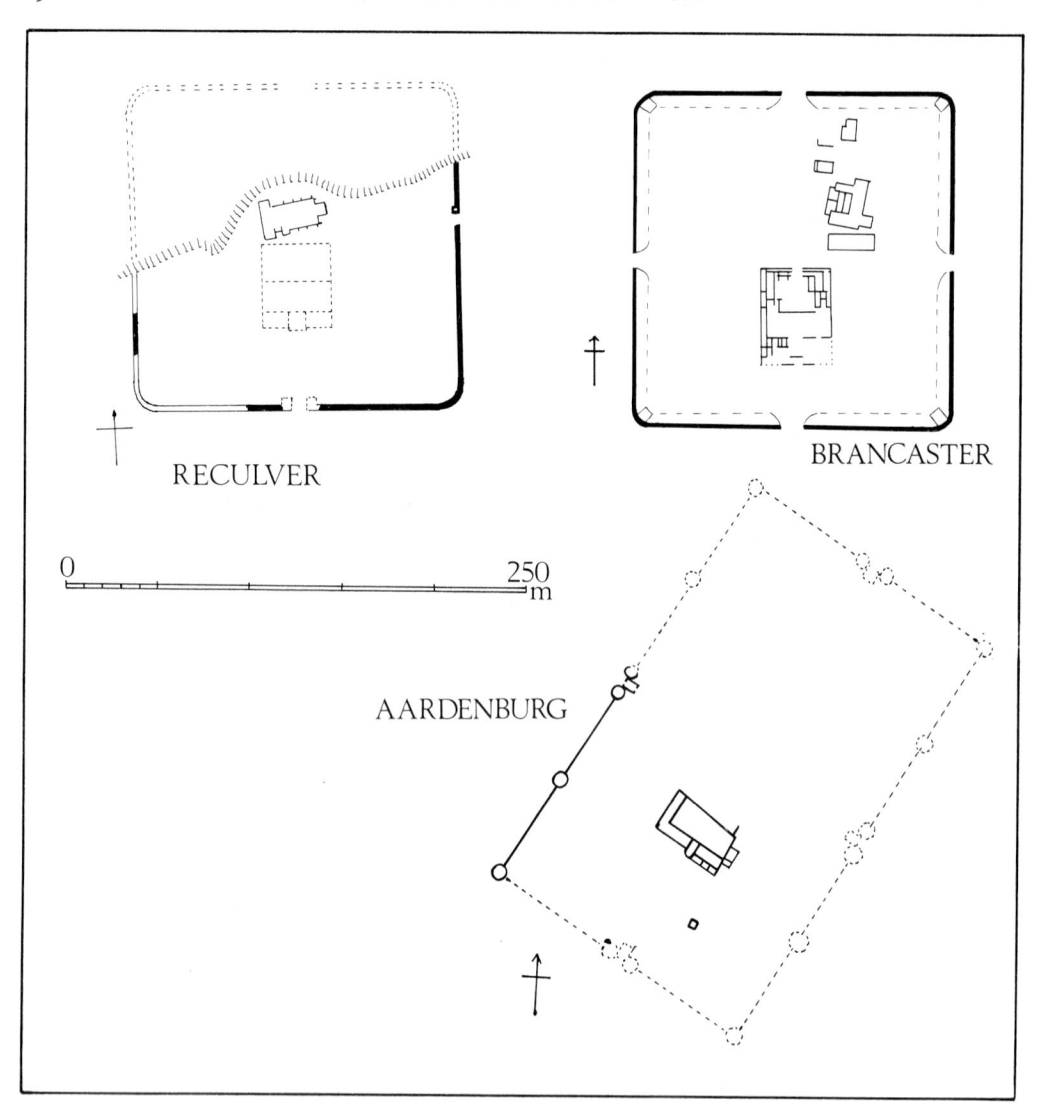

77 Site plans of coastal defences in the early third century (*scale 1:4,000*)

gone out of use at all events by the early 270s. At this period the site was probably abandoned because of the encroaching sea.

Other sites in the area of similar date are less well known, partly because of the high degree of post-Roman alterations in the sea coasts and river channels. No further military installations as such are known, though because of the quality of the remains discovered, a series of deeply buried deposits at the Rhine mouth have been suggested as the sites of *vici* attached to military forts.[9] North of the Rhine mouth, the only known Roman military structure was the enigmatic site of the Brittenburg, now lost in the sea off Katwijk.[10] There is no evidence at present for refurbishment at the known fort and harbour sites on the lower Rhine in the early third century. At Boulogne, remains of a large Roman fort on the Haute Ville hill predate the late Roman and medieval ramparts which now encircle the old town.[11] This, like the *Classis Britannica* fort at Dover,[12] may be of second-century date, but its defensible position, and the fact that it was clasped at the end of the third century within Boulogne's defensive circuit, suggests that it was in use throughout the third century. West of Boulogne, no coastal military installations of early third century date are known.

By about the mid-third century, Roman control of the channel area was on the decline. Two historians who wrote in the 390s, but who had access to earlier material, relate that the coasts of Belgica and Armorica (and by inference, the coasts opposite them in Britain) had become infested with barbarian pirates, specifically named by one of the writers as Franks and Saxons.[13] Various problems, in particular the existence of the Gallic Empire under Postumus and the Tetrici from 258–73, prevented strong action from Rome to redress this situation. A Bagaudic uprising in the early 280s probably also made the difficulties of the area somewhat worse, and it was only in the wake of this uprising that a new admiral of the fleet could be appointed in the area. He received his command at Boulogne, with instructions to draw up a fleet, and to keep the peace in the coastal areas of Armorica and Belgica.[14]

It is rare to be given so much detail about the Roman response to a particular threat, and, were it not for the admiral Carausius' subsequent exploits, later historians would probably not have mentioned the existence of this channel command. Carausius, the story goes, enjoyed considerable success—so much so that stories began to circulate that he was allowing barbarians to raid rich areas, capturing them laden with their spoils on their way home, and refusing either to return the goods to their owners or to surrender them to the imperial treasury. A warrant was issued for his arrest on the charge of treason, and Carausius, trusting both in the popularity he had gained by controlling the seas, and in the skill of his sailors, was declared emperor by his troops. This in effect made northern Gaul and the whole of Britain a new breakaway empire. Other matters for a time claimed the attention of the legitimate emperors, Maximian and Diocletian, and it was not until 291–2, some five years after Carausius' revolt, that the first attempts to dislodge him were made. The plan, a move to reach him by sea from the Rhine mouth, was a failure. The Roman fleet was either defeated in an engagement or forced back by foul weather. It was not until 293 that Carausius' downfall came, in a strike by land on Boulogne, which was suddenly invested and quickly taken. Carausius did not long survive the blow, which now confined him to Britain, despite his control of the channel. His former aide and finance minister Allectus is said to have arranged his assassination. In 296, however, a determined assault by Roman forces gave the British fleet the slip, and landed an assault force in Britain. Allectus' mercenaries were defeated and their leader killed.

Despite the obvious disfavour with which this episode was viewed, in particular by the official Panegyrists, whose speeches in praise of the Imperial house voice a constant denigration of Carausius and Allectus, the episode proves one major point of interest. This is that, given adequate Roman naval presence, the threat posed by Germanic pirates and raiders could be and was effectively contained. Like the *Classis Britannica* in earlier years, the new channel fleet needed

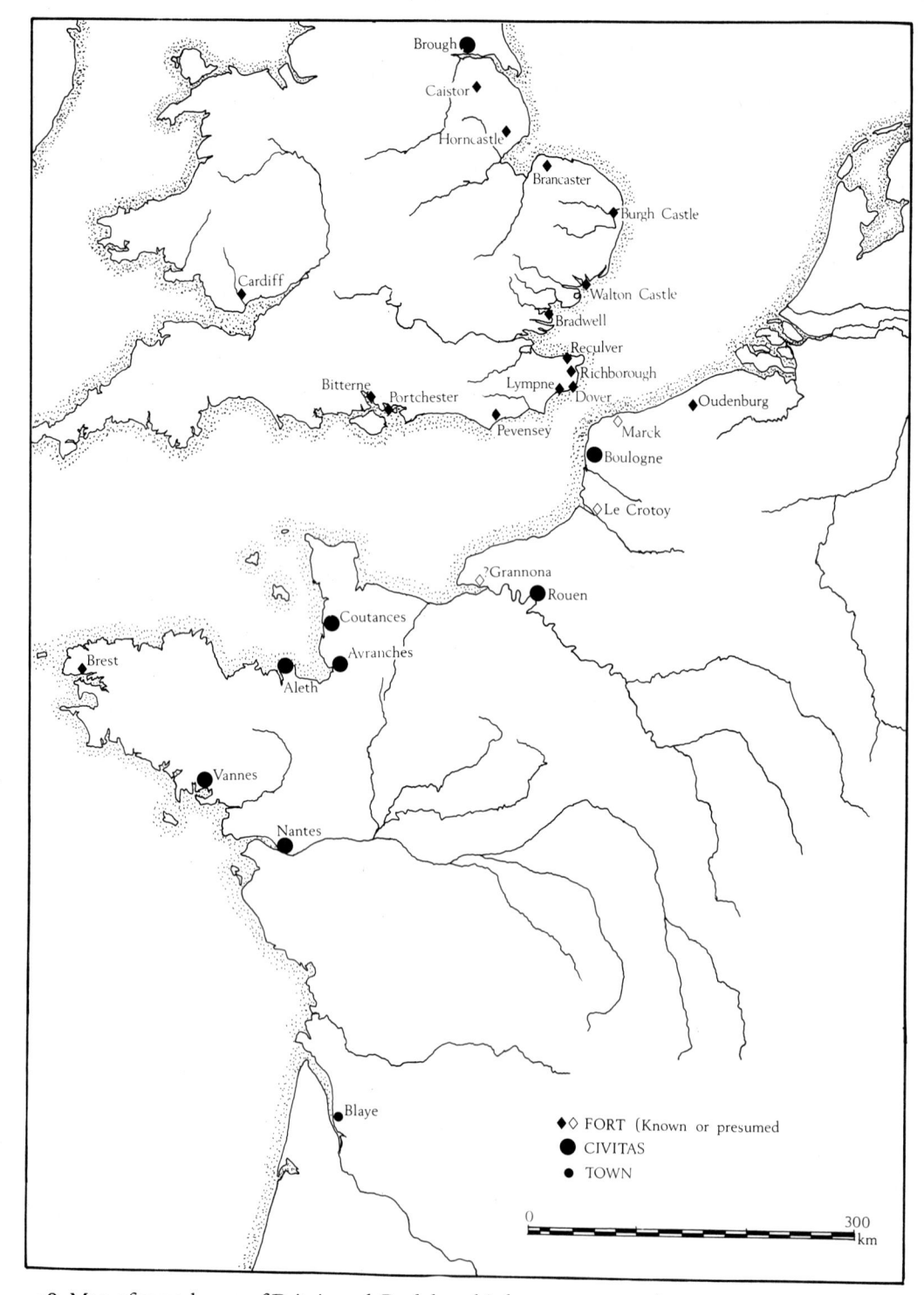

78 Map of coastal areas of Britain and Gaul, late third century onwards

bases, and it thus seems to be no coincidence that at about this period a series of new walled installations were constructed at or near some of the major harbour sites on the East Anglian and south-eastern coastline. Complementary specialist bases may also have been established on the continental shore, but these are by and large still to seek. In Britain, to add to the existing sites at Brancaster and Reculver, bases were established at Burgh Castle, Walton Castle, Bradwell, Richborough, Dover, Lympne and Portchester. On the continent, to add to the command headquarters at Boulogne, new installations were set up at Oudenburg, and at two sites named in a later list in the *Notitia Dignitatum* as Grannona and Marcis, still unlocated.

At a number of the British sites, the defensive walls of these installations survive in a fine state of preservation. Burgh Castle, a site which dominated a large inland estuary forming the mouth of three of the main rivers into the northern part of East Anglia, was constructed with walls of rubble corework faced with split flint and tile-courses.[15] It was an irregular quadrilateral with rounded corners (fig. 79), and the west wall has fallen some 10m down the slope to the estuary, where the Roman harbour presumably lay. External towers, pear-shaped in plan, project from the walls at the two surviving (inland) corners, and at intervals in between. Their facing and construction technique are very similar to that of the walls even though they are not bonded in with the main wall for the first 2.4m of their height (pl. 16). This suggests that they were an addition to the construction work after it had begun, though at an early stage.

Burgh Castle was designed as a walled enclosure with thick, freestanding walls, in a shape which echoes, but does not closely resemble, earlier military forts of 'playing card' type, with rounded corners and internal angle- and interval-turrets. The lack of an earth rampart behind the walls, their extra thickness (3.5m, as opposed to 1.2–1.5m at earlier sites) and the apparent lack of well-substantiated internal angle turrets suggests that the design of this installation is later and more advanced than the Hadrianic standard pattern of British forts. The fact that while construction was in progress external towers were added, possibly as an alteration to the original design, further suggests some refinement and sophistication of defensive architecture was in progress at the time of building.

Objective dating is unfortunately scant. Excavation within the walls has been limited in scope, and has touched on undisturbed Roman layers but rarely. The site was used as a motte-and-bailey castle in the eleventh century, and there were considerable amounts of Middle Saxon occupation material from the site which presage a heavy use in the post Roman period. Coin-finds from the site as a whole ran in a continuous band from the second half of the third to the very beginning of the fifth century, and though there were no defined structures which could be said to belong to the period 250–75. This period, within which the first appreciable numbers of coin finds fall, may be assumed to be the construction date of the defended site. Excavation close to the interior of the walls has shown that these areas were predominantly in use in the Constantinian period, as shown by a massive number of coin-finds (probably largely a dispersed hoard) and pottery types of comparable date from that area. There were no signs of any rebuilding or refurbishment of the defensive walls throughout the period of use.

Further south along the coast, the site of Walton Castle[16] lay at a point which dominated and controlled the access to a further three rivers leading into the East Anglian hinterland. It has now fallen into the sea off Felixstowe, though drawings made during the course of the eighteenth century show that it looked very similar to Burgh Castle: its walls were probably of rubble concrete, faced with split flint and tile. The next fort southwards, too, at Bradwell,[17] barely survives above ground level. It had rounded corners similar to Burgh Castle, with projecting bastions at the corners and at intervals (fig. 79). From the fragment of walling which remains, built of rubble concrete with a tile and septaria facing, little of significance about the building technique can be deduced. Archaeological examination of the site has revealed little of substance.

Of the Kentish sites, the most impressive is Richborough,[18] where the rectangular walled circuit occupies the summit of a low hill, probably a peninsula in Roman times (fig. 79). The late Roman walls were built on the site of an earlier earth fort, itself thrown up in the early third century to provide a lookout post against barbarian inroads. This fort was levelled, and the triple ditches which had surrounded it filled in, in order to provide the platform on which the stone walls were constructed. Substantial portions of those walls still survive, even though much of the eastern side of the fort has fallen through erosion by the River Stour.

The walls were built of rubble concrete faced with small ashlar and double tile-courses. Midway along each of the surviving sides was a gate, posterns to north and south, and the main gate to the west. This was a single portalled entrance between a pair of rectangular projecting towers which incorporated guard-chambers. The north gate was a postern concealed as a right-angled passage through the wall within a rectangular tower, and there were indications from excavation that the south gate was similar. Midway between the angles and the gates lay hollow rectangular projecting towers, while at the two surviving corners there were three-quarter round projecting solid bastions. Double ditches surrounded the walls, and were interrupted only where the main road to the site from the west entered the fort.

The construction date of the fort can be pinned down with some exactitude. Since the demolition of the earth fort and the filling of its triple ditches occurred in order for the stone fort to be built on the site in its place, coins and pottery finds in the fill of these ditches must pre-date the fort construction. At one point, however, a wall foundation, never apparently used, for an east wall was found to overlie the filled-up inner earth-fort ditch. No coins later than 273–4 were found in the ditch fill, in the foundation itself, or in a well shaft dug through the unused foundation and relatively quickly filled up with rubbish. The earth fort was therefore levelled by this date, and the commencement of the construction of the stone fort can be assigned to 275 and after. There are signs from detailed consideration of the masonry of the fort walls that the period of construction may have taken some time, perhaps as much as ten years, with the fort being redesigned after an initial constructional phase. One of the corner bastions is an addition, the east wall was clearly moved further eastwards, the north wall is of

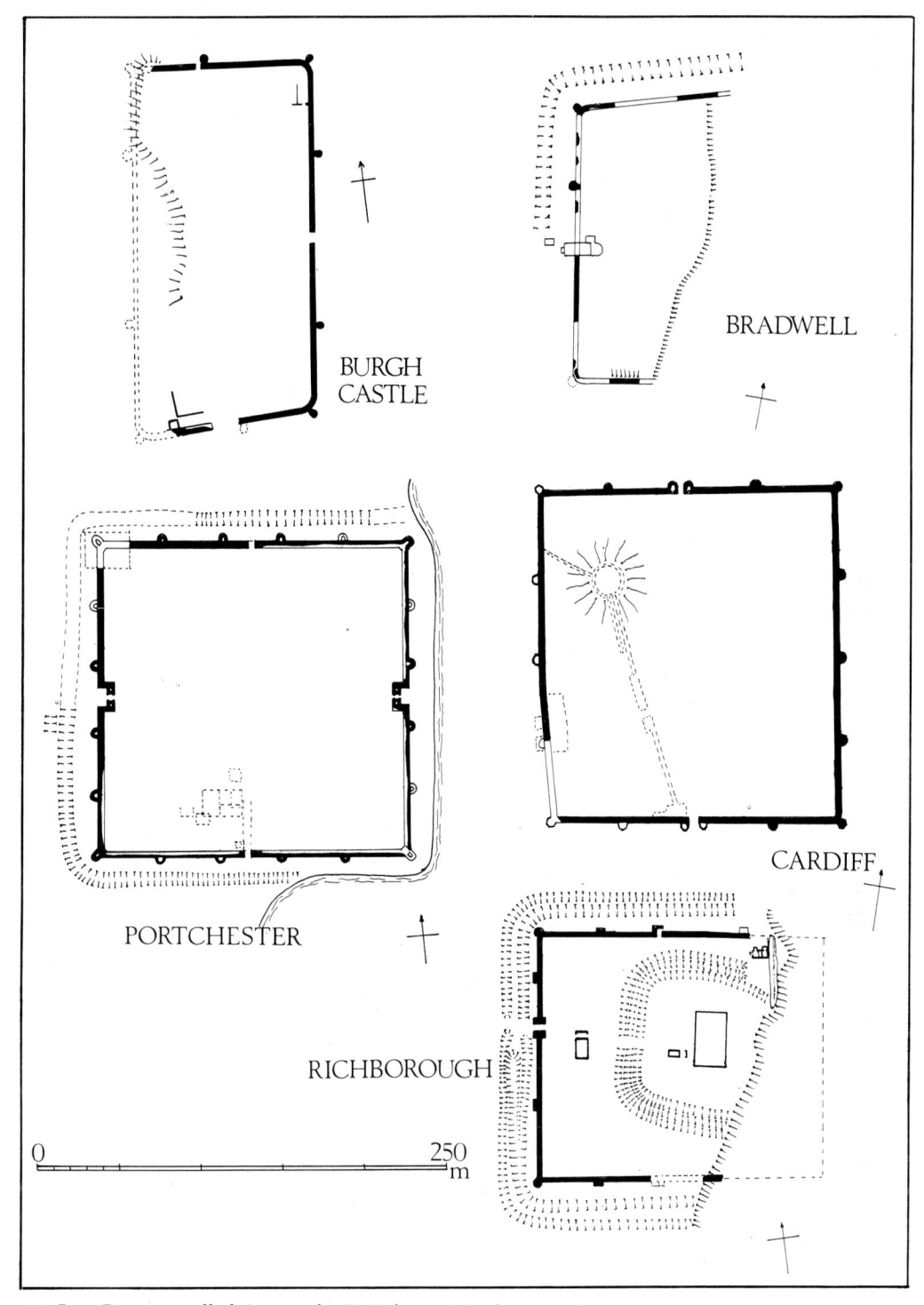

79 Late Roman walled sites on the British coast (*scale 1:4,000*)

different build from the west and south, and the original design catered for only a single ditch, later replaced by two. It may be plausibly suggested, therefore, that the fort was built in the decade 276–85, as part of the preparations then made against pirate raids on Britain and Gaul from German tribes in the coastal areas of Free Germany.

At Dover,[19] where the sites of the Roman naval installations at all periods have been long sought but only recently located, the early Roman fort of the *Classis Britannica* and the later successor have now been shown to lie on the west bank of the estuary of the River Dour, under the modern town. By the late Roman period, the *Classis Britannica* fort, of normal 'playing-card' design, in operation for much of the second century, had clearly gone out of use, though its remains must have been still substantial. After the mid-third century a site slightly nearer the river was chosen for a small irregular fortification with walls 2m thick, backed by an earth rampart (fig. 80). At intervals semicircular bastions, of different build from the walls, were added. No gates are known, and the portions of the west and south sides which have been traced were flanked by a ditch.

West of the channel straits, at Lympne,[20] lay the next fort in the chain. It lies on a slope, now landlocked, but originally dominating a harbour at the estuary of a series of channels, originally part of the river complex leading into the heartlands of Kent. The fort appears now to have been pentagonal in shape (fig. 80), though the landslides to which the hillside is subject have pushed the surviving walls considerably out of their original line and, according to a recent suggestion based on limited archaeological examination, it may have originally been rectangular. Such fragments of walling as still stand show that the fort was built of rubble concrete with a facing of grey ashlar and double tile-courses. A gate, formed by a single passageway flanked by a pair of half-round projecting towers, lay to the south-east. Further solid semicircular bastions projected at intervals along the walls.

Two further Roman installations closely linked with the defence of the shore and harbours lay on the south coast of England, west of the immediate area of the channel straits. At Pevensey,[21] some 50km west of Lympne, a fort of irregular oval plan (fig. 80) now landlocked, lay on the low peninsula of solid ground amid a marshy estuary which probably afforded an adequate harbour in Roman times. The fort walls, at 8–9m high, are considerably higher than any others so far described within this series, and survive round approximately two thirds of the circuit. The foundations, of crushed chalk and pebbles, incorporate timber framing and were pinned to the ground by vertical stakes. Above this the wall was built of rubble concrete with an exterior facing of small ashlar interspersed with bonding courses, sometimes of tile, and more often of large flat stone. At irregular intervals large U-shaped solid bastions projected from the walls, but these were concentrated in the areas near the gates.

The main gate, at the western extremity, lay recessed between a pair of large bastions. It was formed by a pair of rectangular guard towers flanking a single entrance passage, 3m wide. Excavation has shown that beneath these gate-towers the wall foundation was laid continuously. In effect, the gate was a single

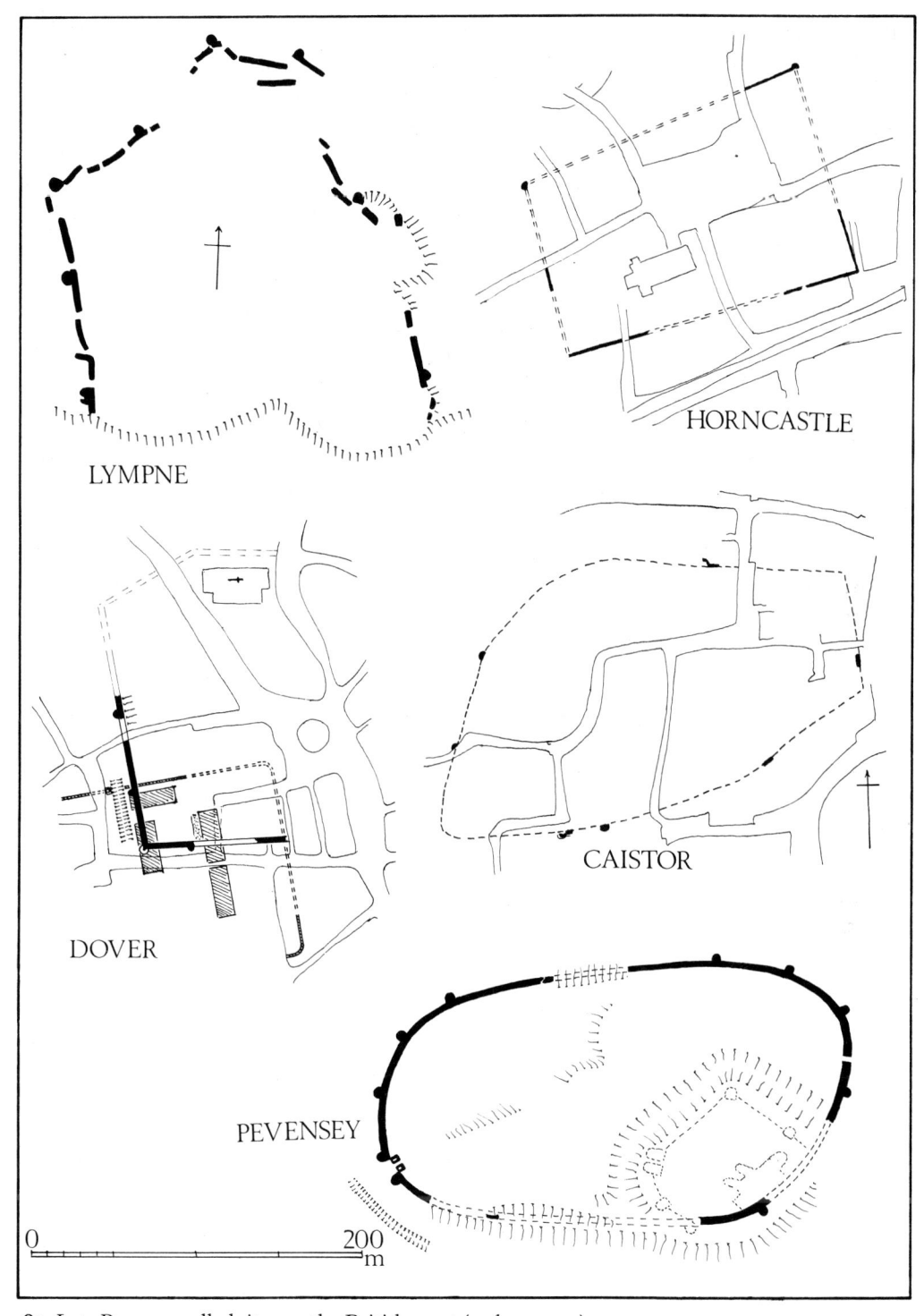

80 Late Roman walled sites on the British coast (*scale 1:4,000*)

passageway given weighty protection by the addition of a massive pair of flanking towers. Two further gates, one a single opening in the wall half-way between a pair of bastions, and the other a smaller, S-shaped passage through the wall, are also known.

The walls, like parts of those at Richborough, were built in sections by separate building-gangs. Subtle changes in the wall-facing, and alterations and disruptions in the levels of the bonding-courses show clearly the points at which the building-gangs failed to make their work correspond. There appears to have been no consistency over the lengths of work built in this way, since the measured distances vary from 15m to over 45m, averaging only around 26–30m.

The last site, Portchester,[22] lay 75km west of Pevensey, and lies almost at water-level in Portsmouth harbour. The fort was nearly square, and the Roman walls have suffered considerably from medieval rebuilding and refurbishment (fig. 79). Excavation has shown that the walls were built on a foundation of rubble and clay capped with crushed chalk, within which a pattern of timber beams was inserted. The wall of rubble concrete, with a facing of small ashlar (where it survives) and bonding courses of tile or flat stone, was largely refaced in the medieval period, and as often as not, diminished by half its width through the removal of the interior face.

Semicircular bastions, at present hollow, but possibly originally solid, projected at regular intervals of 30m from the walls. A gate was placed midway along each of the sides. Two, to east and west, were simple posterns, and excavation of the site of the western one has shown that at one phase at least there was a wooden guard-turret placed just inside the entrance passage. The Watergate and the Landgate to south and north, however, underwent considerable alteration in the medieval period. In the Roman period, both gates were set back within the fort from the line of the enclosing walls, and consisted of a pair of square guard-turrets flanking a single entrance passage. The regular spacing of semicircular bastions, six to a side, was not interrupted to give any added protection to these gates.

Despite the fact that much of the site had been disturbed by periods of occupation and cultivation in post-Roman times in an almost unbroken sequence up to the present day, there were some traces of Roman timber buildings, wells, pits and streets. The construction phases of this fort were dated after 261 by the finds of coins in foundation levels, and the initial stages of organisation of the interior of the fort, which involved piling earth against the inside of the wall to level the interior, are dated after 286–7 by finds of coins within the levelling material. Apart from these definite dated instances, information about the use of the site in the later Roman period was mainly derived from the groups of pottery found associated in wells or pits, and from the relative frequency or scarcity of coin-finds of any particular period.

From study of the numbers of coins lost under separate emperors, it appears possible that the main phases of activity at the fort were in the years 285–90, 300–25, and from 330 onwards, with noticeable occupation gaps in 290–300 and 325–30. The latest coins found on the site date from about 390, but there were so few after 367, that there can hardly have been any continued military presence after that date. Not all these coins need have been dropped by the military, though the presence of large numbers of coins at fort sites is usually taken to be a sign of the presence of troops. It is a problem to decide which phases of life in the fort can be said, on this evidence, to give proof of military occupation. At one period at least, the fort interior appears to have become rather squalid, and there was a large open space apparently without buildings which was possibly used for keeping animals.

At few of these sites is there objective evidence for their period of use, though coin-finds either from excavation or from chance discoveries are almost exclusively of late third or fourth century date. Interior buildings are known or suggested at several sites: at Portchester, Richborough and probably Burgh Castle, there were timber buildings, probably used as barracks, but their plan and layout in all cases is imperfectly known. At Reculver, retained in use into the latter part of the third century and beyond, excavations have shown that the layout was similar to Roman forts of earlier days, with a central stone built *principia*, and other buildings, including a bath-block, spaced out within the fort. Traces of a similar layout and design can also be seen from aerial photographs taken of Brancaster, which was also still in use in the later Roman period. Masonry buildings elsewhere have been but rarely found. At Lympne, traces of two, a bath-block and a range of rooms somewhat reminiscent of part of a *principia*, were discovered in the nineteenth century, while at Richborough, from almost the total excavated area within the fort, the excavations identified only four masonry structures which belonged with it: roughly centrally there lay a large rectangular building, assumed to be the *principia*. In the north-east corner lay a small bath-block, and elsewhere there were two small rectangular buildings, of unknown purpose, fronted by a portico.

Elsewhere in Britain, fort defences of similar design to these are few. Only at Cardiff and Lancaster,[23] where a high hill dominating the River Lune, the site of earlier forts, was defended no earlier than the second quarter of the fourth century, with a fort with thick, freestanding walls, is anything similar known. Other known or supposed fort sites of late Roman date, such as Piercebridge, Elslack or Newton Kyme,[24] retained the old-style playing-card shape, and on Britain's northern frontier, despite considerable alterations in the internal arrangements of Hadrian's Wall forts, the defences were never updated to take account of this new fashion. At Cardiff,[25] there are traces of an alteration in the plan similar to that at Burgh Castle. The walls enclosed an almost rectangular area and were built of a core of river gravel and pebbles faced with a rather coarse small blockwork, with lacing courses of flat limestone (fig. 79). The external towers were five-sided, as were also the hollow gate towers, and all were bonded into the wall. In plan, however, the circuit wall described the pattern of a fort with tightly rounded corners, backed by an earth rampart. To this, projecting towers were added at an early stage, since their plinth courses were not bonded with the wall foundation. The north gate was also an insertion; the wall plinth and foundation course continued uninterrupted underneath the twin guard chambers. Dating evidence from excavations carried out at the turn of the century, suggests a date when the transition between traditional and new fort design was being made, is a strong possibility. Cardiff and Lancaster, and a succession of smaller forts in the western areas of Britain, were probably intended to safeguard the coastline against incursions from Irish tribes, in a fashion similar to the defences against Franks and Saxons on the eastern coast.

Purpose-built forts on the continental side of the channel to correspond to this British cover are few. The major site was at Oudenburg,[26] near Bruges, where a rectangular stone fort with corner bastions replaced earlier earth-and-timber installations (fig. 56, p 147). The exact date of this change has not been precisely defined, for evidence for the date comes not so much from work on the defences and ramparts themselves, which have been examined only in isolated areas, but from consideration of the periods of use of a succession of cemeteries used by those who have lived within the fort. Some limited excavation, however, has been possible inside the fort. Traces of a masonry building lying roughly at its centre were located, and fragmentary remains of timber buildings, possibly barracks, were also discovered to lie parallel to the fort axis. A combination of slight archaeological indications suggested that the stone fort may have been built in the mid fourth century, though it may be somewhat earlier.

Two separate cemeteries were found. The earlier, not comprehensively studied, lay south of the site of the forts, and contained graves deposited by and large in the first half of the fourth century. The second cemetery contained 216 inhumations, with a marked preponderance of adult males. More than half of the graves had some grave goods, with many brooches and buckles but no weapons, suggesting that this was not a Germanic group of irregular soldiers who might bury their dead according to the fashion customary beyond the Rhine, but a garrison of regular soldiery. Overall dating suggests that this cemetery was in use in the second half of the fourth century, and may have belonged with the stone built fort.

Other sites on the continental side of the channel are less easily distinguished. Near the mouth of the Old Rhine lay the Brittenburg.[27] This site, lost to the sea during the course of the seventeenth century, is recorded in a number of contemporary drawings, of which the plan (fig. 56, p 147) forms the main consensus. It has been seen as the site of an auxiliary fort of early imperial date, belonging with others of the early period, while other opinions have varied from fortified granary and late Roman fort to Carolingian or medieval castle. As in much of the remainder of the lowest reaches of the Rhine, there is no known late Roman material from the site. Although the Rhine or one of its tributary rivers in this area, perhaps the Waal, must have been used as the late Roman frontier and to carry late Roman commerce, including cornships from Britain to the Rhine land armies,[28] the course of this frontier is not known.

South and west of Oudenburg on the coast, specifically military installations comparable to Oudenburg or the British sites so far described have as yet eluded discovery. A site of prime importance, now as earlier, was Boulogne,[29] but the circuit of late Roman walls here belonged as much to the city itself as to any purpose built military fort, despite the presence of earlier forts or fortresses discovered underneath it. Boulogne was clearly of considerable importance to the channel command at all periods. It was here that Carausius was instructed to make his headquarters in 285–6, and it probably remained a natural base for the channel command for most of the fourth century. West of Boulogne there were no forts, even on rivers as obvious and affording such ready access to the interior of the province as the Somme and the Seine. Near St Malo, at Aleth, dominating the Rance, is there a small, irregular late Roman post (fig. 27, p 87), lying next to the town.[30] Such excavation and analysis as this site has received suggests that it dates only from the mid-fourth century.

Coastal defences in the Britanny area[31] were concentrated on towns or cities with potential for late Roman use as harbours rather than purpose-built forts. Indeed, the main areas of concern to late Roman military authorities were the eastern facing coastlines of Britain and the area actually at the Channel straits itself, centering on the axis Dover–Boulogne. In later lists in the *Notitia Dignitatum*,[32] two commanders of frontier forces on the Gallic coastline, by then a necessity, had troops partly billeted on existing cities and towns, and partly on installations which appear to be forts. Strangely, Boulogne was not mentioned, but the cities pressed into service numbered Nantes, Vannes, Coutances, Aleth, Avranches, and Rouen. Of the sites more likely to be forts, Blaye,[33] on the Garonne estuary, and Brest,[34] where one wall of a late Roman bastioned fort with tile-courses and typical wall construction has been diagnosed, are the only two located. The other four must have lain near the channel straits by virtue of the commanders under whose charge they lay.[35] One, called Portus Aepatiaci, and another with a garrison of sailors, clearly lay in a strategic naval position. The others, Grannona and Marcis, both held garrisons which might well have been in post longer than the others, and thus could have been forts which belonged to the late third century series. Their position at the head of their respective commanders' lists, and the description they are given—'on the Saxon Shore'— echoing the command of that name given a separate chapter in the *Notitia* (see below, pp 211–213), suggests that they may, if they are located, have more in common with the British sites. These could be the forts apparently missing from the earlier series to bring the Continental defences more in line with those in Britain.

The known defences on both sides of the channel have many features in common. The most obvious are the style of building and the apparent dating attested from archaeological material. In terms of siting, however, the chosen sites are all similar, and rather at variance with the type of defensive position normally chosen for late Roman fort sites elsewhere in the Roman frontier provinces. Such sites would normally make the best possible use of available high ground, but while satisfying this criterion as far as they are able, these coastal installations owe their siting more to the presence of an available harbour or to a series of river or estuarine inlets which required control than to immediate concern for the defensive potential of any particular site. The purpose of this is clear: the forts were intended to defend the harbours and offer protection from sea-raiders. They can thus be claimed to have formed a system designed to answer a specific problem of raiding. This was a development, in rather more organised form, of the scattered series of defensive installations of second- and early third-century date down the eastern and southern coastlines.

Because of this uniformity of purpose and their close geographical siting, some attempt can be made to assess the typological and chronological development of these defences. Because of its overall shape, with rounded corners, and the original supposition that rectangular turrets lay within the curve of the angle, Burgh Castle has been thought in the past to be one of the first of this sequence of new forts. There is no evidence, however, for the presence of an earthen rampart behind the walls at Burgh Castle, and other early features, such as the existence of an angle turret, still require further substantiation. Considered objectively, the similarity in plan between Burgh Castle and a normal playing-card style fort like Brancaster, is more apparent than real, for the only real point of similarity was the rounded corners. The only clear sign of alteration in the design of the fort was the addition of the externally projecting towers, and all the evidence suggests that this was a very early addition to the building programme, and certainly completed at the same time as the upper portion of the fort walls. The positioning of bastions on the exterior of the rounded corners, in a position where they could not enfilade the walls suggests that they were an early modification and addition to the fort's design made by builders unfamiliar with the new style of architecture.

Modifications similar to this can be seen to have been made to the original plan of the fort at Richborough, where one of the corner towers was a later addition to the rectangular corner. At Dover, too, some of the external towers were added after the construction of the main part of the walls of the Saxon Shore fort, and were of different construction. Here the fort wall was narrower than those of all the other Saxon Shore forts, and the external towers had tile-courses while the walls did not. The presence of an earth rampart behind the walls, their narrowness, and the addition of the external towers show that this fort too can possibly be dated relatively early within this transitional stage.

Such details as these may also have a chronological significance, as much as the general style of building. Forts in Britain seem to have had no tile-bonding courses until the late third century, though the style was known in town walls before this date. Dover, with its internal earth rampart and walls without tile-courses, is perhaps earliest in the typological sequence. Perhaps its near contemporary was Burgh Castle, differing only in that the type of construction with tiles and split flints or septaria was the general mode used in East Anglia, whereas in Kent, ragstone was the more normal medium. Dover and Canterbury, constructed in the latter part of the third century, were probably contemporary constructions with Burgh Castle and the walls of Caistor by Norwich and signify the advent in Britain of an entirely new type of fortification, modelled to varying degrees on the style which was then becoming current in Gaul.

It is clear that the greatest impetus given to the Gallic town-wall building programme was a result of the barbarian raids of the middle and later third century, and probably owed much to the efforts of Probus immediately after the greatest of the raids in 276.[36] It seems most unlikely that the forts in Britain were spontaneously built in a new style heralding a revised conception of military defences. The fact that at Burgh Castle and Dover traces of the old style had prevailed and that modifications had to be made suggests that Britain was still some way behind in military development when the scheme was first instituted, and it is equally clear, from such dating evidence as there is, that this was happening in a mid- to late third-century context. A scheme for frontier defence, both along the Rhine and the Channel, would have been the result of planning by an able general such as Aurelian or his self-styled 'successor' of some years afterwards, Probus, who may in some measure have brought some of the schemes of Aurelian to fruition.

These coastal forts, therefore, were probably built in the period 270–85, and form part of longer-term planning against pirates which Aurelian and Probus may have initiated, but found impossible to implement fully before both being assassinated. Thus, when the newer style of fortification was becoming current in various places on the Continent, the forts in Britain, started in an older style, were recognised as being out of date, and were quickly brought up to date by the addition of external towers. There is less evidence among Gallic walls for a transitional stage in construction than in Britain, and it seems only sensible to assume that the newest ideas spread only at a late stage to Britain. The island in any case must have been something of a backwater, especially while the seas which formed its only tenuous link with the Continent were prey to barbarian pirates, making communication more hazardous still.

Although this may have been in broad outline the original design and strategy for these fortifications, there are only the slightest of non-archaeological sources to add substance to our historical understanding of it. The references to German pirates infesting the seas in the 280s and the appointment of Carausius to hold a naval command at Boulogne as a policing measure are virtually the only traces in the literary record of Roman military concern.[37] It has been claimed that the coastal fortifications in Britain were built by Carausius to attempt to secure his position after his usurpation, but these arguments do not carry conviction.[38] Even if mention of the coastal fortifications is implicit in the story of Carausius' usurpation and the eventual downfall and defeat of his empire, all mention of them is lacking throughout the fourth century. Only within the lists of commands in the *Notitia Dignitatum,* at the very end of that century, is mention made of an officer, the *Comes Litoris Saxonici* (Count of the Saxon Shore) who held under his command the garrisons of nine forts, clearly identifiable with those round the southern and eastern coastlines of Britain.[39] This, the only clear reference to these coastal installations, is of limited use. It refers to a single point in time, and provides a name of the Roman command which they then formed. Most important, however, it shows that the screen of fortifications did form a command, and that this was regarded as a frontier in late Roman times.[40]

The fact that only nine garrisons at nine forts are mentioned in the *Notitia* does not necessarily mean that those nine were always the complete complement of the coastal screen in the fourth century. Other places in the eastern and southern coastal zone can lay some claim to relevance within such a defensive pattern. At Brough-on-Humber,[41] originally an earth-and-timber fort, the fortifications were rebuilt in stone towards the end of the third century. These underwent additional modifications during the course of the fourth century by the construction of externally projecting gate- and interval-towers, though no systematic excavation within these defences has shown what they were designed to protect.

Two further sites of relevance lie on or near the east coast. At Caistor,[42] a constricted circuit of walls built on the late Roman pattern (fig. 80, p 205), with solid walls of substantial thickness protected by projecting bastions lay at a considerable height above the River Ancholme. Further south, at Horncastle,[43] a quadrilateral circuit of walls of similar build and pattern, backed by a rampart of sand, enclosed an area of 3.2ha (fig. 80, p 205). Despite the lack of archaeological dating evidence, and the complete lack of contemporary known Roman structures inside these walls, the site has close links with river routes to the sea, then much closer than now, and may well have had relevance to the coastal defensive screen.

Finally, a defended site at Bitterne[44] near Southampton, may have provided defensive cover for the fleet in the south, adding an extra post, its construction dated after 369 from coins found within the wall. At Bitterne, a triangular promontory of land bounded on two sides by the River Itchen was cut off by a strong wall and ditch. Traces of buildings are known within this area, but their purpose and precise date are not clear.

Among the coastal installations in Britain belonging to the late Roman period is a series of towers which lay principally on the north-eastern coastline. These sites are well spaced out, and though normally considered as 'signal stations', are not well adapted for signalling either to one another, or to some inland site, except when weather conditions were fair to perfect. In form, the known sites are similar (fig. 81). A central tower, its upper floors supported on pillars, stood within a small rectangular walled courtyard, with a projecting tower at each corner. This was further protected by a ditch. Though clearly sited to be as close to the sea as possible, and thus to answer in some way a sea-borne menace, their tactical use to the late Roman military authorities is unclear. It may be that they protected harbours or small landing places, and were themselves serviced by the late Roman fleet. A site similar in plan seems to have existed on Alderney, at Longy Bay. The late Roman date of these installations is assured by coin and pottery finds, and by an inscription found at Ravenscar which is normally assigned a date late in the fourth century. The close similarity between these sites, and, for example, Asperden (fig. 53, p. 140) is a further indication of date.

At various times during the course of the fourth century, all these sites may have played their part within any scheme for coastal defence.[45] Pevensey, too, has been claimed as a fourth century addition to the coastal screen on the basis of coin evidence found underneath one of the projecting bastions.[46] Thus, although instituted informally perhaps in the early decades of the third century, and formalised in the late third century by the construction of the new series of British

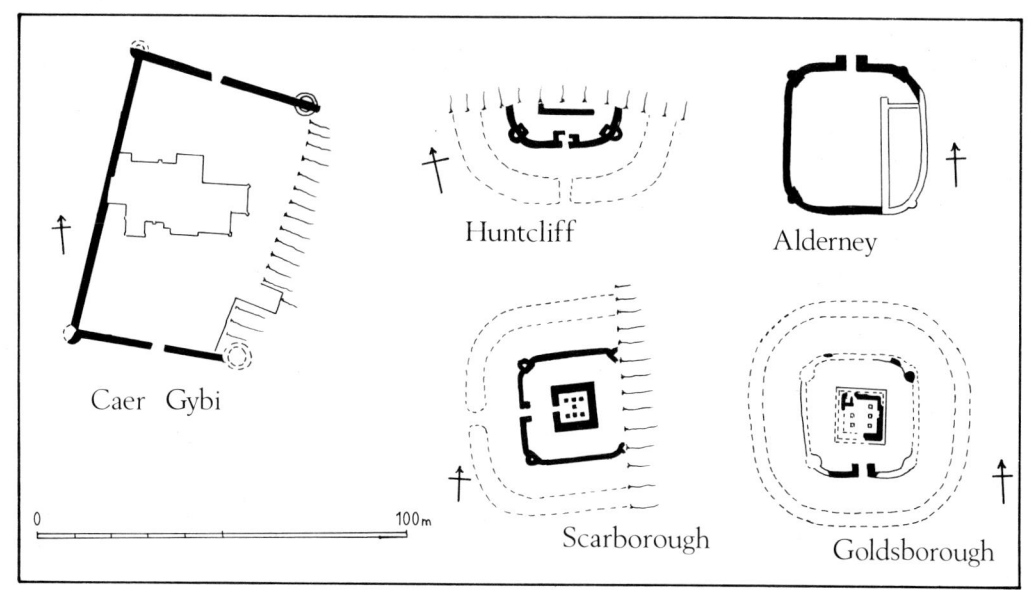

81 Late Roman signal stations on the British coast (*scale 1:2,000*)

forts in particular echoing the style of contemporary Gallic city walls, the coastal defences must be seen as a continuously evolving screen of forts and posts, developed in response to changing needs and circumstances.

Information in the *Notitia*, therefore, which portrays three separate commands in the channel area, must be taken to apply to the very end of the fourth century, and even ten years earlier the situation could have been quite different. There are distinct signs, however, that the command of the Count of the Saxon Shore once embraced not only the forts in Britain, but also at least two of those (Grannona and Marcis) listed under the commanders on the continental shorelines.[47] The date at which this command as such was instituted, and the reason for its holder's status as Count are still to seek, but it is likely that the Count is the successor of a series of naval commanders, of whom Carausius was the most notorious, whose concern it was to hold the forts in the exposed areas of Britain and the Continent. At a later stage, there may have been a back-up commander appointed to the western Gallic coastal area, whose command was formed out of pre-existing towns and cities for the most part, to provide extra cover should the Dover–Boulogne axis at the channel straits fail to prevent regular penetration by the Franks and Saxons. Still later, with the Roman hold on Britain more tenuous than ever, the responsibility for channel defence was fragmented further, with the front line Continental defences too transferred to (now) two coastal commands in Gaul, and the formerly wider ranging Count of the Saxon Shore now confined to Britain.

This complex situation perhaps shows in microcosm how one might expect Roman tactical thinking about one of their most awkward problems of security to change over many years of changing conditions and practice in the field. Other frontiers were perhaps less volatile and afforded a better prospect of a steady response to a known threat. Nevertheless, these threats did not always come in the same pressure point nor from the same direction, and the Roman tactical response in each case may show shifts of emphasis.

9

The Defence of Italy and Spain

One of the main concerns of the late Roman authorities was to protect the approaches to Italy. A series of fortifications lay in the valleys and mountain passes to prevent penetration by barbarians across the Alps. The weight of evidence suggests a concentration of forces in the eastern approaches, but some protection seems to have been given to the western part of Italy: walled towns and posting stations along routes in the Alpes Maritimae were probably intended in part to divert the attention of any bands of raiders who ventured thus far. Cities deep in the Alpine foothills like Grenoble and Die, well protected by wall-circuits which have survived, or others, such as Gap, Embrun and Digne, where the former existence of late Roman wall-circuits is recorded, formed a first line of approach to the passes.[1]

In Italy itself, at the head of most of the passes there stood a strongly walled city, the majority of which had been established in Augustan times. North of these, to supplement this cordon, there is some evidence that there were chains of late Roman watchtowers controlling the access to the valley of the Po. At Castelseprio, for example, there was a series of such towers and perhaps a defended hill-top.[2] Other towers lay at Lecco, near Como, and at Coccaglia on the road from Brixia to Bergamo.[3] A further series of towers is known in the Lario district.[4] Such defensive arrangements, still only poorly known, perhaps provided a forerunner for the later, more advanced, Byzantine frontier-systems in Italy north of Aquileia and Friuli.[5]

The Rhine valley and its associated tributaries in the provinces of Maxima Sequanorum and Raetia also held a formidable series of fortifications—cities, forts and outposts sited to prevent the enemy progressing further than was absolutely necessary towards Italy. Some of these sites were of a clearly military nature. Most obvious were the walled forts at Schaan[6] and Wilten[7] (though perhaps both of these are better described as fortified granaries) and the walled sites at Zirl,[8] Chur[9] and Mauthen.[10] The passes are also dotted with hill-top fortifications, such as those at Muciaster, Schlössle, Frastafeders, Horwa and Stellfeder.[11] Some of these were doubtless intended to serve as refuges for particular villa estates or settlements in the area, but were too scattered ever to have formed a comprehensive defensive system. Militarily speaking they formed, at best, a loose cordon to control and watch over the pass roads.

The importance of the passes through the Julian Alps in north-east Italy for the control of traffic was recognised from the time of Augustus onwards. During the course of its history, this part of Italy was threatened several times by different enemies. As early as the time of the Marcomannic wars in the late second century some of the foundations may have been laid for the defensive scheme, though the systematic provision of permanent forts and frontier posts had not yet been planned.[12]

In the late Roman period, the approach roads to Italy in this region were controlled by a series of small forts and posts connected to barrier walls. High mountains, deep forest, and otherwise difficult terrain made the provision of a continuous barrier wall, or the definition of a frontier line unnecessary. Ancient sources describe the system of fortifications as *claustra*, implying a series of barriers shutting off the approaches.[13]

The walls and towers which block only the more accessible of the passes normally ran across the valley floor from peak to peak. The traveller had to pass through a single entrance through these walls which was normally controlled by a fort or small guard-post. The walls control the two major routes which gave access to Italy. One was the main road from Ljubljana to Aquileia, controlled by three lines of barrier walls and small posts, each one supervised by a central larger fort. The other route led along the coastline, from Rijeka, where there was a further series of walls defended by towers, to Trieste and Aquileia. Rijeka itself was a small fort-cum-town acting as the main control post for this route.[14] Wherever they have been found, the walls were almost uniformly 1.8m thick and built of rubble concrete with square towers spaced at unequal intervals attached to the rear face.[15]

In places, these walls were supplemented by additional forts. At Nadleski Hrib there was a fort built in an earlier style, since it had gateways with claviculae. Another, still unexplored, lay at Sveta Marjeta and a further site is known at Polhov Gradec, to the north of Vrhnika.[16] The major road, however, and the one which was most comprehensively protected, was the route from Emona (Ljubljana) to Aquileia. Besides three forts, at Aidovščina, the Hrušica, and Vrhnika, the road was controlled by at least three separate barrier walls, and two further watchtowers or fortlets.

The key to the system, the Hrušica, lay near the head of the pass.[17] The fort enclosed an irregular oval, but within this there were two separate defended areas (fig. 83). The upper portion, separated from the lower by a strong wall and towers with a single-portalled gateway, may have been primarily intended as a refuge. Few traces of occupation have been found in limited excavation in the interior. The main road ran through the lower portion of the fort, entering from the east by a single-portalled gateway flanked by polygonal projecting towers. Further projecting towers, all of irregular shape, lay at the corners.

Limited excavation within the fort has produced little dating evidence. An

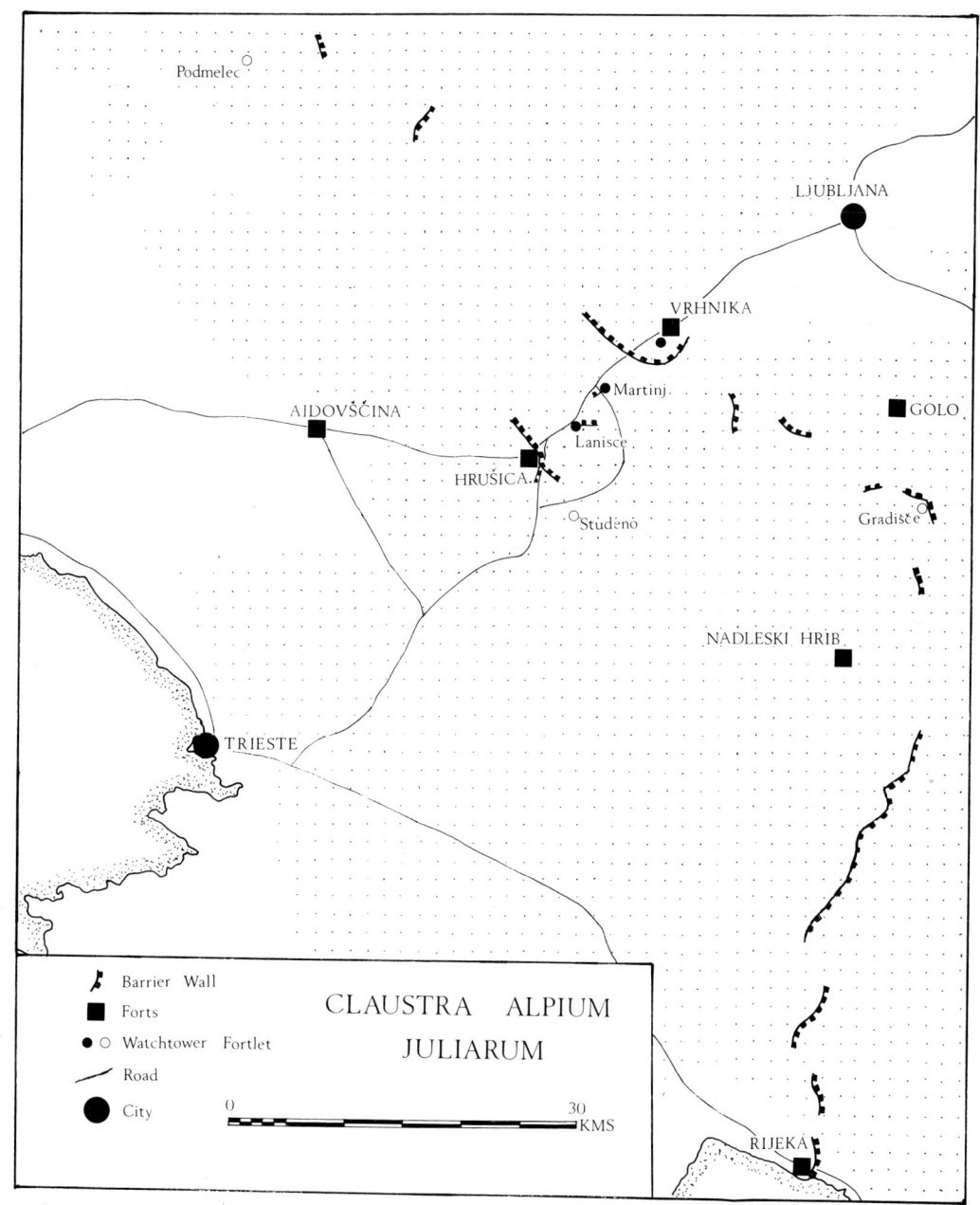

82 Map of the area of the Julian Alps in the late Roman period

occupation level, thought to predate the construction of the fort, contained a coin of 314–7. A scatter of material of fourth-century date from the site suggested a construction date in or near 320, and abandonment in about 400.

The easternmost fort of the whole system lay in front of the main series of barrier walls at Vrhnika.[18] It covered a pentagonal area of just over 2.5ha (fig. 83). Round external towers projected at the corners, and at irregular intervals there were hollow rectangular towers which projected externally. Excavations early this century have never been published, and much of the detailed dating of the structures is thus hypothetical. Finds from the site spanned the late first to the fourth century, but the defensive architecture of this fort aligns it with others of late Roman date.

The fort at Aidovščina (fig. 83) is often compared typologically with Jünkerath (see fig. 60, p. 156) and others of the Rhineland forts.[19] It has thus been interpreted as one of the series of Constantinian posts strengthening the road system along with Vrhnika and Rijeka. Even though Aidovščina was typologically very similar to some of the Rhineland fortifications, it was not necessarily contemporaneous— Aidovščina, Rijeka, and Vrhnika could equally well be Diocletianic foundations. They are not, however, likely to have been constructed later than Constantine.[20]

A further pair of small forts lay on this main approach road to Italy (fig. 83). One was at Lanišče, where a tower, 20m square, was attached to a short length of barrier wall.[21] Coin-finds dated its use to the late fourth century onwards. It did not actually span the main road, but was set in position to watch carefully over it. Other buildings nearby provided evidence for a period of occupation beginning earlier in the fourth century.

At Martinj there was a second small fort, part of which has fallen over a steep drop. It was possibly originally pentagonal, but three sides only now remain. Finds of early Roman material show that this post was in use throughout the imperial period, but the main series of finds dated from the second half of the third century onwards.[22]

The defences of Aquileia had something in common with the Hrušica, as did those of Pula. At Aquileia there was a city wall built with many reused stone blocks, usually attributed to the mid or later third century. This was probably built earlier than the reign of Aurelian, signifying a period closer to the middle of the third century when the area was under some pressure.[23] As the main military headquarters in the area, Aquileia must have been one of the better defended cities. The other walled cities in the area, Concordia, Tergeste, Friuli and Ad Tricensimum, may form a virtually contemporary group, built as a concerted campaign in the latter part of the third century.

Some of the ancillary passes which formed small but difficult routes into Italy also were defended by similar small forts and stretches of barrier wall. Few are known, and even fewer have been explored and excavated. Some of these posts probably date from a period earlier than the late third century.[24] Others may be either military posts or hill-top refuges in which the area abounds.[25] Most of these are re-used hill-forts of an earlier period and several have produced finds of pre-Roman date. For many such hill-top sites there is little recorded evidence beyond their names: detailed archaeological evidence for their settlement or use in the later Roman period has rarely been published. These hill-top sites can have had little purpose apart from that of refuges or supply bases.[26] They do not command any of the passes. One or two of the hill-top sites in the area, for example that at Aidovski Gradec,[27] appear to have had a more positive function to protect some specific site which was of particular importance.

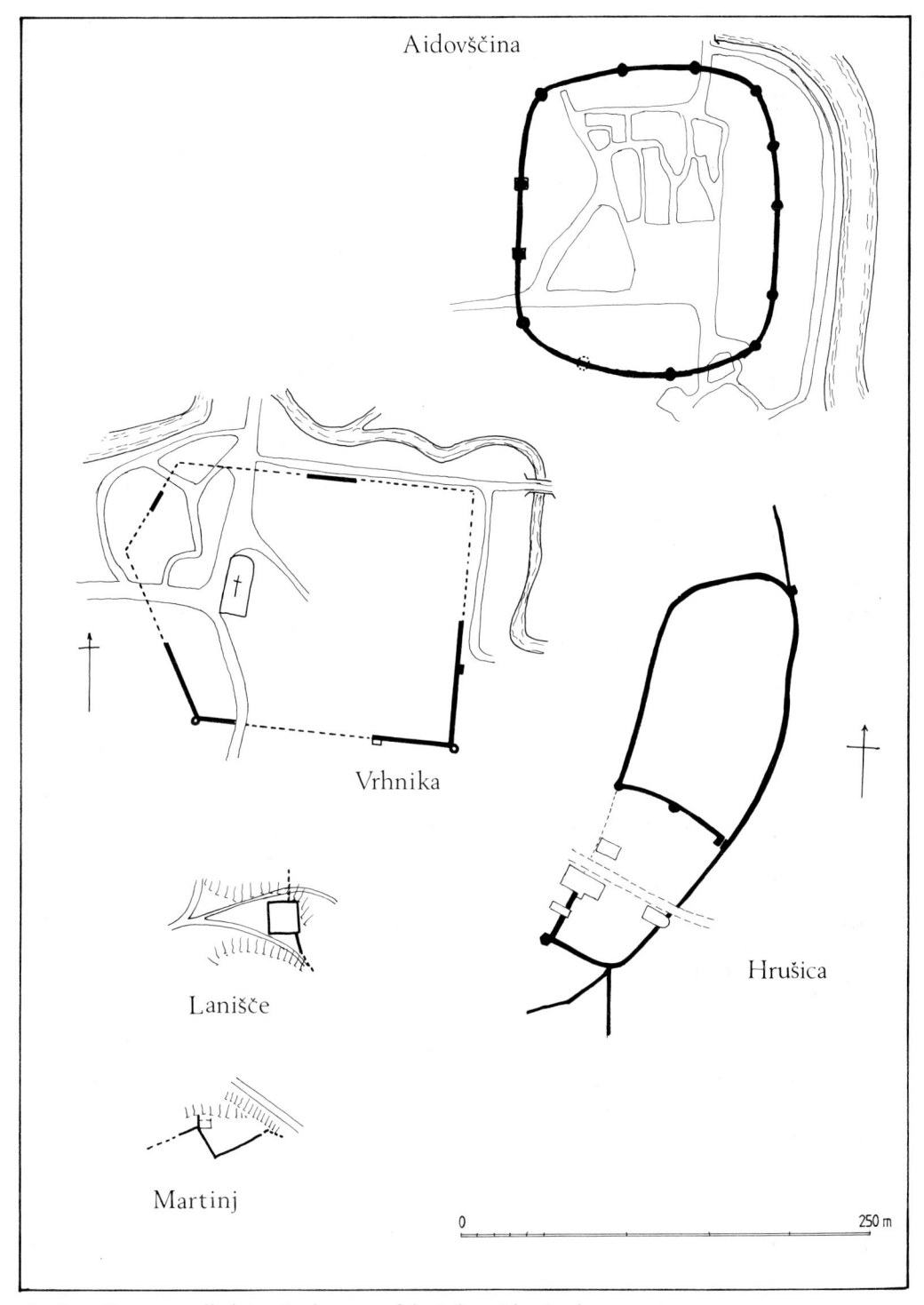

83 Late Roman walled sites in the area of the Julian Alps (*scale 1:4,000*)

The immediate defences of the *claustra* were completed with a series of smaller watchtowers and tiny forts set up at good viewpoints over the approach roads. Two of those still unexplored lie at Grcarevec, on the road from Logatec to Kalce and Planina and at Polhov Gradec.[28] Above Vrhnika itself lay the small fort of Turnovšce—a tower about 20m square which commands the view of the plain of Ljubljana and the approaches to Vrhnika. The fort at Golo may also have been similarly intended; it had an equally fine view over a different part of the Ljubljana plain. Watchtowers complementing the system may lie under the Lombardic defences in the hills above Friuli.[29]

A possible extension to the *claustra* lay in the valleys and passes to the north of Ljubljana, in Kärnten. The Loibl pass, carrying the main route from Ljubljana to Virunum, was controlled by a fort at Kranj (Karnium). In addition, the approaches from Austria were controlled by the hill-top forts of Falkenburg and Karnburg.[30] West of this, in the Alpes Carnicae, there was a similar series of forts at the approaches to the Arnoldstein and the Plöcken passes. Here lay Tscheltnigkögel, Draschitz, and the peculiarly shaped stronghold of Hoischügel (fig. 90, p. 241), claimed to have been constructed as a post for *beneficiarii* at the beginning of the third or even at the end of the second century.[31] Further up the pass at Rattendorf, lay another barrier wall, while the hill-top sites at the Gurina and Mauthen at the head of the pass provided some deeper cover. These sites did not provide a comprehensive defensive system for the area, but they were vantage points which would have been especially useful to guard the roads and to provide some security for the inhabitants of the valley slopes.

The early period of the fourth century was in general a period of peace and growth for this area, as for the majority of the west. Building flourished under Diocletian at Virunum, and villas in the area reached a new peak of prosperity. On the whole such towns and villas remained undefended: even the so-called defended villa at Vranje has now been shown to be a hill-top church site without defences.[32] The main disturbances in the area came in the period when the conflict between Julian and Constantius II was at its height. The main arena of the struggle for power was the main route down the Sava between Aquileia and Serdica. Thus the refuge posts in this area and along this thoroughfare may have been used not only at times of threat of barbarian invasion, but also when warring armies were on the move.

Literary evidence gives no clue as to the precise date of the defensive system in the Julian Alps or of its constituent elements. There is no historical mention of the defences in the mountains until the fourth century. As late as 238, there is still mention of the *protensio Italiae* in inscriptions, used to refer to this extension of Italy behind Aquileia.[33] In that same year, Maximinus crossed the Julian Alps without any trouble at all: if there had been any significant barriers in his way, they might have been expected to be held against his progress. The sources record that Maximinus was particularly anxious about his approach to Italy, and expected to be ambushed.[34] It is not until the struggle more than a century later between Magnentius and Constantius II that there is any reference to the *claustra* defences. Ad Pirum, the fort at the head of the pass, became the scene of bitter fighting: Constantius besieged it and it finally fell in 352. Julian, in describing the event, called the site an 'old fort'.[35] Ammianus also mentioned the *claustra* in 361, when Julian besieged two of Constantius' legions at Aquileia: he feared that Constantius would have cut off his approach to Sirmium.[36] In 374-5, the Quadi and Marcomanni broke through the barriers in the Alps.[37]

In 394 Theodosius gained control of the passes of the Julian Alps by treachery.[38] The system of barricades was clearly still of enough importance to cause a sizeable obstacle to an invader or assailant attempting to reach Italy from the east. It was not until 401 and 409 that the passes were breached by Alaric, who marched from Emona to Aquileia without encountering any resistance.[39] Even later, the situation once more restored, the *claustra* appears to have been again in use, for the Codex Justinianus issued instructions in 443 for the disposition of a *limes*, probably but not necessarily that of the Julian Alps.[40] Though the series of fortifications in the area is mentioned from time to time[41] the control of the passes seems no longer to have been a vital or effective bar to progress along the roads.

When the Danube frontier collapsed in the reign of Honorius, there was consequently no last ditch stand at the *Claustra Alpium Juliarum*. The provinces of Noricum and Dalmatia were now more open than they had ever previously been. Salona itself, despite its sheltered position deep in Dalmatia, came under pressure, and large portions of the province were abandoned to their fate.[42] Whether intended as such or not, the Claustra appears never to have had to double as a *limes*: it was apparently an elaborate longstop controlling the access routes to Italy, but never really formed an effective barrier.

Problems of a different type were met at the other end of the Western empire in the diocese of Hispania. The claim has often been made that the troops listed in the *Notitia* as part of the garrison of Spain, formed a sort of frontier in Northern and Western parts against potential threats from the Basques, Asturians and Cantabrians.[43]

These arrangements refer specifically to the stationing of *limitanei* (frontier or static troops) in six named towns or cities, including León and Lugo, whose walls are described above.[44] To supplement this static force, the *Comes Hispaniarum* also had a mobile force of troops.[45] The *Notitia*, however, does not reveal where these troops were stationed, nor the period at which this *Comes* held his force in Spain. The mobile force is attested otherwise under the emperor Honorius,[46] but may not have been in existence much earlier than the late fourth century.

As a complement to these regular forces, archaeologists have also suggested that a number of fortified sites in the Duero valley and the cemeteries containing men buried with their weapons in the same region belonged to a more irregular group of Germanic mercenaries or *laeti*, whose task was also to strengthen the defences of Spain's heartland against potential trouble from the north.[47] Although the cemeteries discovered did indeed contain men buried with grave-goods very similar to those of other, north Gaulish, cemeteries normally considered to contain *laeti*, the argument which seeks to define the men as Germanic federates and to suggest that they were posted here, as it were, on frontier duty, goes beyond what the evidence seems immediately to support.

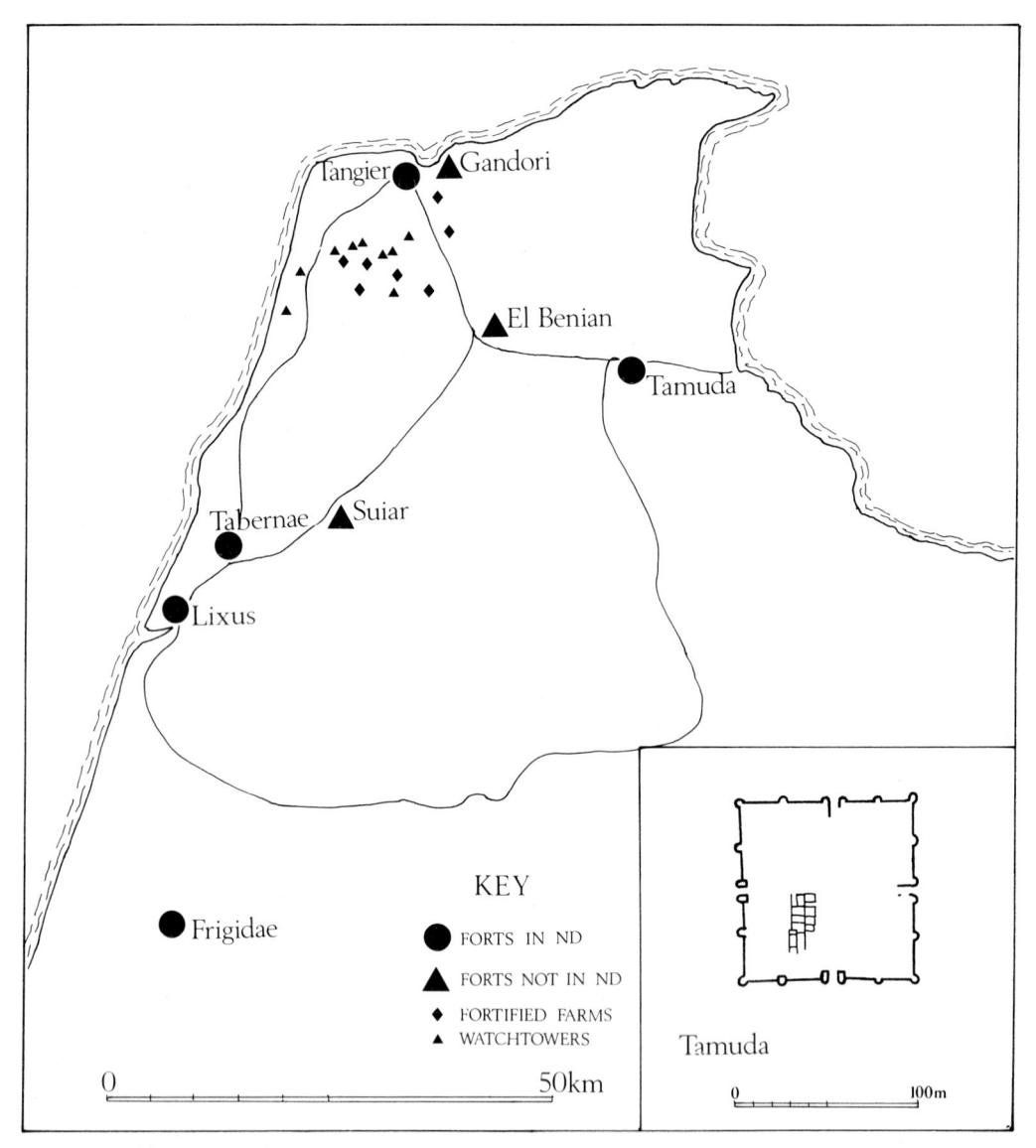

84 Map of late Roman Morocco, with (inset) plan of Tamuda (*scale 1:4,000*)

Whether one wishes to call this defensive arrangement a frontier system or not, it is clear that the Roman defence of Spain did not depend upon a tightly held and rigidly patrolled frontier line, but rather was a series of rearward towns with garrison troops, stationed to respond if required. This appears to be much the same system of precautionary defence as is also seen in Northern Italy, where, according to the *Notitia*,[48] the major towns were the headquarters of *praefecti laetorum* whose task it was presumably to protect the towns and their important installations from the pressing threat of barbarian invasion. The threat to Spain appears to have been more internal than external, and to that extent, the word 'frontier' appears not really to be apposite here.

There was a province within the Spanish diocese, however, in the late third and fourth centuries where frontier defence between Roman and non-Roman was a necessity. Towards the end of the third century, the former large province of *Mauretania Tingitana*, which had covered much of the coastal strip of Africa, was overrun by barbarian pressure, and a large tract of land between the former provinces of Tingitana and Caesarensis was given up. Mauretania Tingitana now shrank to a tiny size, comprising only a small bridgehead of African territory opposite the southernmost tip of Spain. The exact date of this abandonment of territory is not clear, but it may have been no later than about 290, and occasioned by a large, and well documented, revolt of African tribes which occurred in 289.[49] From now on, Tingitana no longer had a separate identity: in the list of Roman provinces known as the Verona list dating from 297,[50] it now was one of the seven provinces which made up late Roman Spain.

The *Dux Tingitanae*, however, the frontier commander in this area, appears in the *Notitia* in his own right, with troops based on seven sites, four of which may reasonably be equated with known archaeological sites, and three of which are unknown.[51] At Lixus[52] (Aulucos in the *Notitia*), the most important part of the flourishing early imperial city was destroyed soon after the mid-third century, never to be rebuilt. The richest part of the site was abandoned, and the nucleus of the later town now moved to the *forum* area, where a late Roman fortification, two sides of which are known, was constructed. The walls were 3m thick, with projecting rectangular bastions.

The site at Tetuan,[53] known to the Romans as Tamuda, was originally an extensive unwalled town of the first and second centuries. In the third century a new near-rectangular fortification was erected on the site (fig. 84). It was small —a mere 80 by 80m—and occupied the highest ground available within the town area. The walls were narrow for a fortification of this date (only 1m thick), contained much reused material, and were further protected by a series of frequent semicircular projecting towers of secondary build.

Tabernae[54] is the name normally applied to a site some 10km north-east of Lixus, where selective excavation has taken place. The remains appear to be those of a square fort, similar to Tamuda, and slightly larger. This main fortification was surrounded by the remains of the *vicus* buildings which may have given it its Roman name, and these, too, appear to be encircled by a further defensive wall. The fort occupied the site of previous Roman buildings which had been destroyed in the 260s.

The fourth site mentioned in the *Notitia*, Friglas,[55] may refer to a site also known as Frigidae, where a rectangular fort of similar type, 100 by 75m in extent, has been located but not excavated.

Although the other three named sites in the *Notitia* cannot be immediately linked to sites on the ground (and it must be said that some of the identifications above are doubtful), there are three other known late Roman defended sites in the province. The least explored is at Suiar,[56] which appears from field-work to be similar to the sites at Tamuda and Tabernae. It has produced a hoard of coins of mid third century date. At El Benian,[57] a site of parallelogram shape, measuring

183 by 140m, and occupying a slight hillock above a major road junction within the province, limited excavation has suggested the presence of a similar fort, with narrow walls built of large blocks of stone, and finds dating from the late third and fourth centuries. Finally, a small *burgus* at Gandori[58] protected the immediate surrounds of Tangier and its bay. The site was probably placed to guard the nearby tile-kilns, and was a mere 26 by 29m, with a single gate flanked by U-shaped towers. At the corners of the walls, which nowhere reached a thickness of more than 1m, were projecting three-quarters round towers.

It is remarkable how different these Moroccan fortifications were from late Roman sites in Spain. The size of the *castella* and their usual rectangular shape recall the small road posts and forts of the Arabian or Syrian frontiers[59] rather than the massive defences of northern Spain. The Moroccan forts showed none of the signs of anxiety displayed by the European fortifications—enfiladed gateways, massive walls, and walls with frequent bastions to keep an enemy at arm's length. These small posts were those of garrison troops moved into the centre of the province, and not the Spanish or Gallic style 'acropolis' into which the populace might crowd at the first sign of danger.

European tribesmen could never have been a threat to Morocco, even though Victor says that the Germans in 260 sailed for Africa after the sack of Tarraco,[60] perhaps after being hounded out of Spain by the Roman army in hot pursuit. There is some record of destruction in the coastal zone of Morocco at about this date, and an inscription recording victory over a group of *barbari* who had broken into Tamuda, at this time a small impoverished township. The coastal zone, then, of Morocco was the one to suffer most in the raids by Germans, but the damage done by a small band of inexperienced sailors was in all probability slight. The same sort of response to the fear of the German invaders which prompted such massive town defences in Spain was not elicited here.

It is far more likely that the construction of this series of defences, and the more serious cases of destruction noted at Lixus and other sites, were connected with the earlier of the two uprisings by Mauretanian tribes, who in 253 rose up and caused havoc in Mauretania Caesariensis.[61] This revolt must have had repercussions elsewhere, and since the 'terminus post quem' of destruction on most of the sites in the zone of Mauretania Tingitana is about 250, it seems likely that the initial construction of this series of small forts for garrisoning the interior of the province may date from this incident. Even if the German raid exposed the insecurity of the province of Tingitana, it was from the tribes of the south and east that trouble was expected, and came, in 289, resulting in the separation of the two halves of Mauretania.

Unlike Spain, Morocco was a frontier province. In this small area, defence of a frontier line to the south had to be linked with the defence of the region's industries and of the communications system which kept them alive. In Spain, the mines in the north-west and the routes of access to them were protected with the main concentrations of fortifications. So in Morocco, the *garum* industry,[62] and the routes of access through to Spain received fortifications to reinforce the frontier protection. A number of fortified block-houses on inland routes in the province were probably used, like *centenaria* on other African frontiers, by farmer-settlers. Tingitana was thus unusual in that it had a network of regular forts as well as a system of fortified farmhouses, but this made defence doubly effective.[63] The system of defence grew up, little by little, as a series of garrison posts protecting important areas. No longer were the troops on the frontier in a strictly militarised zone: in Tingitana, the frontier zone was now spread throughout the province.

10

Local and Rural Protection: Hill-top Defences

The defence of towns and cities and the securing of frontiers were two areas of particular concern for the late Roman world faced by barbarian incursion. The new fortifications erected to protect these, by and large the official establishments of Roman power, did little or nothing to improve the lot of ordinary provincial men and women. Dwellers in some of the major cities of the empire would still be lucky, despite the proliferation of new walls and defences, to own a house within their protection. Those who lived in the suburbs would still, at the hint of danger or the approaching path of a barbarian raid, have to cluster for personal security within the tightly constricted circuits of city walls. There they might feel safe, but their property could still be ransacked by marauders.

It was clearly in the frontier zones of the empire that the threat of invasion and consequent loss was at its most imminent. Here of course there were towns, *civitates*, forts, with *vici* nestling for protection round them, and a variety of walled centres, ranging from road posts to grain stores. In times of danger, however, there was no guarantee that local civilians would be welcomed within the walls of all of these for protection. There was therefore a need to provide more security for provincial folk who lived and farmed in the frontier areas. What was perhaps more important was the removal of livestock and produce from the path of an invading band. Such measures were vital to sustain the livelihood of those who farmed the frontier areas, often probably at little more than subsistence level, and by removing easy sources of food supply, to ensure that barbarian raiding did not increase in intensity.

A further form of fortification thus became increasingly more common to supplement the protection afforded by the official establishments. These took a variety of forms, but normally consisted in the provision of defences round a convenient hill, either enclosing its summit entirely with walls, or, more commonly, defending the easily accessible side of a hill spur with walls, ramparts or ditches, and relying on natural defences for the rest of the circuit. In essence, such defences are late Roman hill-forts, and, as often as not, they occupied the site of Iron Age predecessors. Refortification or refurbishment of the ramparts was often of course necessary. This followed a variety of forms, and occasionally incorporated a degree of sophistication surprising for such local and rural sites. Normally they lay well away from main roads, and deep in the heart of the countryside.[1]

The areas where these sites are most common are the Rhineland (parts of the Roman provinces of Belgica and Germania I), the upper Rhine[2] (Maxima Sequanorum and the Alpine foothills) the Austrian valleys[3] and the passes leading to Italy across the Julian Alps.[4] Isolated sites, however, are known from several of the late Roman western provinces, and the geographical concentration of such sites does no more than follow the availability of suitable terrain as well as the scale of threat from barbarian raiders across the frontiers. A substantial number of sites of late Roman hill-top refuges have been identified as such from chance finds made on the site, unconfirmed by excavation. An exhaustive look at the evidence is perhaps overdue, but it will be necessary here to concentrate only on those sites which have been sufficiently studied. Total excavation of a hill-top site is a rare event, and perhaps a daunting task, for which the rewards are liable to be scant.[5] If used as a refuge only, such a site is liable not to have contained any buildings of permanence, but to be more in the nature of a camp-site and stock-enclosure.

Those sites which have been studied, however, show late Roman defended hill-forts to be as multifarious and perhaps as multi-purpose as contemporary fortifications of other types. It will be convenient first to present the archaeological evidence from some of the type-sites, and to discuss their relevance to other, less well documented sites in their neighbourhood.

In the Rhineland area, of the 100 or so known or suspected late Roman hill-top refuges, less than a dozen have been examined systematically. Much of this work has been concentrated in the northern part of the area lying in a wide circle round Trier—in the thickly wooded and hilly areas of the Meuse valley and its tributary rivers in Belgium.[6] The countryside is ideal in this area for such sites. In Roman times, the roads and to a lesser extent the larger rivers formed probably the only easy communication links between major settlements, like Trier and Metz and the Rhine frontier forts and towns. The roads, by and large, were policed by a series of small posts to protect them. Once off the main routes, particularly in the thickly wooded areas of southern Belgium and Luxembourg, and the Pfalzland, progress will have been extremely difficult and hazardous. The majority of these hill-top refuge sites lie well away from main roads, and are often places already naturally well defended, which, by the addition of a limited amount of defence, could be made virtually impregnable in late Roman times.

The Cheslain d'Ortho[7] (fig. 85) is typical of these sites. In a valley of the River Ourthe, a narrow triangular spur of hill was ringed with defences in the late Roman period. On the south side, the base of the triangle, an unmortared wall of schist and limestone of crude construction blocked off the main access route. Limited excavation has revealed three constructional phases: the earliest had been an earthen rampart reinforced by wooden towers. This was replaced, and in parts crested, by the stone wall with towers. In a final phase the defences were reinforced by bigger quadrangular towers. At the northern entrance there was a

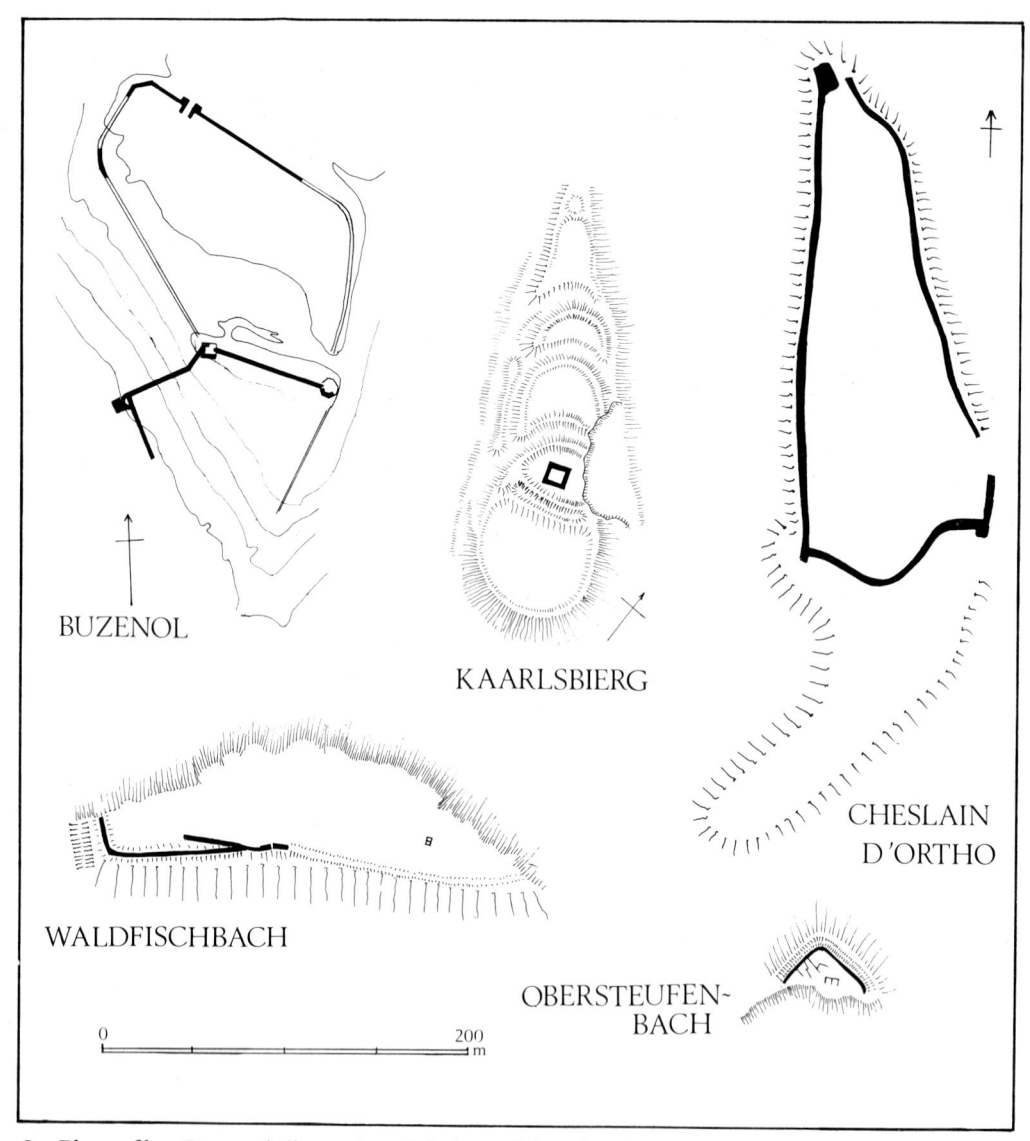

85 Plans of late Roman hill-top sites, Belgica and the Rhineland (*scale 1:4,000*)

large advanced bastion. The wall incorporates reused stone, and pottery finds from the site include Argonne sigillata, the common fourth-century pottery of the region. No buildings within the defended area have been identified.

Other, similar, sites which have been studied to a greater or lesser extent in the same region are the Château des Fées at Bertrix, the Roche à Lomme at Dourbes, and the Vieux Château at Mont-Sommerain. Brief descriptions of all of them will suffice to show both the variety and the similarity of such defensive arrangements. At Bertrix,[8] a promontory site with a steep slope towards the north was chosen for defence (fig. 86). The earliest defences seem to have been of wood, but these were converted to stone at a later stage. These walls enclosed an irregular area no

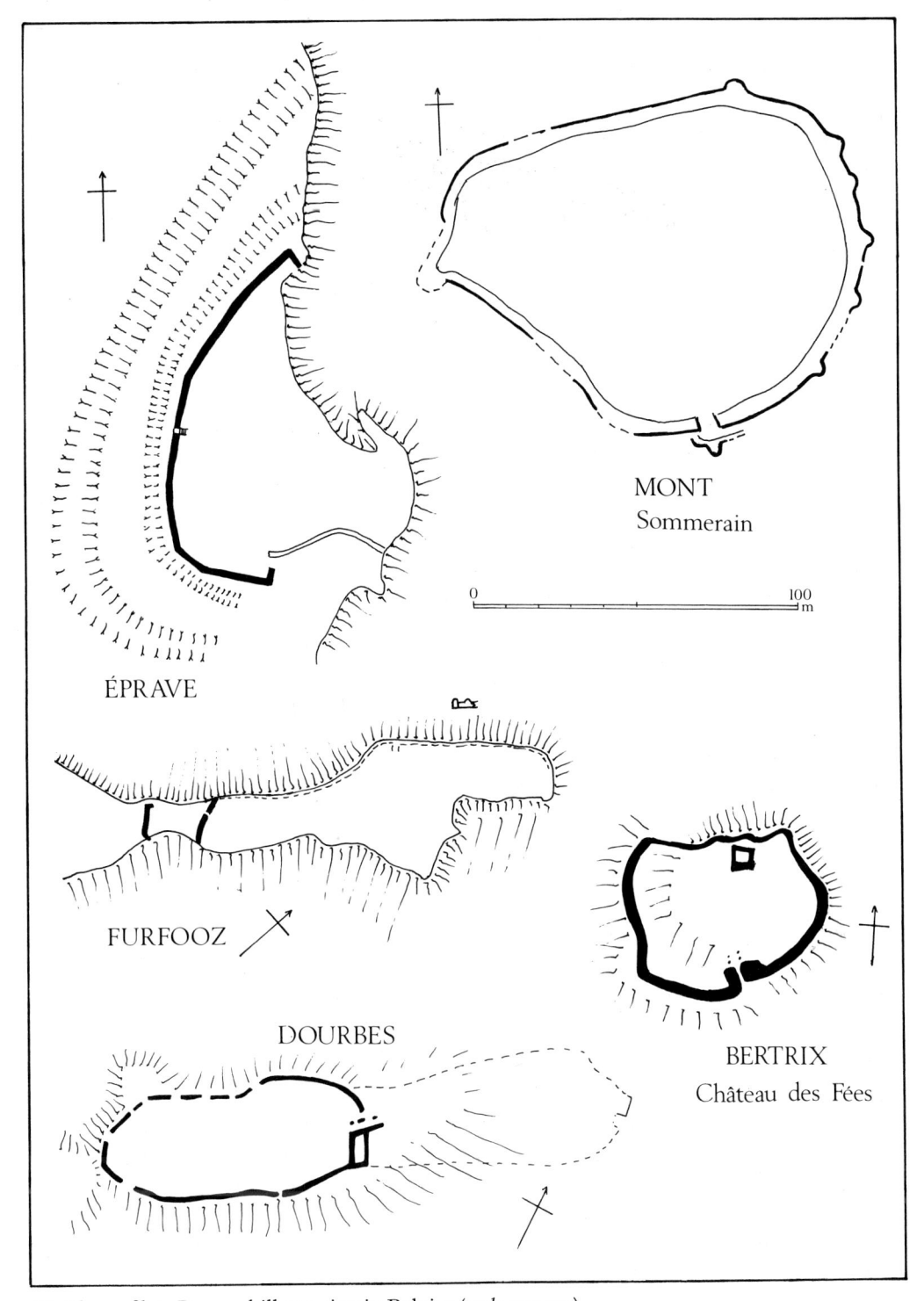

86 Plans of late Roman hill-top sites in Belgica (*scale 1:2,000*)

bigger than 60 by 45m: on the southern and western flanks they were upwards of 3m thick, but to the east and north, where the area was naturally better defended, the wall was much narrower. The entrance, a narrow defile between a pair of towers formed by the thickening of the enclosure wall, lay to the south-east. A small square tower occupied the summit of the hill: this was probably a Roman watchtower, but its remains were incorporated into a medieval structure. A Roman period of occupation at the site is confirmed by the discovery of appreciable amounts of Roman pottery and tiles.

The site at Dourbes[9] occupies an elgonated hill-top plateau above the river Lomme, which is also naturally well defended (fig. 87). The major defences are on the eastern narrow side of the hill, and comprise a ditch and a double wall. The defended area is some 20 by 70m, and the entrance is on the east side only. Although the site was occupied in the medieval period, a significant number of Roman finds prove its use in the late third and fourth centuries. The coin series from the site runs from the end of the third to the beginning of the fifth century, and the finds seem to substantiate two periods of use in the fourth century.

At the Vieux Château at Mont-Sommerain,[10] a roughly circular defended area was well adapted to the terrain (fig. 87). An area of roughly 1ha was enclosed by an earthen bank on stone foundations, including a drystone front revetment. The eastern and southern sides were protected by semicircular towers, and the gate lay on the south side, and was fronted by a further semicircular tower, after the manner of a *titulum* in earlier Roman marching-camp defences. Occupation of a temporary nature in the mid and late Roman periods is suggested by a scatter of pottery finds from the material forming the earth bank.

The four sites so far described are typical of many, and show the range of types of fortification of this nature which sprang up in the later Roman period. The natural topography was normally assisted by defences sufficient to provide adequate protection. This might mean, therefore, either a stone wall, mortared or drystone, or an earthen bank. Where the hill's natural defences made approach difficult, it was sufficient to provide a protected enclosure by a simple barrier wall across a narrow portion of the hill. Otherwise, the whole hill-top might need encircling with defences. In all four cases so far described, the sites were fortified, so far as one can judge, for the first time in the Roman period. Where prehistoric defences had encircled a hill, as in several sites in the same region, it was normally necessary at a later period only to heighten or to refurbish the existing defences in some way.

The variety of types of construction is a potential source of confusion between prehistoric, late Roman, and even medieval forms. At Sommerain, for example, the earthen banks incorporated a certain amount of timber strapping, and they also were consolidated by a facing of large stones. The timber strapping finds late Roman parallels elsewhere—notably in the more formal architecture of the fortress at Strasbourg or in the Saxon Shore forts.[11] The more irregular earth and stone construction finds parallels at other hill-top sites, such as Jemelle[12] and Rouveroy.[13] The projecting towers at Sommerain, however, place the defences, even if no other finds confirmed it, in the late Roman period.

Not all sites necessarily have such distinctive features, and in the absence of research or excavation which can definitely establish the late Roman date of the defences, it is often difficult to be certain whether it was actually used in late Roman times for defence. At a number of sites where no excavation has taken place, a combination of earthwork defences (even though they may in fact conceal stonework) and a scatter of Roman finds, mainly coins, are the only indication of late Roman re-use. Hence a number of sites marked on the maps (fig. 87, 88) are possible rather than definitely established. By no means an insignificant number of these 'possible' sites have produced coin-hoards of the later third century, for example Baileux, Lompret, Macquenoise, Junglinster and Wellesweiler.[15] It is worth remembering, however, that in late Roman times an isolated hill-top with surviving prehistoric defences might be used as a temple site. Without excavation and confirmation of late Roman refurbishment of the defences, it is difficult to tell such sites apart.

The problems of the phases of use of these defences within the late Roman period are not lessened by the character of the sites themselves. By and large their evident use was as temporary refuges for men, beasts or produce in the face of barbarian threat or invasion. Some served small communities—for example the fortified site above the settlement at Williers-Chameleux,[16] or possibly the Katzenberg near Mayen, a small settlement on the road from Trier to Koblenz.[17] Others probably served villas,[18] while many more do not seem to have lain near any established settlement. They testify, however, to the relatively dense occupation of the countryside in the late Roman period, and to the value of the produce and beasts it supported. Essentially, such sites are 'safes', into which animals could be driven on the hoof to escape their capture by barbarians. The sites were not impregnable to a determined foe by any means, but they first had to be located, and in many cases their remoteness and difficulty of access would make this a task not worth pursuing for a marauding band intent on plunder and richer pickings elsewhere.

One can expect, therefore, at best temporary or transitory use of such sites; possibly they would never need to be used at all. They thus would lack permanent buildings, and excavation of the interior is potentially unrewarding. At a number of sites, the only building discovered to be of late Roman date is a small square tower—an example can be seen at Bertrix, but there are others for example, at Katzenberg, Bonnert, and Kaasselbierg.[19] Not all sites would have such a watchtower—indeed, its presence might attract unwelcome visitors. Sites which had them (if indeed these small square buildings are towers, and not, for example, sheep folds or small byres) were possibly therefore a little more martial in aspect, and may even have formed part of an official communications chain. In the case of some of the Swiss sites, this seems definitely to be the case (see p. 189): it is less easy to substantiate in the Rhineland area.

A late Roman date for these hill-top sites is occasionally confirmed by the very sophistication of the defences. At Sommerain, the circuit of wall had projecting towers and a gate protected by a semicircular tower which fit well with late Roman defensive practice. The gate at Bertrix too, was of late Roman form.

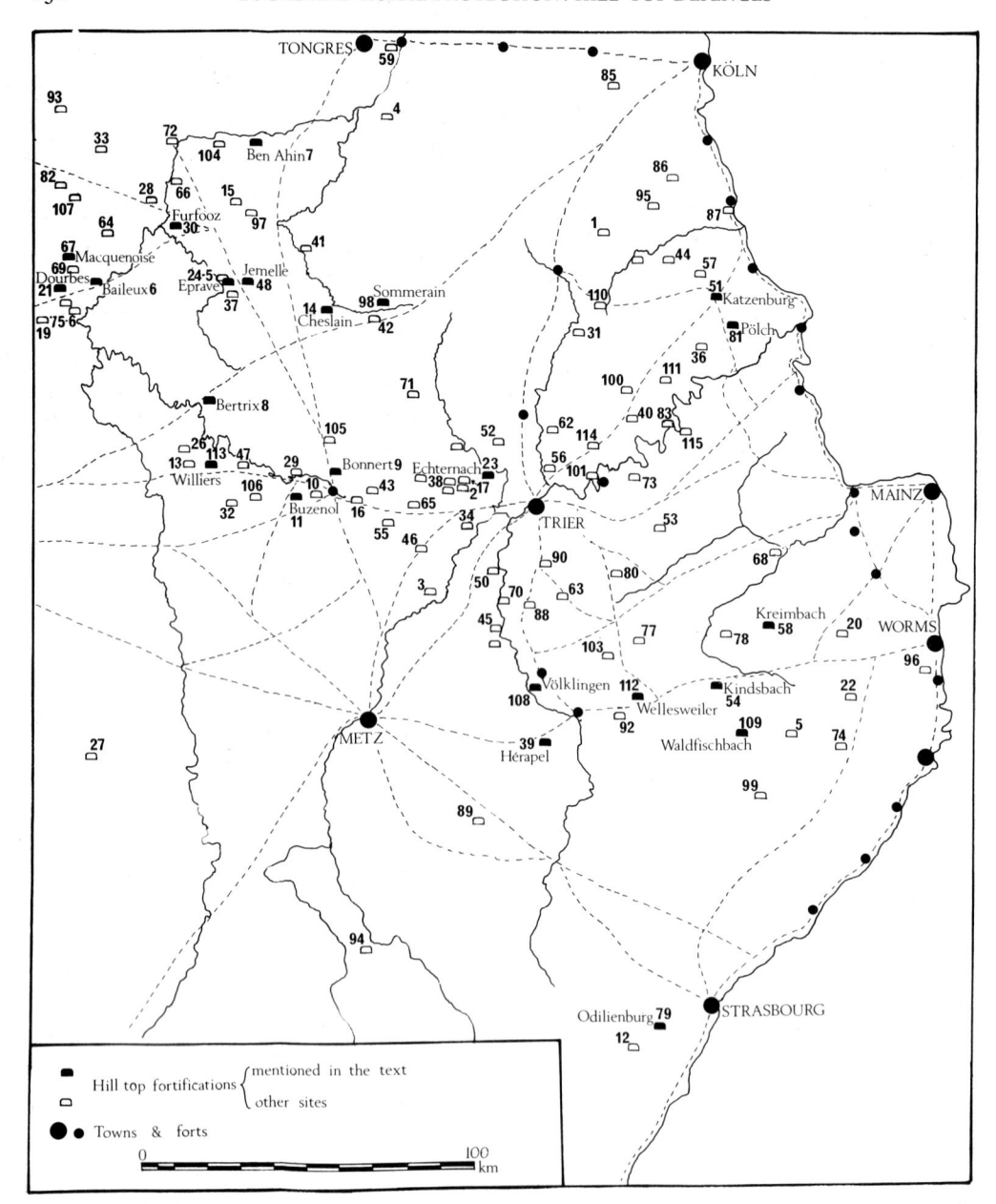

87 Distribution map of hill-top sites defended in the late Roman period in the Rhineland and Belgium. The numbered sites refer to the list of sites in Appendix 3, pp 280–286

Where the defences were stone built, their defensive arrangements were often more explicit. At Buzenol,[20] a roughly triangular site was enclosed by a rampart and ditch, crested by a wall 1.2m thick. The wall incorporated a large number of reused sculpted blocks—mainly tombstones, found at one spot in the defensive wall of the inner enclosure. The site was split into two separate enclosed areas, on the site of a prehistoric hill-fort. The upper enclosure was surrounded by walls which incorporated square and U-shaped towers. A lower additional enclosure contained a gate-tower. Perhaps the most surprising example of this form of sophistication is at Völklingen.[21] At this site an elevated promontory was cut off by a bank and ditch to enclose a roughly triangular area about 100m by 70m. The bank was 6m wide, and stands 1.2m high. A gap roughly midway along this rampart was found on excavation to have contained an elaborate stone gateway, with a pair of semicircular projecting towers flanking a wide rectangular building which must have contained an entrance-passage.

In the Pfalzland area, west of Mainz, finds from some of the more remote sites, for example the Heidelsberg near Waldfischbach, or the Heidenburg at Kreimbach, are perhaps more significant. At the Heidelsberg,[22] natural defences were used to protect all but the western side of the plateau (fig. 85): on the western side, two building periods have been identified, the second of which was formed of a drystone wall with casemate buildings, and was dated to the beginning of the fourth century. Much of the wall was built of re-used stones, a number of them tombstones, and one inscription found recorded a *saltuarius*, showing a man who was clearly an imperial forester portrayed with a small handaxe. Buried within the walled enclosure was a large hoard of ironwork, including agricultural equipment. At Kreimbach,[23] a walled hill-top enclosed about a hectare: there were gates to east and west, and the curtain wall had three towers. The coin series runs from 260–360: the site has also produced a large hoard of ironwork, including a fretsaw, scythes, blacksmith's tools—anvils, hammers and branding irons. A similar hoard of material was found at Kindsbach[24] in the same area. These large finds confirm the essentially rural nature of these sites, and their burial for safety within this protected area suggests both the importance of such equipment for raiders who might carry it off, and the security expected of the defended sites themselves.

At Kreimbach, another small-find was a chip-carved belt-buckle. A number of hill-top sites have produced chip-carved equipment: stray finds of such material have been found at Pölch, at the Hérapel, and at Williers among others, but several sites have cemeteries nearby which contained a large proportion of men buried in association with this equipment.[25] The type site for this, since the excavation of its cemetery in the nineteenth century, is Furfooz.[26] The fortified site here consists of a rocky promontory of irregular shape, elevated above the River Ourthe (fig. 87). There were two main periods during which the hill-top was used in late Roman times, corresponding in the main with structural phases recognised within the development of the defences. In the first period, (270–350), a wall 1.15m thick, and carefully built of small blockwork, spanned the narrowest part of the south-western approach to the plateau. The wall was 16.5m long, and at its northern end incorporated flanking defence for a simple passageway at the cliff edge affording entrance to the defended area. A bath-house lay outside the defended area, and probably also belongs to this period.

The second phase of defensive arrangements belongs to the period 370–420. The original blocking wall was retained, but a second, parallel to it and 30m within the defended plateau, was added. This wall, 2.15–3.1m thick, and with a simple gate 2.5m wide set at its northern end, incorporated material from the by now ruined baths, whose site was now occupied by a cemetery containing graves of men who wore late Roman 'Germanic' style military apparel.

The richness of the finds from the cemetery leads to the conclusion that this site was garrisoned by the military rather than served as a refuge. The first phase of fortification of the site may have been carried out by a foreign auxiliary troop unit stationed for some undisclosed reason at this rural spot. As further evidence for this view, one must take the bath-block, the relative richness of pottery and coin-finds from the site, and the overall conception of construction of the defensive wall into account. The gap in occupation in the mid-fourth century was probably due to the effects of Germanic invasions in 354–5, and it has been suggested that in the late fourth century the site was occupied by a group of Germanic warriors whose dead were interred in the cemetery. A small coin-hoard with terminal coins of Valentinian III (425–55) shows that the site was still in use in the middle of the fifth century, but the date of its abandonment by the Germans has not been established.

At few of the hill-top sites in the Rhineland area is there a similar combination of fortifications and cemetery as at Furfooz. At Polch there is a fortified hill-top site with a cemetery at its foot, graves from which have produced comparable material to that of Furfooz.[27] Other sites with associated material of this kind are Völklingen, notable for the sophistication of its gate, and the Belgian sites of Éprave (fig. 86),[28] Ben Ahin[29] and Dourbes[30] where the relative richness of the finds, the well made construction of the walls and the overall time-span of the finds suggest a long and fairly constant occupation of the site rather than a short stay. Only at Éprave is there an associated cemetery.

The presence of this material and the cemeteries with men buried in items of military apparel suggest that some of these hill-top forts were more permanent than mere refuges, and that they held garrisons of soldiers. Apart from the forts actually on the line of the River Rhine, the only soldiery mentioned in this area in the *Notitia Dignitatum* is a list of *praefecti laetorum* stationed at various places in the territories of cities[31] in northern and eastern Gaul: the locations of these prefects vary from Orléans, Bayeux or Le Mans in the west through Arras, Reims and Noyon, to 'near Tongres' in Germania. A similar list of *praefecti Sarmatorum* appears to be more widespread about Gaul[32]—at Poitiers, Langres, and further to the south. Such men were in charge of what seems to have been a back-up force to protect the country regions: the sorts of units of troops or mercenaries under their command can only be guessed at. It has been suggested that hill-top sites like Furfooz, Éprave and the others where a military presence is to be deduced from the associated finds are the locations of garrisons of such *laeti*.[33] The date of the *Notitia*, however, which is more than a century later than the beginning of settlement at some of these sites, is a crucial problem, for its relevance to the particular sites identified on the ground cannot be established. The question whether sites such as Furfooz were occupied by *laeti* rather than troops of supposedly more regular type is at present unanswerable,[34] not least until the difference between *laeti* and regular troops has been archaeologically better defined. Suffice it to note a possible military presence at some of the hill-top sites—itself a matter of some consequence since it provides evidence once more for the considerably altered nature of the military in the late third and fourth centuries and also provides a context for some, but not all, of the sophistication of design apparent in the hill-top sites.

A further possibility with regard to some of these sites must be raised, if only at some later stage to be positively discounted. Discussion about *laeti* which has for long centred on these hill-top sites has concentrated on the twin aspects of settlement by Germanic tribes and the dual role undertaken by such tribesmen in the later Roman period as settlers and soldiers if need be. It is possible, however, that some of the hill-top sites where some permanence of occupation can be established are the homes of settled tribes within the empire, whether strictly speaking they should be known as *laeti* or not. Research on contemporary Germanic settlement in free Germany has identified several hill-top sites similar to the fortified or re-fortified hill-tops within the empire.[35] At the Glauberg,[36] the prehistoric defences of a hill-fort on an elongated plateau, enclosing about 2ha, were capped in the later Germanic Iron Age (roughly the late Roman period) with a new drystone wall 1.5m thick. Occupation of the site at this date is attested by remains of a series of rooms with late Roman finds, including Argonne sigillata and coins. Occupation at a similar date has now been attested at the Runder Berg near Urach and at the Gelbe Burg near Gunzenhausen, where occupation was apparently nearly continuous from the second century to the ninth or tenth century.[37] A number of other possible sites of similar settlement have been identified without any corroborative archaeological work. Clearly hill-fort style settlement was still known in the area (though not all free German settlement was of this type) and may have spilled over into areas within the Roman world if settlement by German tribes was allowed there. In fact, areas of a high concentration of late Roman hill-top sites might be seen as protectorates, devoted to the settlement of German tribes: in particular the Pfalzland area, ringed with forts such as Alzey, Bad Kreutznach, Pachten and Saarbrücken, might seem a likely candidate for such a view. Much still depends, however, on our imperfect understanding of the variety of forms that Germanic or barbarian settlement within the Roman empire might take.

The overall map (fig. 87) shows a provisional picture of the number of known or suspected sites of late Roman defended hill-tops in the Rhineland area, and is accompanied by a list of names and a brief bibliography (pp. 280–286). This shows no more than the present state of research into the subject, and may reflect an unbalanced overall view of the spread of such sites because of uneven effort in different areas put into identification and survey of the material. A brief résumé such as this is also unsatisfactory in that it can only deal in the broadest terms with the best known of the sites. Others which perhaps deserve fuller mention and discussion, for example Echternach,[38] a remarkable 'hill-top' site which, though late Roman, appears to have much in common with medieval shell-keeps, or the fort on the Odilienburg[39]—a vast sprawling defended site of several hundred hectares—have to be passed by.

Further up the Rhine, in present-day Switzerland and Bavaria, such sites are not so thickly concentrated, possibly because field work has so far failed to identify them but more probably because the barbarian threat to this area may have seemed less than that in the frontier zone between Mainz and Köln, and because the routes followed by barbarian raiders on their way through to Italy were closely defined by the nature of the terrain. These roads were studded with forts, and the prospect of potential raiders straying off these well known paths to any great extent was not serious. In the more mountainous regions, the pattern of hill-top refuge sites took on a different aspect altogether. Here many small settlements, villas or townships in the valley bottoms selected and defended a prominent hill often quite close to their homes, and made of it a visible citadel, martial in outlook, and clearly designed to act as a deterrent to would-be attackers.

There is a group of fortifications, however, in the Jura which corresponds with the Rhineland sites. This includes a number of sites which, though defended, incorporated watchtowers within their enclosure, and form a possible chain of signalling stations. These lay at Mandacher Egg, Portiflüh, Renggen,[41] and the site which has received most attention, the Sturmenkopf near Wahlen. The summit of this hill was enclosed in late Roman times by defensive walls (fig. 89), protecting a tower whose remains have been located and excavated. A scatter of Roman tiles and late Roman pottery and coin-finds support this identification.

Apart from this series of sites, there are in Switzerland and Bavaria a number of more isolated hill-top sites which seem to have acted merely as hill-top refuges after the same pattern as those in the Rhineland.[42] In describing just the three of these which have been the subject of relatively detailed examination, it is necessary naturally to miss out a great deal of the detail. It must be said that the majority of 'possible' defended hill-top sites in this region rely, like many of their Rhineland counterparts, on the suggestion of defences on the hill and a slight scatter of Roman material from the summit. This might have other explanations than the presence of a late Roman refuge.

The first of the three sites meriting closer attention is the Wittnauer Horn, near the Rhine Frontier east of Basel.[43] The site lay at some distance from main Roman roads, and occupied a steep-sided plateau of long narrow triangular form (fig. 89). The site was occupied in the Hallstatt period, but in late Roman times the crest of the Hallstatt bank was strengthened by the addition of a single wall, 1.5m thick. The coin series from the site is long and impressive, the majority of the coins belonging to the second half of the third and the first half of the fourth century. The excavator defined two main periods of use. A coin of 222 lay under the wall. It was suggested that the main occupation in 250–300 lay in the interior of the defended area, which was partially examined, while, in the period 300–50, the area immediately inside the wall was used. The form of the actual barrier wall at the Wittanuer Horn has occasioned some debate. The narrow base of the triangular plateau was walled off with a single, virtually straight wall, only about 100m long. At the northen end of the wall, a large projecting square tower protected the access route to the site which lay along the northern edge of the hill.

88 Distribution map of hill-top sites defended in the late Roman period in the Upper Rhine and Danube areas. The numbered sites refer to the list of sites in Appendix 3, pp 286–290

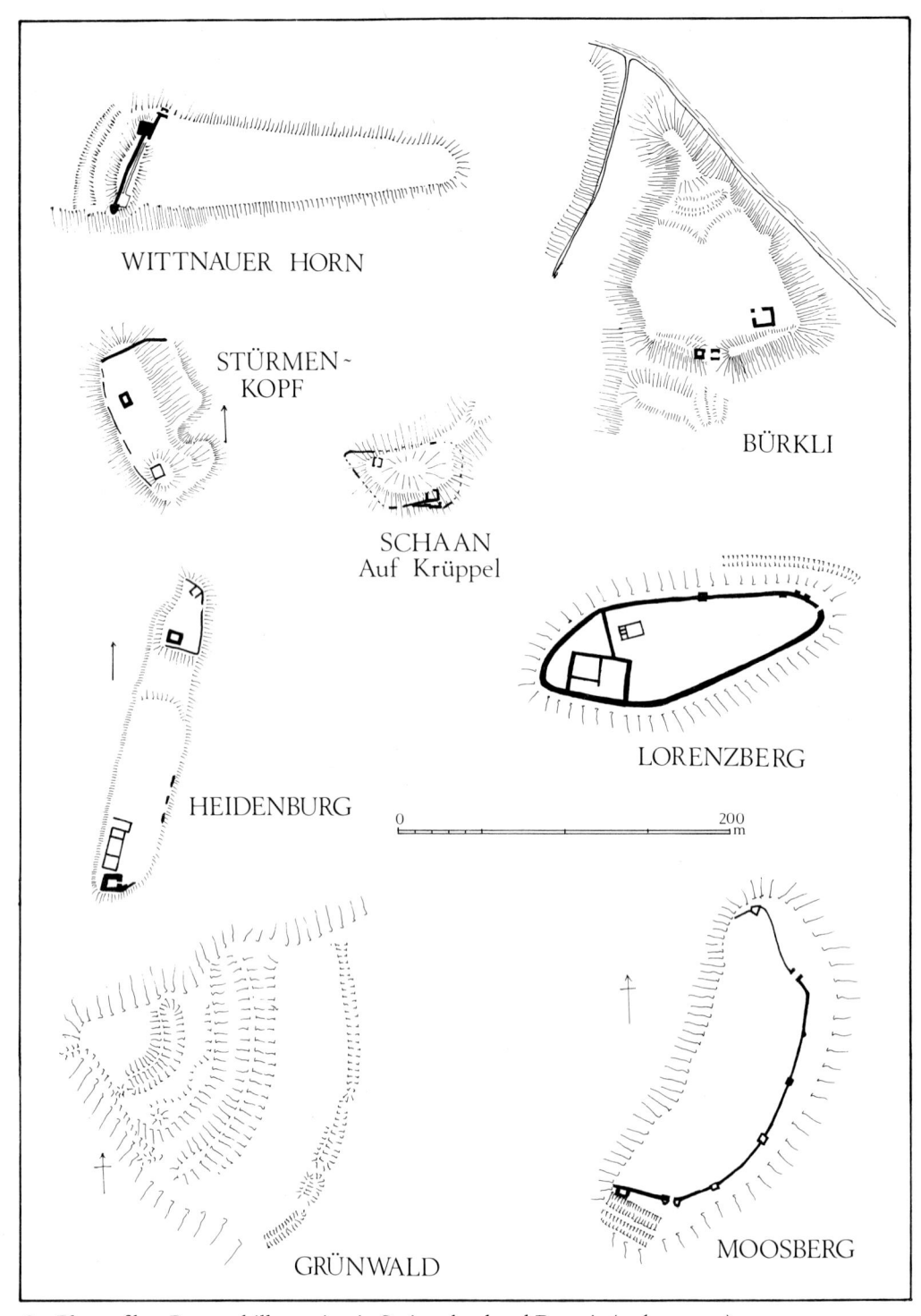

89 Plans of late Roman hill-top sites in Switzerland and Bavaria (*scale 1:4,000*)

From the back of this tower a thinner wall ran down to the gate-passage itself. This took the form of a simple passageway through a single gate tower. This small wall and gate adjoining the large projecting bastion was thought by the excavator to be contemporary with the rest of the defences, but it has been claimed as a medieval addition.[44] In fact there seems no need to postulate a medieval use of the site, which the finds otherwise hardly support, for this form of gateway is found in other late Roman fortifications nearby.[45] Doubts raised at the time of its discovery about whether so apparently 'eastern' looking a gateway could really be late Roman have in the intervening years been dispelled by the discovery of other, equally outlandish looking, forms of late Roman defensive architecture (see pp. 49–50).

The other two late Roman refuge-sites which have been examined both lie in Bavaria, and are the Lorenzberg near Epfach and the Moosberg near Murnau. At the Lorenzberg,[46] a roughly oval plateau in a bend of the river Lech was re-occupied at a date within the decade 260–70, and given a complete circuit of stone defences soon afterwards (fig. 89). Occupation within the walled area seems to have been in timber buildings which were destroyed probably in Alemannic raiding in or around 350. After this destruction, a new stone building was put up at one end of the site. It may have been a store-house or a *mutatio* housing travellers for the Imperial postal system. The site retained its civilian character until the 370s, but the first evidence for military use comes soon after this, when a cemetery with at least 33 military graves came into use nearby. No trace of accommodation for these troops has been found within the defences, however, and the only additional stone building known at the site has been identified as a church, dating from the last years of the fourth century. Surprisingly, the presence of troops at the site, confirmed by the cemetery and by finds of military equipment from the site itself, resulted in no further improvements to the defences: the form of the gate (at the north-western tip of the site) is not known, and the western flank is protected by a pair of rectangular projecting towers (little more than buttresses) and a large square solid tower which stands astride the wall.

The settlement on the Moosberg[47] was different in character, however. The hill has been entirely quarried away, but a complete series of excavations was carried out prior to its destruction and did much to reveal the full plan (fig. 89). The site occupied a long hill-top: no defence was required down a steep slope to the north-west, and the defences were correspondingly weighted round the southern and eastern flanks. The defensive wall, enclosing a crescent shaped area, was of slight thickness (usually less than 1m), and was protected by a series of square and semicircular towers. The southern flank was provided with a pair of defensive ditches, and the main south gate had a pair of projecting towers. The north-eastern gate, however, was of 'Andernach' type, with a pair of solid rectangular projecting towers. The excavation record suggested a series of timber buildings within the enclosure, some in the form of freestanding post-built halls along the main road through the site, and others arranged, casemate fashion, along the interior of the south-eastern curtain wall in particular. Two phases of post-hole construction were recognised in the southern part of the site: it was suggested that the first phase of use came in the late third century, followed by more permanent buildings towards the middle of the fourth.

Parallels for such hill-top refuges can be found in other provinces—in Dalmatia, or in southern Gaul.[49] In the Alpine valleys however, as already noted, the pattern is rather different.[50] In Noricum, towards the end of the fourth century, many towns and settlements had moved, often permanently, to their hill-top locations. A major town, like Virunum, once the capital of the province of Noricum Mediterraneum, appears to have been abandoned by the end of the fourth century, in favour of the town of Teurnia, itself a site somewhat similar to a fortified hill-top.[51] In the region of Virunum itself, a number of hill-top settlements took on some permanency of form at the same date: there were sites like the Kirchhügel at Karnburg[52] or the Grazerkögel.[53] How much earlier such sites had been in use as temporary refuges is not certain, but they are unlikely suddenly to have sprung into being around 400. Only a few such sites are known in the north of the province, near the *limes*, and the major concentration of refuge sites lies in eastern Noricum: a town like Aguntum (Stribach), partially walled in the late second century, had a refuge nearby at Lavant, which, like the site at Virunum, had become the main settlement centre by the mid-sixth century.[54] At this site, like many others in the region—for example Duel,[55] the Hemmaburg,[56] the Hoischügel (fig. 90)[57] and the Grazerkögel—one of the most important buildings on the site is the church, and this clearly formed a nucleus of continuity into the sixth century and beyond.

Closer to Italy, in the passes leading from present-day Switzerland and Austria, a number of hill-top sites took on a more strategic aspect. The possibility that they formed part of the frontier system in the Julian Alps, either as look-out posts, or by holding a specific troop garrison, has already been suggested.[58] At the Duel and at Hoischügel, buildings of military style have been recorded. How far this military role can also be supported for the large numbers of hill-top sites claimed to lie in the Julian Alps region—many of them scarcely supported by investigation in the field—is unclear. Few sites outside the Austrian valleys have been thoroughly excavated: only those at Schaan (fig. 89),[59] at the Gurina,[60] and at Vranje[61] have been examined in recent years.

A final series of sites of this type lies in Spain. In the valleys of tributary rivers of the Duero, a number of hill-top sites, very similar to those of the Rhineland have been located. One of the most obvious examples of this type of fortification is Inestrillas,[62] where a barrier wall cuts off the most easily accessible side of a steep hill. The wall is defended with semi-circular towers, and the gateway is deeply recessed between a similar pair. Others are less well known, and lie at Suellacabras,[63] which is a fortification of a similar type, while there are rustic style forts at Yecla de Yeltes, Lumbrales and Iruena.[64] At Monte Cilda, near Palencia, there is another hill-top fortification.[65] Finally, there is mention of a late Imperial fort at La Calzada, near Salamanca.[66] Several of these hill-top sites have associated cemeteries of apparently military type.[67] Some may have been for Laetic style soldier-settlers, though there is no mention of such settlers in Spain in the lists of the *Notitia*.

LOCAL AND RURAL PROTECTION: HILL-TOP DEFENCES

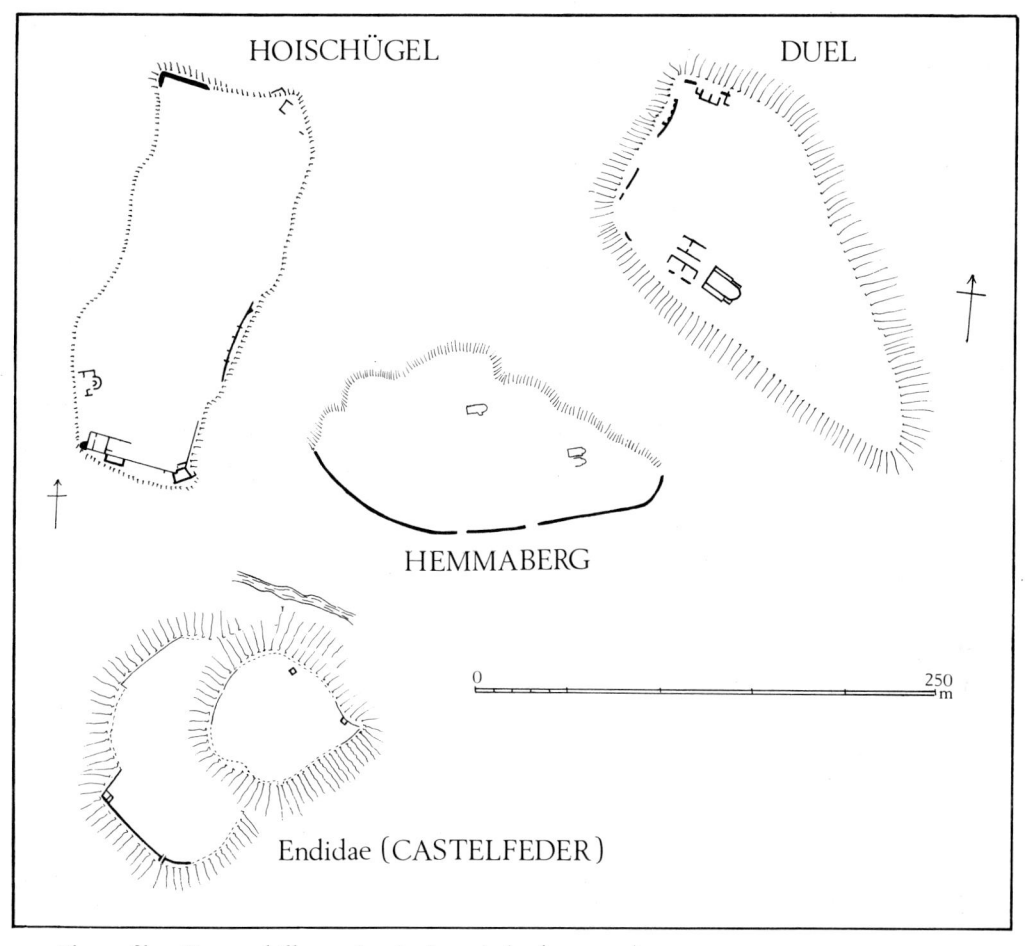

90 Plans of late Roman hill-top sites in Austria (*scale 1:4,000*)

One province where the lack of such late Roman hill-top sites is surprising is Britain. In Britain there is perhaps a denser concentration of walled settlements than in other areas of the Roman world, but the whole island could be regarded as a frontier province. A number of sites, particularly in the West country, have produced evidence of late Roman usage, though this is normally no more than a scatter of occupation material which may presage no actual refortification of the hill, or some use other than defensive.[69] In the late Roman period specifically, one of the few sites claimed to have been redefended was Highdown, near Worthing, Sussex.[70] This prehistoric hill-fort was redefended, possibly in the mid-late fourth century. An early Saxon cemetery lay nearby. The use of selected hill-forts further west and in Wales is attested rather more securely, but cannot be claimed to form a specifically late Roman reaction to threatened invasion. The redeployment of a power base from the defensible hill-top was the attraction.[71]

The use of a defensible hill-top to provide a private or countryside refuge in the late Roman period is merely one aspect of the general move towards the selection of naturally suitable locations for protection.[72] Various types and classes of fortification now occupied sites similar to these hill-tops, whether this represented a late Roman re-use of a former prehistoric site or not. Towns or cities in particular which made use of such a strongpoint included Kempten, Lausanne, Geneva, Verdun or Vermand. Trends in fort design since the mid-third century had been leading in a similar direction. On the Rhine, the site at the Qualburg or the Valkhof Hill at Nijmegen are good examples. Elsewhere, especially where a new frontier had to be commissioned, as in the series of posts linking the Rhine, Iller and Danube, the forts, though clearly military, made the same use of high ground. The general similarity between military establishments and hill-top fortifications makes for difficulty of modern-day definition: sites like Füssen[73] or Kuchl[74] or even Furfooz and Völklingen might as well be civilian or military in purpose. The ambiguity of use of such places is well illustrated by the Lorenzberg, where a site which appears to have been originally fortified as a refuge because of local need, probably held a garrison of troops at a later stage.

The construction of defensive walls at many of the hill-top sites discussed in this chapter, however, cannot properly be laid at the door of the military. Provision of defence at this level—for *vicus*, villa or country-holding—was a matter for self-help. Prime agents in such provision were the local *potentiores*, occupying a position rather similar to a medieval lord of the manor, to whom the local populace would be bound by domestic, financial and social loyalties. As their part in this, the local population could expect protection against, among other things, excessive taxation, central governmental interference, and barbarian raids.[75] The countryside refuges are a tangible proof of the effectiveness of this latter aspect. The other two forms of harassment have left little archaeological trace.

The literary record is dotted sparsely with references to such men and the fortifications they provided. The clearest example is in the south of France near Sisteron,[76] where a rock-cut inscription at the entrance to a narrow defile proclaims in badly spelt Latin that in the place known as Theopolis, walls and gates were provided to form a private defence for common use. The inscription was proudly set up by Cl. Postumus Dardanus to show his keenness for everyone's safety and his public devotion. The provision of this enclosed and fortified space for his family and retainers was clearly seen by the *patronus* as a public-spirited gesture, the nature of which he was keen for all to see. The majority of other men recorded as having made similar provision seem to have intended their fortifications as private fortresses, much after the fashion of the rather aggressively possessive inscriptions which adorn some of the small African farm-cum-blockhouses for the mid-second century onwards.[77] From Sidonius Apollinaris we learn of his friend Aper, who was apparently spoiled for choice of which of his hill-top forts to spend his time in;[78] or of Pontius Leontius, whose *burgus* is described in poetic detail and sounds like a veritable palace.[79] At a slightly later period, there is mention of Marcellinus, bishop of Milan, who died in 536, building a fortification on an island in Lake Como,[80] and of Theodoric, who drew into his fortification of Verucca (in the South Tirol) the possessors of neighbouring estates for protection.[81]

The case of Leontius shows that the defended hill-top as a private fortification

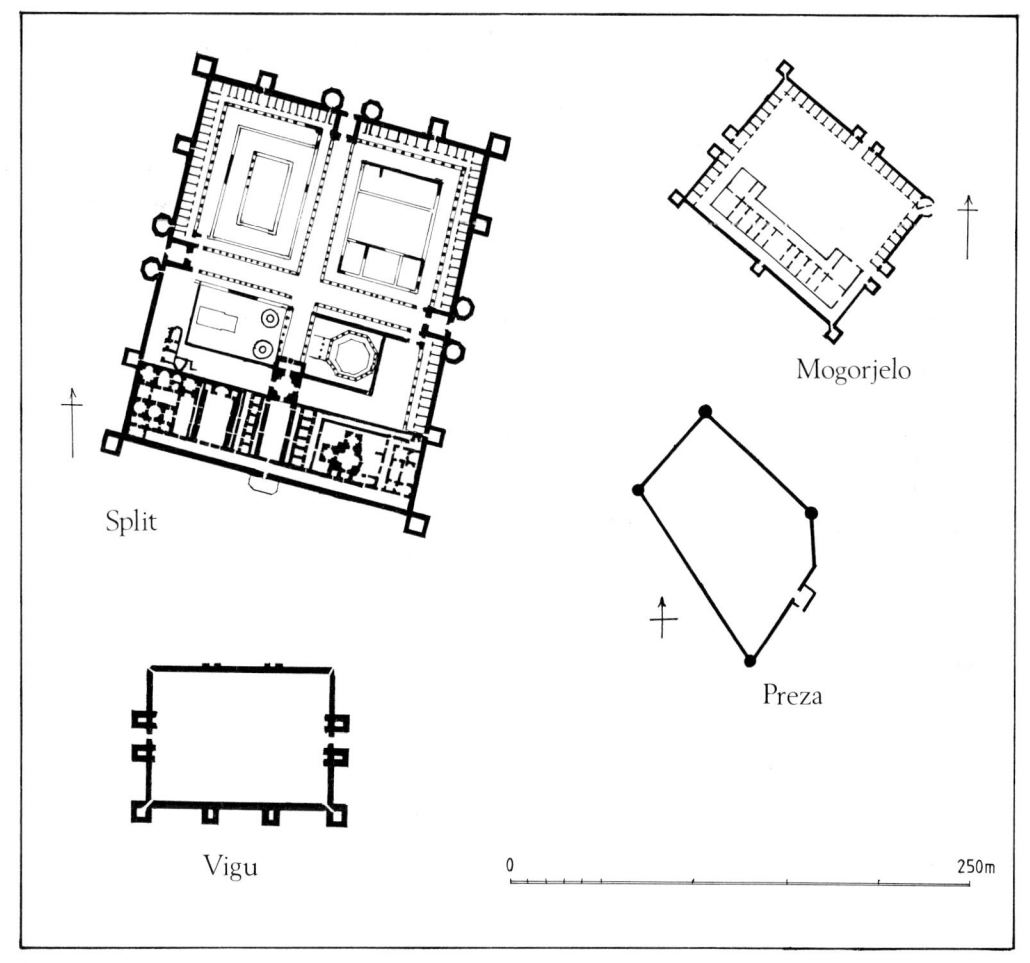

91 Late Roman walled sites in Dalmatia (*scale 1:4,000*)

was only a short step removed from the defended villa. Villas which merit this description have occasionally been found,[82] particularly in the frontier provinces of the empire, but such buildings are less common than one might suppose. Of the best documented sites, one lies at Pfalzel, in the Rhineland,[83] but perhaps the most impressive examples of the type are at Mogorjelo,[84] Gamzigrad[85] and, of course, Split (fig. 91), the fort-cum-palace which Diocletian built for his retirement in 305.[86] These hardly qualify as 'private' fortifications except in the sense of being the private residence of the emperors or their deputies. The martial aspect of some of the Rhineland sites is emphasised by Ausonius: villas with towers protrude into his picture of rural rusticity on the Moselle.[87]

The few literary references to the provision of refuges include at least one provided by a bishop—Marcellinus of Milan. It has long been recognised that a substantial number of such sites may have an ecclesiastical origin. Particularly in the east—the Tirol area and parts of Dalmatia (Jugoslavia)—many of the sites contain churches, and they therefore possibly formed the seats for bishops in the later Roman period.[88] Some hint of this can be gleaned from the *Notitia Galliarum*, where in several of the provinces a number of *castra*, supplementing the *civitates*, have been suggested to be alternative seats for bishops whose home church was under considerable threat of trouble or invasion.[89] Such sites were, for example, Horbourg or Breisach in Maxima Sequanorum, Macon or Chalon-sur-Saône in Lugdunensis I or Bigorre in Novempopulana. A further literary example, Horbourg or Breisach in Maxima Sequanorum, Mâcon or Chalon-sur-Fortunatus at the end of the sixth century.[90] It is important to realise that the late Roman use of such sites as Horbourg and Breisach as *castra* for bishops dates from well after the abandonment of the forts for military purposes. A number of other forts are known to have contained late Roman churches which in many cases now occupy the sites of medieval and modern successors:[91] the use of disused fort sites by the ecclesiastical authorities in the fifth and sixth centuries may have been relatively common. In the eastern region, sites of similar type at Lavant, (near Lienz),[92] the Georgenberg (near Michelsdorf),[93] and Vranje (near Sevnica)[94] have been relatively well studied. Not all such sites which contained churches will necessarily have been the seats of bishops, but the possibility is worth consideration.

As with other types of late Roman fortification, therefore, the hill-top refuge, with its defences of greater or less sophistication, portrays a range of probable and potential uses which pose considerable problems of interpretation. This brief résumé, which looks merely at a few of the wealth of sites of this type, serves to display the variety of responses to the pressures of the late Roman barbarian invasions which lie concealed in out-of-the way and seemingly unimportant places. From this can be gauged the depth of concern and fear caused, even among the peasantry of the Roman world, by the barbarian onslaughts of the late third and fourth century.

II

Late Roman Frontier Policy

It is important in attempting some overall view of late Roman fortifications to see them not only as a part of the military history, but also as contributory factors to the social and economic background of the late Roman world as a whole. At best, the study of these defences will chart a social reaction—the growth of an awareness of the dangerous presence of unknown outsiders threatening the peace of the empire. At worst, it will show the strictly military concern for weak points in a defensive chain.

Though the majority of the fortifications included within this study were clearly put to military purposes, there are many also which could be seen as forming part of a local reaction—a 'grass roots' response to the growing threat of invasion. Where this is so, there is a potent source for gauging the social unrest brought about by the unsettled times. Coin-hoards can possibly be misinterpreted, but the fortification of a hill-top to keep one's family and flocks safe against a potential or actual invasion is positive proof of fear.

To a lesser extent, the construction of city walls offers very much the same inference. Here, perhaps, reaction was more officially controlled. The construction of city defences was often undertaken by troops and the money used to provide pay and materials, where these were not simply requisitioned from local people, came out of local funds or even imperial grants. Unfortunately little about the mechanics of Roman local government, still less about who could have made the recommendation that cities defend themselves, is known. In some cases, certainly, a check on the defences was ordered by the emperor, but there is no reason to believe that it might not also sometimes have been the local decurions, who, worried about the security of their city, applied spontaneously to the emperor to allow them to do something positive about it. It was often these same decurions, who, with lands in the area, also saw to the safety of their own retainers, and had refuges built for the times when danger threatened.

There was a heightened awareness, too, of the presence of the barbarian beyond the Roman frontier. The frontier was now no longer a thin line along the Rhine and Danube separating Romans from outsiders. Nowhere was safe, nor was the line itself adequately defended, for towns and cities were now closed up against the unsettled times. Perhaps the military frontier itself, planned not as a barricade to separate the Romans from the barbarians, but as a method of controlling their intercourse, was as watertight as it had ever been. The pressure from outside, however, had increased and the effect can be clearly seen on the Roman citizens who built protective walls for cities many miles from a recognised Roman frontier line.

Rome now attempted all the more to control and curtail barbarian activities by negotiation, since she was increasingly unable to achieve this by military power. In addition to trying to buy peace by paying tribute to tribes immediately outside the empire, it was also part of imperial policy to grant some of the more fractious tribes the responsibility of vacant land within the empire's bounds. Though this was potentially a dangerous exposure of Roman citizens to the fickleness of these tribal groups, the policy was aimed at encouraging these tribesmen to put vacant spaces normally near the frontiers to good use by farming them and by resisting any further invader who came to rob them of their hard won privilege to help protect the empire. Occasionally, the attempt misfired, and these farmer-settlers attacked Roman cities,[1] but on the whole the scheme was a success. Invasion often had as its root cause the deprivation of lands and consequent starvation. This forced tribes to forage further afield for supplies. Vacant spaces inside the empire were ideally populated by just such tribesmen, provided that they could be kept under the strictest control. This meant posting them as near the frontier and its network of forts and roadposts as possible. Thus the network of small posts on roads near the frontiers could as easily be used for the policing of these areas as the safe reception of *annona* for military use.

The barbarian's view of his treatment by the Roman authorities is not revealed in any unbiased ancient source. Roman writers put into the mouths of their barbarian antiheroes words which suited the Roman picture of the barbarian temperament. Ammianus Marcellinus talked in stereotyped terms of the barbarians' stupidity when confronted by walls.[2] Faced by a Roman Caesar in battle, the barbarian rabble could never hope to succeed. Themistius, describing the fort which Valens built in Moesia, related that it was built to inspire terror into the barbarian beholders.[3] But not all such Roman operations inspired so much terror: under severe pressure, Valentinian had to withdraw his troops from Mons Piri, a post he was attempting to establish in barbarian territory north of the Rhine.[4] Indeed, considering the resistance put up by most of the *barbari* to the Roman presence on their territory, it is surprising how many such posts are known, both beyond the Danube in Czechoslovakia and also in Sarmatian territory beyond Pannonia.

Whatever the effect on the watchers outside the empire, the provision of town and city walls within it was a great boost to Roman morale. As propaganda value in establishing new settled conditions in a time of change and uncertainty, the provision of walls was to constitute a 'defence for all time'—a sentiment found echoed several times in various forms, notably from Thrace in the period of the Tetrarchy.[5] Other dedications tell of *burgi* which were set up for the same reasons.[6] But another inscription, again from a *burgus*, spotlights one of the problems of frontier studies: this site was called 'Commerce', the reason for its establishment.[7] Clearly, as well as a barrier to keep the more undesirable outsiders from entering the empire, the frontier was a legitimate channel for trading, and was thus a gateway through which all traffic had to pass. The severance of trade could cause grave problems to barbarians who depended for their livelihood on an efficient and thriving contract with their Roman market.[8]

For a glimpse of condition on the frontiers, albeit some decades after official Roman withdrawal from the Danubian area, it is possible to turn to Eugippius' life of St Severinus.[9] The only garrisons in the area left to Severinus were a small contingent under a *tribunus* at Favianis and the *numerus* of troops who garrisoned Batavis. These men were still paid, but only irregularly, because transport to and from Ravenna was risky, and when the money failed to arrive the posts were often abandoned. The garrison of Batavis held out against the barbarians and attempted to send messengers to collect their pay from Italy, but this mission was ambushed on its way there by bandits. Only 40 men were left to guard the fort against a barbarian attack. The garrison of Favianis was so weak that it was unequal to the task of dealing with brigands—until granted miraculous strengthening by Severinus. Most fortifications were now in a state of decay.

Towns in Noricum were thrown back in great measure onto their own resources for defence, and this was not always totally successful. Comagenae (Tulln) had a treaty with the barbarians who thereby became masters of the town, and Severinus was able to reverse the position. The neighbouring tribe, the Rugii, held the towns more and more under their influence: they were now settled north of the Danube, and eventually the *oppida* of Noricum became their tribute-paying subjects. Communications were easy enough, for roads were still in good repair, but brigandage was evidently rife. There was no obvious way of preventing brigandage by tribes with whom the remaining Romans did not have an agreement, and in this respect the Goths were the main culprits. Trade continued unabated on a weekly market basis, attended both by Rugii and Romans.

Though Severinus' day should not be taken as typical of the later Roman period as a whole, it does exaggeratedly portray the sorts of pressures which were operative in the late Roman period, and which, surprisingly enough, had not yet been strong enough completely to break down the Roman order. The movements of Severinus, however, cannot be regarded as any more than an *ad hoc* response to the situations as they occurred. In an earlier age, when the response to barbarian threat might be expected to be more calculated, it should be in a limited sense possible to detect some stirrings of frontier policy. It has always been difficult, however, to correlate events happening on different frontiers, even where there is adequate literary testimony to roughly simultaneous campaigns at different ends of the Roman world. Only when there is a record of an emperor on campaign to give a near-contemporary impression of the policy being followed by the emperor concerned can we attempt to correlate fortifications built with particular historical events. Even so, the pitfalls are great, and in attempting to

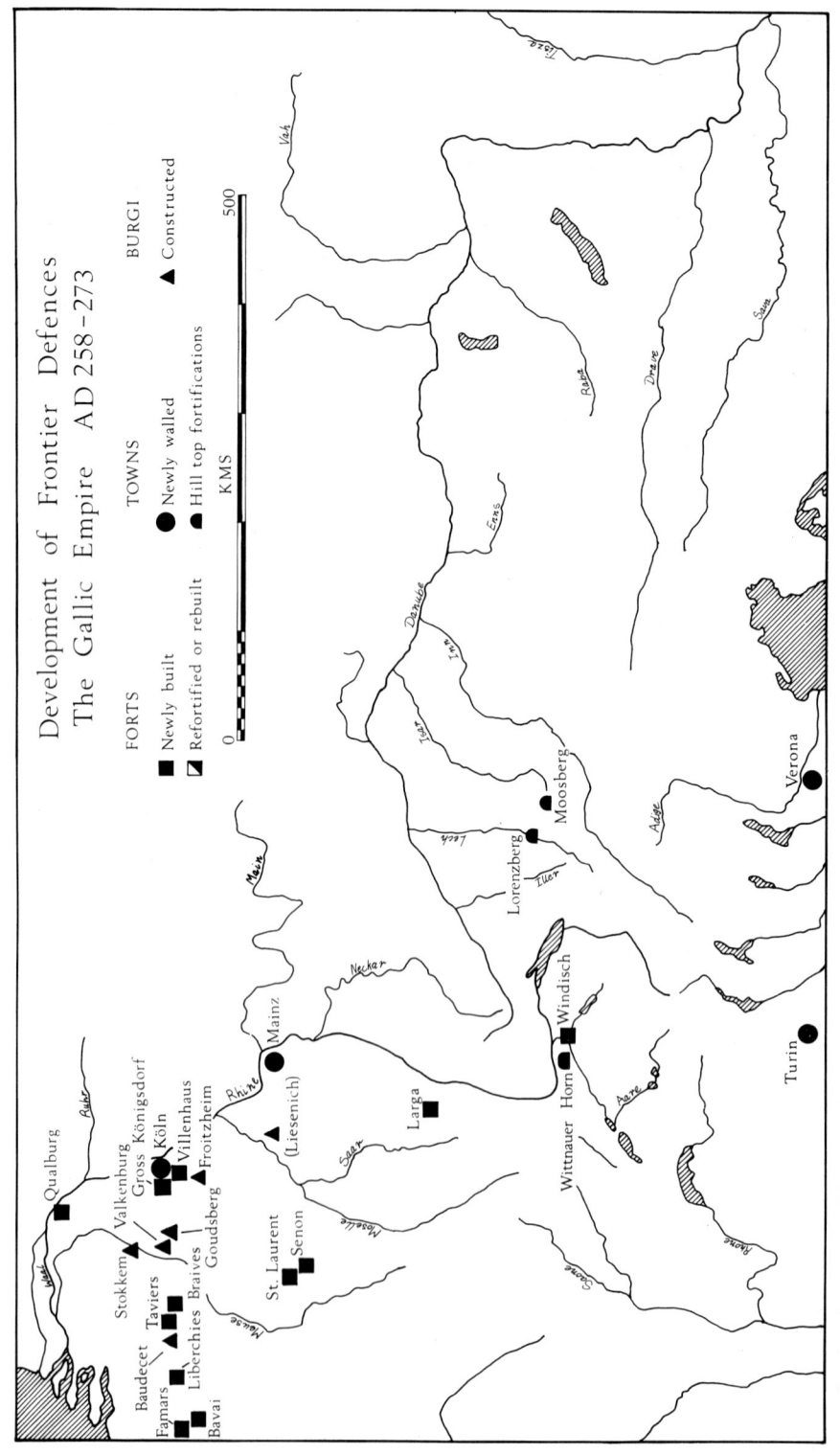

92 The development of frontier defences under the Gallic Empire

give as wide a basis as possible to the study of what the late Roman emperors were trying to do, it is necessary to turn not only to the western sites, but also to any useful comparative dates and campaigns in the east.

The distribution of sites with new defences built during the period of the Gallic Empire concentrated on the area round Köln and Bavai. Most of the small posts on the road from Köln to Bavai had an early earth-and-timber phase of defences. Bavai and Famars, too, both built in stone, had walls built in two phases. The earlier phase at both sites, later brought up to a standard late Roman thickness, belongs to a period when makeshift walls were hurriedly thrown up under Postumus and his successors.

The main defensive development of this period, however, was the appearance, all over the empire, of local hill-top refuges. Three, the Wittnauer Horn, the Moosberg, and the Lorenzberg bei Epfach, have been dated by archaeological means to this period, but many others which have produced few, if any, finds were probably their contemporaries. Coins of the later third century from such sites in Gaul suggest an occupation then, and two have produced hoards with terminal coins of Postumus.[10] It is likely that not only these, but a great number of comparable sites in Noricum, Raetia and Pannonia were defended at the same period. Elsewhere, and even as far away as Thrace, similar fortifications on remote hilltops were set up, particularly in the mountainous area of Rhodope, in the south-west of the province.[11]

Fortifications on the upper Danube itself received little attention from Gallienus. His main energies were concentrated on Thrace, where many of the larger towns were walled, and the fortifications of those which had been ravaged in the invasions of the late 240s and 250s were repaired. Epigraphic evidence shows that at Philippopolis (Plovdiv),[12] Serdica (Sofia)[13] and at Montana (Michaelovgrad)[14] repairs and rebuilding were undertaken, striking confirmation of the reported activities of Gallienus' architects Cleodamus and Athenaeus who made a tour of inspection to ensure that this part of Thrace was adequately defended.[15] Between them, Gallienus and Diocletian saw in great measure to the protection of this area, and it is therefore difficult to assign the city walls not dated by inscriptional testimony to the reign of one or the other emperor.

Despite his fame in beginning the walls of Rome, Aurelian appears to have done little constructional work which can definitely be assigned to his reign either on the frontiers or inland. According to Gregory of Tours, Aurelian was responsible for the construction of the walls of Dijon.[16] The small road-posts of Anse, Tournus, Mâcon and Beaune may also form part of the same strategic plan. Apart from these suggestions, Aurelian is also said to have been responsible for repairs to forts on the lower Danube.[17]

From his record in the literary and epigraphic sources, Probus fares little better. He completed Aurelian's task of building the walls of Rome and 'restored' the Gallic cities after the raids of 276.[18] The main area of frontier construction under Probus was in the small area of Raetia most at risk during the period which saw the fall of the German frontier east of the Rhine. On this new frontier line linking the Danube to the Rhine the small forts of Isny and Goldberg were now built.

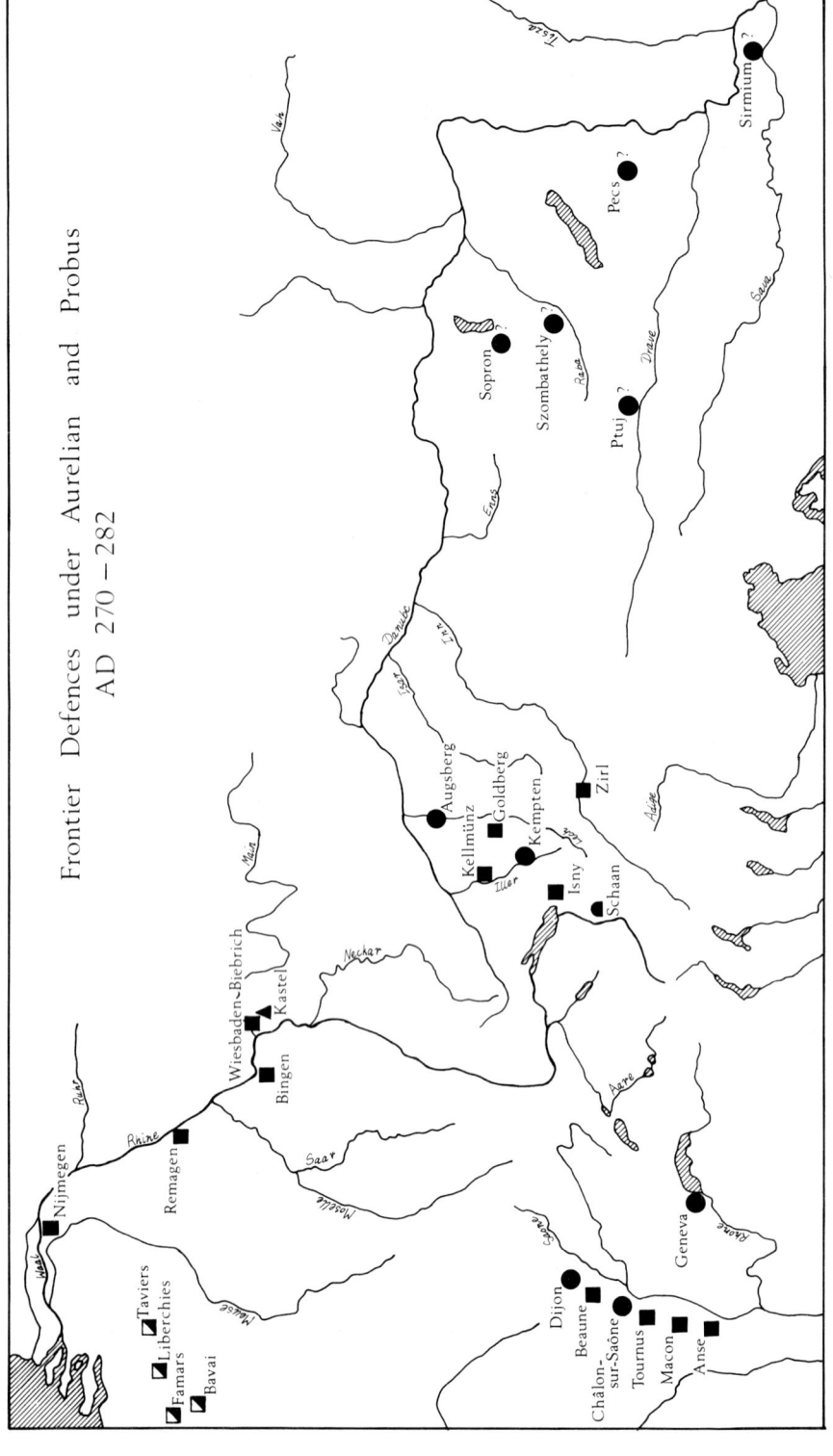

93 The development of frontier defences under Aurelian and Probus (270–82). For key symbols see fig 92, p 248

Few other sites are closely dated, but there was probably other activity in the area to back up this weak point of the frontier.

The largest part of the task facing Probus was to begin the fortification of many of the previously unwalled Gallic cities. Discovery within the foundation courses at Rennes and Nantes of milestones of the Gallic emperors dating from 273-4 suggests that the group of cities in the north-west of Gaul was defended at this period. Similar groups of cities, probably also contemporary, lay in other Gallic provinces. It is scarcely credible that every city in these areas was walled simultaneously, even though the regional groups preserve very much the same general style. The process of ensuring that all Gallic provincial cities had adequate defences probably took several years, only to be completed under the Tetrarchy.

A further defensive scheme, in origin Probus', was the related task of ensuring the safety of the British and Gallic Channel coasts. The seaborne Franks and Saxons who menaced the coastlines of Britain and North Gaul always arrived from the same direction, and in order to penetrate deeply within the empire, had to pass through the Straits of Dover. Therefore, a series of Roman forts and harbours designed to prevent their surprise arrival within the civilian areas of Britain and Gaul was a strategy closely linked to the defence of the civitates of Britain and Gaul as a whole. Thus the 'Saxon Shore' forts came into being—a ring of fortified posts which guarded the major estuaries into Britain and Gaul on their exposed coastlines north-east of the Straits, and by their presence compelled the Saxons to run the gamut of the concentrated forces at the Straits themselves. Provided that Saxons were known to be in the area, the Romans in their own separate fleets could give chase and prevent the Saxons from making the most use of the surprise attacks which were their main weapon.[19]

The Saxon Shore forts and the defence of this quarter of the empire went hand in hand with the protection of the civilian areas of Gaul and Britain. This well explains the otherwise rather awkward similarity between the defences of the Saxon Shore forts and the new walls of the Gallic cities. Planned by Probus as part of the scheme for the protection of the north and western parts of Gaul as much as for Britain, the Saxon Shore forts can be seen as far less of a 'nationalistic' phenomenon peculiar to Britain as has often been thought. The episode of Carausius, therefore, has spotlighted past attention on the Saxon Shore forts, wrongly suggesting that they are somehow connected with the defence of Britain alone, whether against Maximian and Constantius, or against the Saxon raiders. In fact, the defence of the Roman empire at this stage of her history can rarely be so compartmentalised. In this case, the scheme of defence embraced Britain and Gaul, highlighting the interdependence of provinces as they faced up to the barbarians outside.

Though Probus spent little of his time in Gaul, it is surprising that the majority of fortifications which can be assigned to his reign are found in the west. The only epigraphic source to commemorate city walls built in his reign in the east comes from Bostra[20] and even this is of uncertain date. It is possible that Probus was more active in Pannonia, and he may have seen to the defence of the major towns

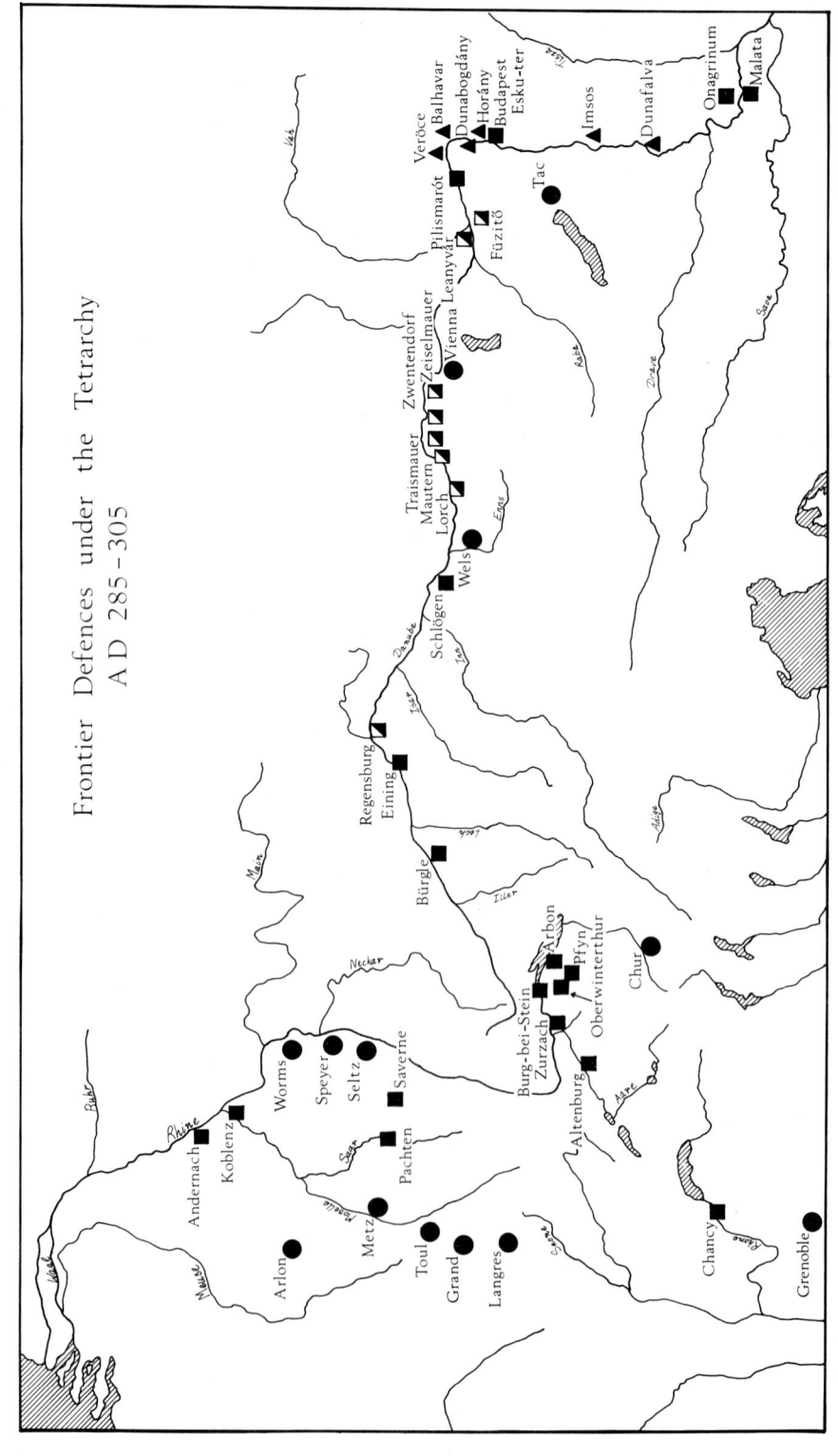

94 The development of frontier defences under Diocletian and the Tetrarchy (285–305). For key symbols see fig 92, p **111**

like Pécs (Sopianae), Sopron (Scarbantia), Szombáthely (Savaria) and Sremska Mitrovica (Sirmium).

The reign of Diocletian and the Tetrarchy was marked by great activity, reflected from many sources, in building forts on the frontier and city walls inland. Several historians recorded this impression of the reign, referring to the main river frontiers of the Rhine, Danube and Euphrates.[21] Malalas even recorded that Diocletian established a chain of new commanders called *Duces* in the forts of the Euphrates frontier.[22]

This impression of frontier activity is well borne out by the map of new defences in the west (fig. 94). The only western sites to have produced inscriptional evidence for their construction at this time are Burg-bei-Stein am Rhein,[23] Oberwinterthur[24] and Grenoble,[25] but many other towns and forts can be assigned to this date on archaeological grounds. To Diocletian and his colleagues can be credited the completion of Probus' building projects in Gaul. The defences of cities in Belgica I and II and Germanica II in particular were probably completed at this period, and their construction went hand in hand with the strengthening of the frontier in the areas where more defence was needed. In Maxima Sequanorum, too, there was a group of small forts, which can be assigned, like Burg-bei-Stein, to a Diocletianic reconstruction. On the Danube, the reconstruction of Regensburg is usually thought to have occurred in the last years of the third century, and may be associated with contemporary rebuilding of the same type at Vienna and at a group of other forts in the area, including Schlögen and Passau.

Even if all the improvement to the defences of Pannonian and Norican forts, which included the addition of projecting towers of fan- and U-shape, cannot be dated to the reign of Diocletian, it is likely that this innovation was now introduced. A Diocletianic date, however, can be assigned to some of the smaller forlets and landing-stations on the 'barbarian' bank of the Danube in the region of Aquincum. These may be associated with the 'Devil's Dyke' fortifications east of the Danube bend, which suggest the maintenance of Roman control in Sarmatian territory. A further indication of Tetrarchic influence on fort building and improvement is the relatively common incidence within the Pannonian area of the names Jupiter (Iovia) or Hercules (Herculia), Diocletian and Maximian's adopted tutelary deities.[26] In addition, one fort was given the name 'Constantia', which must refer to Constantius, the Tetrarchic Caesar, or the later emperor Constantius II.

Farther down the Danube, the cities of Transmarisca[27] and Durostorum[28] received new walls at this period, while at Tomi[29] one of the gates was rebuilt. Further building on the frontiers is attested in the east: Diocletian had at least three forts on the eastern *limes* built, according to Procopius,[30] one of which may have been Cercusium, fortified by Constantius II in his campaigns, but originally built by Diocletian.[31] A fort in Egypt (Manfalut—Hieraconpolis) has also yielded an inscription of 288[32] and Procopius mentions another fort on the Nile which was built at the same time.[33]

Archaeological discoveries in the desert frontiers of the east suggest that a fort-

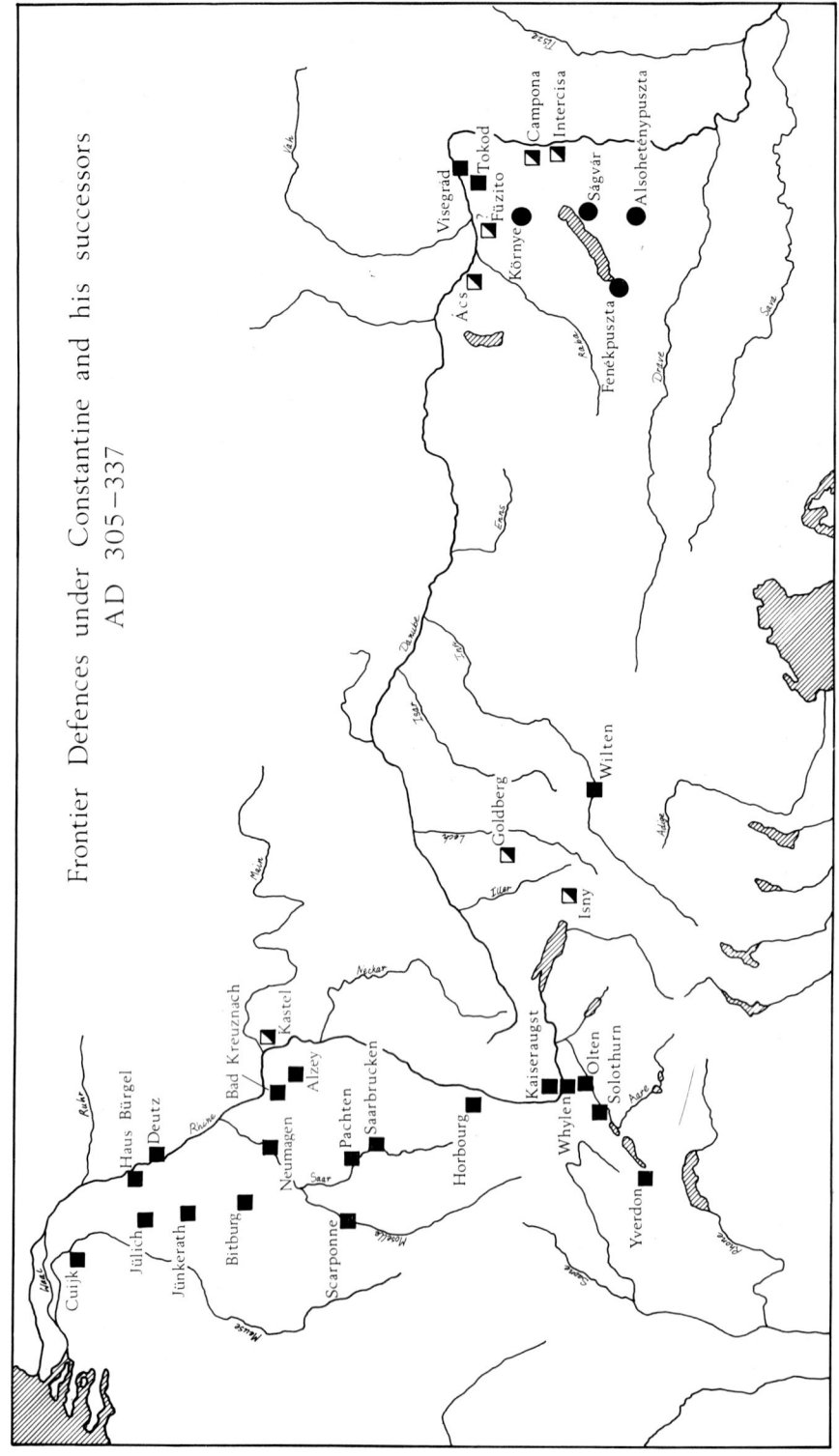

95 The development of frontier defences under Constantine and his successors (305–c350). For key symbols see fig 92, p 248

type in use in the period of the Tetrarchy in these areas was the *quadriburgium*, known also on the African *limites* in Tripolitania and in Israel though it was known at the beginning of the third century (see p. 27). One example of a securely dated fort of this type is Deyr-al-Kahf, in Syria, which has an inscription recording its construction in 306,[34] technically speaking one year too late to belong to Diocletian's reign. Many other sites are in the same mould, for example Han Anneybe, in Syria, and Kasr Bser, in Arabia.[35] Diocletianic building is also attested on the African *limes*, where *centenaria* have yielded inscriptions dating them either to the Diocletianic or Constantinian period. Finally, another inscription from Rapidense in Mauretania attests rebuilding there after barbarian raiding.[36]

The successors of Diocletian thus had an exceptionally strong example to follow, with much of the work of plugging the strategic gaps probably already complete. Zosimus, who compared the policies of the two reigns, accused Constantine of abandoning Diocletian's positive approach to the strengthening of the frontiers and allowing the barbarians unhindered access to Roman territory. He went on to point out that Constantine denuded the frontier of the troops which it most needed by stationing them well back from the front line.[37] His judgement is probably over harsh, for the policy shows well the two features of Constantine's reign most typical of his equally positive approach to the whole problem of frontier management. This was shown in the organisation of the army into its two parts—*limitanei* and *comitatenses*—the one for static defence on the frontier itself, the other for mobility to answer distress calls wherever they might arise. There was also concern to strengthen communications routes near the frontier, and, in effect, provide deeper lying cover in addition to the front line bases.

One feature for which Constantine's reign was better known was the strengthening of more of the bridgehead sites across the Rhine. An inscription, now lost, from Deutz, assigned a Constantinian date to the fort there.[38] The bridgehead fort of Mainz-Kastel appears on the Lyons medallion of Constantinian date. This may indicate its construction then. Recent researches at Kaiseraugst have suggested that the legionary fort there may have been of the same date: it, too, had a bridgehead across the Rhine, at Whylen. Other bridgeheads opposite the more important legionary forts may belong to Constantine's reign rather than Diocletian's: for example the refurbishing of the earlier fort at Leányvár (Celamantia), opposite Szóny (Brigetio). Further east, the construction of a bridge from Oescus to Sucidava across the Danube may have been the occasion for updating the defences of both forts, and the fort of Daphne, opposite Transmarisca, appears certainly to have been a Constantinian foundation. The only town which has produced inscriptional evidence for the reconstruction of the city walls at this period is that of Tropaeum Traiani (Adamklissi). Its walls were completely rebuilt between 315 and 317.[39]

In Pannonia and Noricum, the process of bringing fort defences up to date by the addition of fan-shaped towers at the corner was continued under Constantine. The fort of Visegrád was almost certainly now built. Whether the fan-shaped

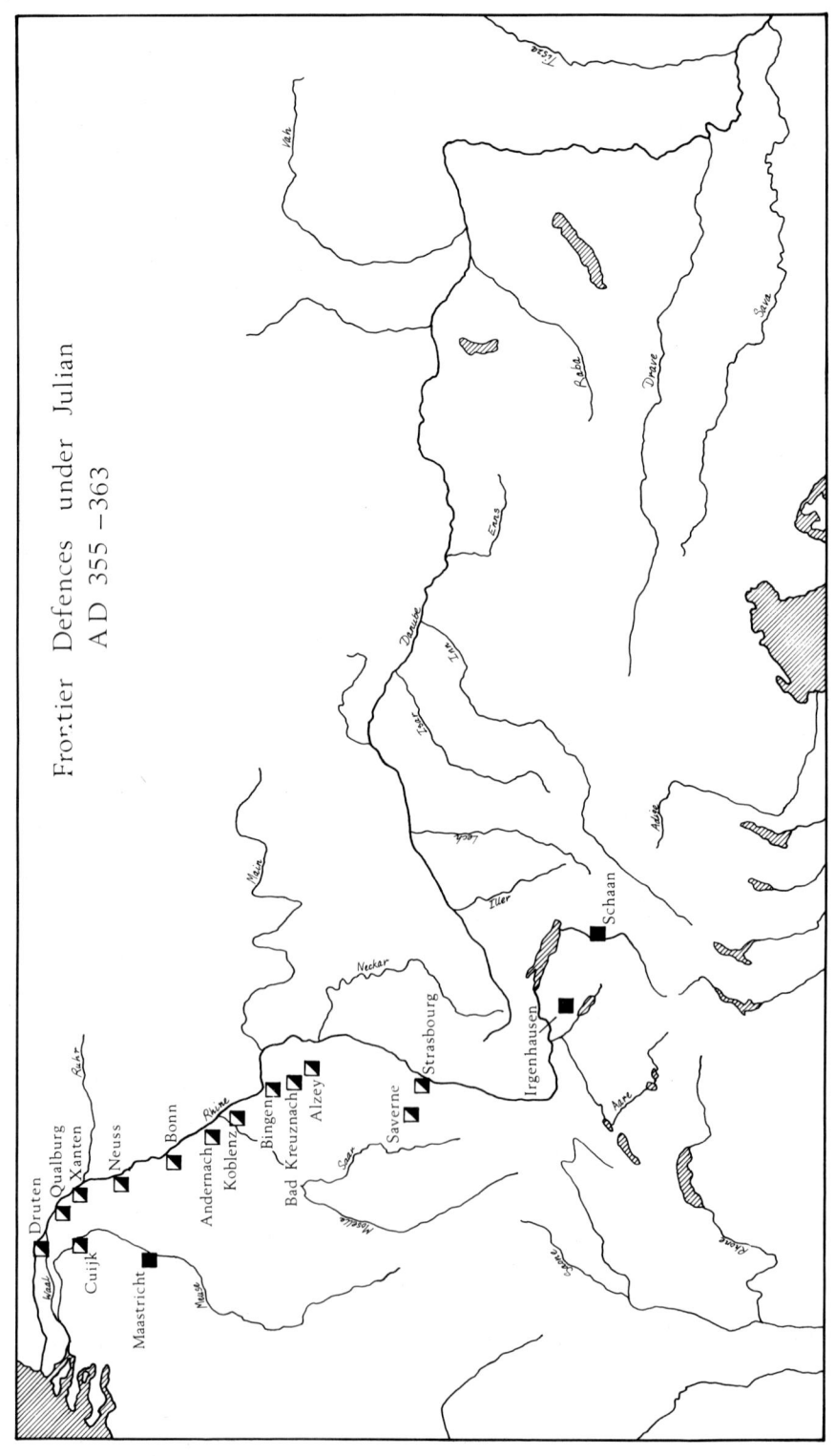

96 The development of frontier defences under Julian (350–3). For key symbols see fig 92, p 248

towers at others of the *limes* forts belong to their reign cannot as yet be determined.

After Constantine's death, there appears to have been no sudden upsurge in the number of new fortifications begun either by Constans or by Constantius II. At Troesmis in 337–40 an inscription records the construction of a fort and an arms depot to keep brigandage under control.[40] Another inscription from the same reign records the recapture of a former fort which had been taken over by 'bandits' who were using it as the base of their operations.[41] Apart from the visit by Constans to Britain,[42] and the possible strengthening of the Saxon Shore frontier by the addition of the fort of Pevensey, there is little trace of any major contribution to the defence of either the western or the eastern empire until the campaigns of Julian, undertaken in response to a more serious bout of barbarian pressure.

The campaigns of Julian are poorly recorded in epigraphic sources, and it is necessary to place heavy reliance on the accounts of Ammianus. Julian's Gallic campaigns were made necessary by the Alemannic invasions of 351–3. Most of the places mentioned in Ammianus' account appear already to have been walled under previous emperors. Many sites, however, needed repairs, among them Saverne (Tres Tabernae),[43] three forts along the banks of the Meuse,[44] various captured sites in lower and upper Germany,[45] and a former Trajanic fort near Mainz which was reoccupied and fortified with *ballistae*.[46]

In the east, Constantius II was no less active at about the same time. To prevent the barbarians from crossing the river he saw to it that the whole of the bank of the Euphrates was fortified with towers and other impediments.[47] Ammianus recorded also that Constantius himself had been responsible for the construction of the walls of Amida and the neighbouring town of Antinonupolis while he was still Caesar.[48] As emperor in 361–3, Julian was equally active in Thrace and along the Danube.[49] The impression gained however from the reading of Ammianus is that neither emperor undertook major schemes to refortify the empire. Both men preferred to patch up and repair existing defences rather than construct new ones.

The last emperors to have undertaken anything like a major refortification particularly along the European frontiers, were Valens and Valentinian. Ammianus recorded that one of Valentinian's major accomplishments was to have fortified the whole of the Rhine[50] and gave instances of the emperor at work in seeing that its defences were adequate. He was present, for example, at Basle when the fort of Robur was being built,[51] and his troops were vigorously opposed when they tried to occupy and establish a base on a hill inside barbarian territory.[52] On the Danube, however, he fared better and was able to establish posts across the river in his concern to fortify the frontiers.[53]

Inscriptional evidence for Valentinianic building of the chain of small towers along both Rhine and Danube is relatively common. In all, five such towers have produced in some cases over-lengthy inscriptions for the size of the towers which were being built. This might suggest that Valentinian was making much of having fortified only a small stretch of the frontier.[54] The number of other towers, however, on the Rhine and the Danube in which tiles stamped with the

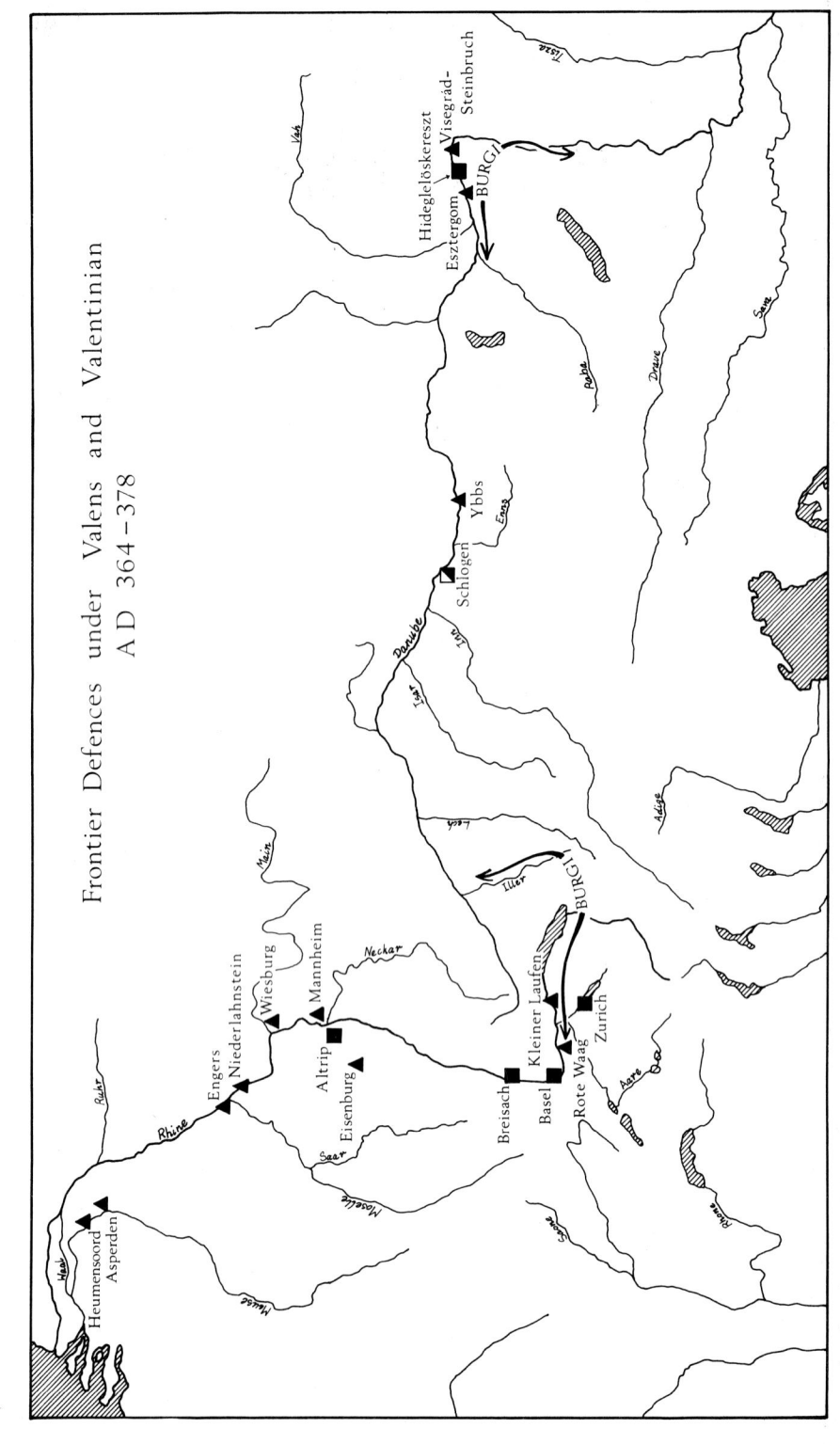

97 The development of frontier defences under Valentinian (364–78). For key symbols see fig 92, p 248

names of Valentinianic military commanders have been found confirms that building work and manning of these posts was common in Valentinian's reign. Although most of the Pannonian landing stations were built earlier, those in the Rhineland—Engers, Mannheim-Neckarau, Rheinbrohl, Niederlahnstein and Zullestein—seem from associated finds to have been built late within the Roman period and probably date from Valentinian's reign. An inscription from the fort of Hidelglelöskerestz on the Danube attests Valentinianic fort building there in addition to the construction of *burgi*.[55] On archaeological grounds the forts of Altrip, Zürich and Breisach have also been suggested as Valentinianic foundations.

Despite this comprehensive activity in all quarters of the empire under the emperors Valens and Valentinian, there in no way appears to be even as complete a plan as Diocletian's to fortify the whole frontier. The chains of *burgi* which are in many ways typical of the Valentinianic period are concentrated only on the upper Rhine and Iller and on a short stretch at the band of the Danube. After the Valentinianic period, there was little further work in building on the frontiers.

In this summary, based on the more securely dated examples of fortification in the later Roman period, some of the trends which ancient writers also recorded are immediately obvious. Only relatively few forts can be dated at all accurately; a number of frontier forts have never been examined even preliminarily, and excavation can rarely give a firm date without some reference to the written sources as well. From this series of maps, the stirrings of frontier policy can be glimpsed. These include the first attempts at an overall policy by Aurelian and Probus, the renewal of the frontiers under Diocletian, the provision of adequate back-up facilities by Constantine, and the plugging of breaches and gaps by Julian, Constantius II and Valentinian.

The development of fortifications during the late Roman period of more than 100 years does not seem to have seen much progress. The first fortifications in the west built in the new style were those of the cities of Gaul and the Saxon Shore forts which can be regarded as an extension of the Gallic building campaign onto the Channel coastline to protect Gaul. These were mostly built or at least started under Probus and Aurelian. The apparent gaucheness of the British Saxon Shore forts is perhaps an indication of the unfamiliarity of this style of architecture as applied to military needs. In Raetia, where the forts of the Danube-Iller-Rhine *limes* can be regarded as their contemporaries, the style of construction was very different, though perhaps, if more survived, the style of construction of forts like Isny would seem no less awkward.

Despite the burst of new fortifications built under Diocletian and Constantine in particular, there are no clear signs of a preferred layout and design which might enable a typological assessment to be made. The fort of *quadriburgium* shape, for example, is attested on several frontiers as a Diocletianic construction. Forts of this shape in the west, however, cannot with complete certainty be assigned to his reign, and at least one, Schaan, is closely dated from coin- and artifact-finds some 50 years later. Conversely, the only two western inscriptions from forts of Diocletianic date come from Burg-bei-Stein and Oberwinterthur, the one a quadrilateral with polygonal towers, the other a very irregular oval with semicircular towers. Despite their striking differences of design, the inscriptions suggest that they were built in the same year.[56] This shows well the dangers and pitfalls of modern attempts at typological dating. Of more use is the suggestion that fortifications within a restricted area were perhaps built according to the same general principles of construction, with uniform tower-types and siting, and standardised sizes of materials. This allows the assumption that fortifications which can be linked to such a regional scheme were contemporary, or very nearly so.

Uniformity of design or of ground plan, then, significant only when found within a restricted area, can seldom be applied to a wider framework to imply contemporaneity of construction. Unfortunately, the defences and in particular the ground plan, have until now been the major factor by which a fort's date is determined. The disposition of internal buildings, probably equally significant, has never been studied in sufficient depth, mainly because a series of stratified levels of different plan showing the various late Roman phases at a fort have not been adequately examined. The study of the plans produced at Drobeta (Turnu Severin) however gives a startling view of the diversity of layouts which may be expected at a single site within the same four walls at different periods.[57] There is no reason, however, why the design of fort interiors should have been any more standardised, or distinctive of any particular late Roman period, than the design of the defences themselves.

In one sense, therefore, late Roman fortifications in the west failed to develop. Within the late Roman period the date of a fort cannot at present be distinguished by the growing complexity of its defences or internal arrangements. City walls also saw little change well into the medieval period. Even hill-top fortifications thrown up rapidly in response to the invasions are not easily distinguished from either Iron Age defences or their Carolingian successors.

For very much the same reason, late Roman fortifications were also a culmination of Roman defensive tactics. Their very lack of uniformity and the triumph of adapting the fort to the terrain led to the discovery and widespread use of a style of fortification with externally projecting towers which was to provide a pattern with slight modification for medieval builders. It was only really superseded in the west by the Italianate towers and the complex artillery fortifications of the late fifteenth and early sixteenth centuries.

The study of later Roman defences throws light on the clash between two separate forms of defensive tactic. On the one hand there was the established Roman method of maintaining strict control of the empire's frontiers by patrolling a visible line with guards who were stationed in permanent posts. On the other, there was a more negative approach which protected main nodal points from an invading force by investing them with the strongest walls the Roman military architects could produce. The pressures of the late third and fourth centuries sealed the downfall of the frontier system from its first spirited defence under Diocletian to its neglect and final abandonment, at least in some areas, under Honorius and Arcadius. Interwoven with this trend was the gradual and corresponding reliance on the static defence, from its beginnings under Aurelian and Probus, to the time, late in the fourth century, when, despite the influx of barbarians, the majority of walled cities were sufficiently well defended and formidable to escape widespread mutilation and survive intact.

Appendix 1 SELECTED GROUPS OF GALLIC CITY WALLS

LUGDUNENSIS I

	Size and shape	Thickness of wall	Size of facing (in mm)	Details of construction	Tower types	Gate types	Dating
LYON	Triangular						
AUTUN	Irregular reduced circuit, 10ha, originally 200	2.5m	110 × 110 blocks	Re-used material in blocking wall between extended town and citadel	Circular, astride wall	Augustan gates, double portal with flanking towers	Augustan circuit, reduced
LANGRES		3.35m				Archway of C2 included in circuit	Built by 300 (Eutropius, *Brev.* 1,9)
CHALON-s-SAÔNE	D-shaped, 11ha	2.5–3.5m	Small blockwork ?tile-courses		Circular, astride wall, 6.8m diam.		
MÂCON	Irregular, *c*10ha						
DIJON	Roughly circular, 11ha	2.4m	Medium-sized blockwork		Circular astride wall	Postern gate protected by bastion	Attributed to Aurelian
BEAUNE	Roughly circular, *c*2ha	up to 5m?	Medium-sized blockwork, not uniform		Circular?		
TOURNUS	D-shaped, 1.5ha	3.9m		Wall said to be constructed between wooden shuttering	Circular astride wall		
ANSE	Roughly oval, 1.5ha	3.5m	Small blockwork, tile-courses		Circular astride wall		

LUGDUNENSIS II

	Size and shape	Thickness of wall	Size of facing (in mm)	Details of construction	Tower types	Gate types	Dating
ROUEN	?Rectangular, c8ha	1.32m?	Blocks; 110 × 120–160; tiles: 420 × 320 × 40, and 370 × 280 × 40		U-shaped, projecting in towers?		
BAYEUX	Rectangular, rounded corners, c9ha		Blocks: 110 × 110; tiles: 310 × 45		U-shaped projecting		
AVRANCHES	Oval?						
EVREUX	Rectangular, rounded corners, 10ha	2.90–4m 3.1m	Blocks: 110 × 100–180; tiles: 370 × 40	Tile-courses go right through wall	U-shaped projecting?		
LISIEUX	Rectangular, rounded corners, 8ha	2.5m	Blocks: 110 × ?; tiles: 180 × 180	Built into side of steep hill	U-shaped projecting	Gate shown on plan protected by 2 U-shaped towers	
COUTANCES	Oval						
LILLEBONNE	Rectangular, 1.5ha	2.5m	Tiles: 340 × 250 × 35	Built onto theatre at foot of steep hill	U-shaped towers?		

LUGDUNENSIS III

	Size and shape	Thickness of wall and facing type	Size of tiles (in mm)	Details of construction	Tower types	Gate types	Dating
TOURS	Rectangular, includes amphitheatre, 9.38ha	4.3m small blockwork, widely spaced double tile-courses	340 × 240 × 50 330 × 410 × ?	Amphitheatre incorporated on S side	¾-round towers at corners, semicirc. elsewhere	2 posterns known, one with simple tiled arch	
LE MANS	Rectangular but odd. c10ha	4m small blockwork, closely spaced triple tile-courses	340 × 240 × 50 320 × 270 × 45	Putlogs 1.5m apart horiz., in tile-courses and above. Patterning in chalk blocks in facing	U-shaped projecting from wall. One tower pentagonal. Windows at rampart walk height	Posterns with tile arches. ?Main gate in re-entrant with projecting towers	*tpq* of early C3
RENNES	Irregular oval, c10ha	3.5m, small blockwork, closely spaced triple tile-courses	390 × 280 × 50 410–490 × 330 × 30	Patterning in chalk blocks in facing. Putlogs 1.75 apart, in tile-courses	No towers known	Postern roofed with single slab of stone and relieving tile arch	*tpq* of 273, from series of milestones
ANGERS	Irregular, oval, c9ha	4m small blockwork, closely spaced triple tile-courses	400 × 280 × 50 300 × 200 × 35		Projecting U-shaped towers, and small rectangular buttresses	Gate has two projecting U-shaped towers	

NANTES	Pentagonal, c18ha	3.8–4.5m small blockwork, closely spaced triple tile-courses	400 × ? × 50 310 × ? × 40	Projecting U-shaped towers 7.5 m in diam. and 30m apart	Gate on plan shown with two projecting U-shaped towers	*tpq* of 275, from two re-used milestones
VANNES	Kite-shaped, c5ha	4m small blockwork, widely spaced triple tile-courses	330 × 30	Projecting U-shaped towers	Putlogs 1.35 apart horiz. and 1.5m vert.	
ALETH	Irregular rectangular?	1.5–1.95m small blockwork, no tiles		Rectangular buttresses	Round or U-shaped towers: single portal	Late C3?
JUBLAINS	Trapezoidal, 1.2ha	4.5m, small blockwork, trible closely spaced tile-courses	400 × 275 × 40	U-shaped and ¾-round towers. One rectangular	Postern gates defended by corner towers	Putlogs 1.4m apart horiz., in tile courses and above
RUBRICAIRE	Square, 0.3ha	2.06m, small blockwork only		Corner towers c 7m square		

LUGDUNENSIS IV (SENONIA)

	Size and shape	Thickness of defences	Size of facing (in mm)	Details of construction	Tower types	Gate types	Dating
SENS	Oval, 45ha	3m	Blocks: 110 × 110–400; tiles: 420 × 280 × 50	Patterning in the facing blocks resported, but not now visible	U-shaped, projecting from wall	Postern with tile-stone arches	*tpq* of 268 from coins in wall-core
AUXERRE	Irregular quadrilateral, 5.5ha	?4m			U-shaped, projecting from wall, with large windows at rampart-walk level	Posterns with tiled arches	
TROYES	Rectangular, c16ha						
ORLÉANS	Rectangular, 22ha		Tiles: 'large'		U-shaped, projecting from wall on plan		
PARIS	Oval, c8ha	2.79–2.5m		On island in R. Seine			
MEAUX	D-shaped ?, 10ha		Blocks: 110 × 110; tiles: 330 × 25 and 140 × 25		U-shaped, projecting		
MELUN	Oval, c3ha	2.35–2.5m	200 × 120	On island in R. Seine			

BELGICA II

	Size and shape	Thickness of wall	Size of tiles (in mm)	Details of construction	Tower types	Gate types	Dating
SOISSONS	Rectangular, 12ha	3–4m	200 × 35, 500 × 400 × 70	Putlogs always in row above tile-courses	Square corner tower. ? U-shaped interval towers	2 gates presumed, but no details known	
SENLIS	Oval, 5.6ha	3.3m	400 × 300 × 25	Putlogs always in row above tile-courses	U-shaped, astride walls. 3m radius and 3m external projection	6 known: 4 defended by single tower 1 by double towers 1 by none	
BEAUVAIS	Polygonal, 10ha	2.5–2.6m	410 × 260 × 30, 570 × 570 × 80	Putlogs in row above tile-courses	U-shaped astride walls, 3m radius. Corner towers c9m square		tpq of 286 from coins found in W foundation
AMIENS	Polygonal, 6ha?	3.6m		Amphitheatre built into circuit. Wall on piled foundation	?Rectangular gate tower, 8m by 5.5m	Gates flank the amphitheatre	tpq of 278 from coins and from C3 well
BOULOGNE	Rectangular, 12.6ha	3.2–3m					
NOYON	Irregular oval, c4ha		460 × 360 × 50, 433 × 325 × 27		Semicircular or round interval towers		
BAVAI	Rectangular, 4ha	Double wall 3m and 1.5m	260 × 360 × 40	Putlogs in row above tile courses. Wall on piled foundation	Semicircular, ¾-round at corners	Gates defended by single towers	
FAMARS	Roughly rectangular, 1.8ha	Double wall, 1.8 or 1.2; 2.3 or 2m			Semicircular, ¾-round at corners		
OUDENBURG	Rectangular, 2.3ha	1.8–2m			Round towers at corners, 9m diam. Octag. at gates		Stone fort late C3 or mid-C4?

AQUITANICA I

	Size and shape	Thickness of wall	Size of facing (in mm)	Details of construction	Tower types	Gate types	Dating
BOURGES	Irregular oval, c26ha	2.5–3m	Blocks: 120 × 180, 90 × 120; tiles: 400 × 30, 260 × 40, 310 × 70	Putlogs 0.8–1.8m apart. Irregular placing	Semicircular projecting, with large windows at rampart walk height	One gate flanked by semicircular towers with moulded pilasters in opus quadratum	
LIMOGES	Roughly circular, 10ha			Re-used material	?Circular tower		

AQUITANICA II

	Size and shape	Thickness of wall	Size of facing (in mm)	Details of construction	Tower types	Gate types	Dating
BORDEAUX	Rectangular 23.4ha	4–5m		Wooden beams in foundation course and wall on piled foundation	Semicircular projecting towers		tpq of 268 from coins in wall
SAINTES	Irregular quadrilateral, c10ha	3m–3.6m		Re-used material in foundations	Semicircular towers		
POITIERS	Irregular, 42ha	5.4m			Semicircular, 25–34m apart		
PÉRIGUEUX	Oval, 5.5ha	4–6m		Re-used material in foundations. Patterning in tile cubes and small blockwork	Semicircular, 5m diam.	Porte de Mars, flanked by two semicircular towers with moulded pilasters	

NOVEMPOPULANA

	Size and shape	Thickness of wall	Size of facing (in mm)	Details of construction	Tower types	Gate types	Dating
AUCH	Oval, 3.6ha		Blocks: 110 × 110, 90 × 110–170; tiles: 200 × 40		Semicircular projecting towers		
DAX	Rectangular, ?12ha	4.25–4.5m	Blocks: 100 × 120; tiles: 280–400 × 40	Timber framing in foundations. Drainage outlets	Semicircular projecting towers, not bonded with wall. 28–30m apart	Postern of large stone blocks with tile arch above	
LECTOURE	Oval						
ST. BERTRAND	Irregular, 4.5ha		Blocks: 110 × 130–300; tiles: 270 × 210 × 40	Tegulae in wall. Drainage outlets	Semicircular projecting, 4.5 diam.	Simple posterns with one defending tower	
ST. LIZIER	Irregular, 3.5ha		Blocks: 110 × 110–240; tiles: 340 × 240 × 40	Drains outlets in walls	Semicircular projecting and rectangular towers		
LESCAR	Irregular triangle, 2.64ha			Re-used material from walls			
BAZAS	Irregular, ?c12ha	2m?					
OLORON	Irregular quadrilateral, ?3ha						
BAYONNE	Irregular, 6ha	3.4–3.5m	large blocks		Semicircular projecting, 40m apart		

Appendix 2 LATE ROMAN WATCHTOWERS *(BURGI)* IN THE RHINE AND DANUBE AREAS

The numbers given to the sites refer to the series of maps, figs 51, 62, 70, and 72

Site	Description	Dating	References
1 HEUMENSOORD	Timber palisade 13m square, 2 periods of use	Late C3 and C4 use: coins and pottery	Bogaers & Rüger, 1974, 81 (see fig. 53, p. 140)
2 ASPERDEN	Outer bastioned wall, 40m square, round tower 15.6m square	Late C4 coins and pottery	Bogaers & Rüger, 1974, 100 (see fig. 53, p. 140)
3 RHEINBERG	Timber tower 4.4m square surrounded by parallelogram ditch 30 × 34m	C1 or C2 site, used into C4	Bogaers & Rüger, 1974, 125 (see fig. 53, p. 140)
4 MOERS-ASBERG	18m square tower built over SE corner of early fort	?Valentinianic	Bogaers & Rüger, 1974, 128 (see fig. 53, p. 140)
5 GRUBBENVORST-LOTTUM	Possible *burgus*	Late Roman finds	Bogaers & Rüger, 1974, 88
6 STOKKEM	Large stone blocks. Possible *burgus*?		Bogaers & Rüger, 1974, 153
7 VALKENBURG	Watchtower on site of a villa, surrounded by bank	C4 date from pottery and coins	Schönberger, 1969
8 HULSBERG-GOUDSBERG	Tower 12.2 × 8.8m in larger enclosure, surrounded by ditch	C3–C4 coins and pottery	Bogaers & Rüger, 1974, 177 (see fig. 53, p. 140)
9 GROSS KÖNIGSDORF	Rectangular earth/timber bank surrounded by ditch 31 × 26m. Interior building of wood	C2–C3 coins, pottery	Bogaers & Rüger, 1974, 157
10 VILLENHAUS	Earth/timber fortlet, 2 periods. Area inside ditch 26 sq m	2 periods, both in late C3	Bogaers & Rüger, 1974, 180 (see fig. 53, p. 140)
11 FROITZHEIM	Group of 3 *burgi*, (a) 55 × 53m, (b) similar, (c) 45 × 30m	Period of use late C3–late C4	Bogaers & Rüger, 1974, 208 *RA* 3, 9f.
12 RÖVENICH			Heimberg, *Ausgrabungen im Rheinland* (1975) 50–1
13 RHEINSROHL	Rectangular tower, with walled courtyard protecting river harbour	C4 date?	Schönberger, 1969
14 ENGERS	Rectangular tower with walled courtyard protecting river harbour	C4 date	*Germania* 30 (1952), 115f (see fig. 54, p. 141)

270

Site	Description	Dating	References
15 NIEDERLAHNSTEIN	Rectangular tower, with walled courtyard protecting river harbour		*NA* 61 (1951), 8 Schönberger, 1969
16 WIESBADEN-BIEBRICH	Rectangular tower	Valentinianic date	*FHess* 4 (1964) 224
17 ZULLESTEIN	Rectangular tower, with walled courtyard protecting river harbour		*AKorr* 3 (1973) 75f (see fig. 54, p. 141)
18 MANNHEIM NECKARAU	Rectangular tower 23 × 16m with walled courtyard		*Germania* 26 (1942) 191f (see fig. 54, p. 141)
19 EISENBERG	Blockhouse 16 × 20m; 6 rooms inside		Sprater, 1929, I, 55f (see fig. 54, p. 141)
20 STRASBOURG ST NICHOLAS	Roughly rectangular, 30 × 22m	Not dated	*CAHA* 132 (1952), 63f (see fig. 54, p. 141)
21 BASEL-BRÜCKENKOPF	13m square, walls 3.95 thick, ¾-round towers at corners	Valentinianic?	*AKorr* 4 (1974), 161f
22 STERNENFELD	8.6m square, walls 1.50m thick		Stehlin, 1957, no. 1
23 AU HARD	Tower 8.5 × 8.65m, walls 1.65m thick		Stehlin, 1957, no. 2
24 WHYLEN	Bridgehead opposite Kaiseraugst. Remains of 3 towers only		Anthes, 1917, 131f
25 PFERRICHGRABEN	11.4 × 11.6m, walls 1.70m thick	Small finds from site of Valentinianic date	Stehlin, 1957, no. 3
26 HEIMENHOLZ	10 × 10m		Stehlin, 1957, no. 4
27 RYBURG	10 × 10m, walls 1.6m thick		Stehlin, 1957, no. 5
28 FAHRGRABEN	9 × 9m, walls 1.5m thick		Stehlin, 1957, no. 6
29 UNTERE WEHREN	9 × 9m, walls 1.6m thick	Small finds from site of late C4 date	Stehlin, 1957, no. 7
30 STELLI	17.5 × 17.5m, walls 2.3–2.4m thick	Small finds of late C4 date	Stehlin, 1957, no. 8
31 UNTER DER HALDE	8.6 × 11.3m, walls 1.4m thick		Stehlin, 1957, no. 9
32 WALLBACH	8.95 × 8.95m, walls 1.5m thick		Stehlin, 1957, no. 10
33 MUMPF	17.5 × 26m, walls 2.5m thick. Site surrounded by ditch and includes a separate bath-house	Coins of Gratian Magnus Maximus and pottery of C2 but mostly C4	Stehlin, 1957, no. 11 (see fig. 65, p. 164)
34 SISSELN	18.5 × 26.5m, walls 3.5m thick	Some fragments of pottery C4 date	Stehlin, 1957, no. 12 (see fig. 65, p. 164)
35 KAISTERBACH	Walls 1.45m thick		Stehlin, 1957, no. 13
36 LAUFENBURG			Stehlin, 1957, no. 14
37 SANDRUTI	Walls 1.55–1.65m thick		Stehlin, 1957, no. 15
38 HAUENSTEINER FAHRE	7 × 7m, walls 1.2–1.5m thick		Stehlin, 1957, no. 16
39 ROTE WAAG		Site discovery of *CIL* XIII 11538 dating construction of *burgus* to 371	Stehlin, 1957, no. 17

Site	Description	Dating	References
40 UNTERES BÜRGLI	17.6 × 17.9, walls 2.2–2.5m thick	Coin finds of C4; pottery seems to be uniformly of C2–early C3 date	Stehlin, 1957, no. 18
41 OBERES BÜRGLI	7.5 × 7.5m, walls 1.25m thick	Coins of Constantine; pottery of C3–4	Stehlin, 1957, no. 19
42 BERNAU			Stehlin, 1957, no. 20
43 JUPPE	9.2 × 9.8m, walls 1.4–1.8m thick	Sparse pottery finds, probably late C3–C4	Stehlin, 1957, no. 21
44 IM SAND	8.2 × 8.2m, walls 1.2–1.3m thick		Stehlin, 1957, no. 22
45 KLEINER LAUFEN	8.05 × 8.05m, walls 1.6–1.8m thick	Inscription *CIL* XIII, 11537, dating to 371, and recording name of *burgus* 'SUMMA RAPIDA'	Stehlin, 1957, no. 23
46 RHEINHEIM	50m square	Square fortlet with projecting rectangular corner towers	Filzinger and others, *Die Römer im Baden-Wurttemburg* (1976), 459f
47 OBERFELD	7 × 7m, walls 1.08–1.59m thick		*JSGU* 15 (1923), 110
48 RECKINGEN			*JSGU* 14 (1922), 87
49 RHEINZELG			*JSGU* 14 (1922), 87
50 MELLIKON	7–8 × 7–8m, walls 1.5m thick		*JSGU* 14 (1922), 88
51 SANDGRABEN	7.9 × 7.9m, walls 1.5m thick		*JSGU* 32 (1940–1), 150; *US* 18 (1954), 10
52 BLEICHE	9 × 9m, walls 1.5m thick		*JSGU* 14 (1922), 88
53 LEBERN	7.05 × 7.8m, walls 1.1m thick		*JSGU* 14 (1922), 88–9
54 HARD	7 × 7.8m, walls 1.6m thick		*JSGU* 14 (1922), 86
55 SCHLOSSBUCK	8.9 × 8.9m, walls 1.7m thick	Finds dating to end of C4	*US* 18 (1954), 10
56 TÖSSECK	8.1 × 8.1m, walls 0.9m thick		*JSGU* 15 (1923), 110
57 EBERSBERG			*JSGU* 15 (1923), 111
58 STRICK	9.92 × 9.92m, walls 1.7m thick	Late Roman coin-finds perhaps show use in Valentinianic times	*US* 18 (1954), 10
59 MANNHAUSEN			*JSGU* 15 (1923), 109
60 LINDENBUCK	13 × 13m, walls?		*JSGU* 18 (1926), 108
61 SCHAARENWEISE	6.3 × 7m		*JSGU* 10 (1917), 75
62 GALGENHOLZ			*BRGK* 8 (1917), 108
63 RATIHARDT			*JSGU* 12 (1919), 119
64 DIESSENHOFEN			*JSGU* 47 (1958–9), 187
65 BURGSTEL	8.7 × 8.7m, walls 2.6 thick		*JSGU* 10 (1917), 75; 29 (1937), 99
66 HÖRBRANZ	11.8 × 12m, walls 1.55m thick	Coin of Theodosius	Garbsch 1967, no. 43
67 GWIGGEN			Garbsch 1967, no. 44
68 HOHENWEILER			Garbsch 1967, no. 45
69 BURGSTALL			Garbsch 1967, no. 46
70 WALDBURG			Garbsch 1967, no. 47

APPENDIX 2

Site	Description	Dating	References
71 UMGANGS			Garbsch 1967, no. 48
72 OPFENBACH			Garbsch 1967, no. 49
73 MELLATZ			Garbsch 1967, no. 50
74 MECKATZ	11.8 × 12m, walls 1.5m thick	Small finds and pottery of Valentinianic date	Garbsch 1967, no. 51
75 HEIMENKIRCH			Garbsch 1967, no. 52
76 DREIHEILIGEN	11.8 × 12m, walls 1.3–1.4m thick	Valentinianic?	Garbsch 1967, no. 53
77 OBERHÄUSER			Garbsch 1967, no. 54
78 NELLENBRUCK			Garbsch 1967, no. 55
79 WENK	11.5 × 11.5m, walls 1.8m thick	Valentinianic?	Garbsch 1967, no. 56
80 BUCHENBERG			Garbsch 1967, no. 57
81 AHEGG	12.2 × 12.2m, walls 1.3–1.4m thick		Garbsch 1967, no. 58
82 STIELINGS		Valentinianic?	Garbsch 1967, no. 59
83 HEISING			Garbsch 1967, no. 60
84 OBERRIED			Garbsch 1967, no. 61
85 HORENSBERG	11.2 × 11.2m, walls 1.8m thick		Garbsch 1967, no. 62
86 WALDEGG			Garbsch 1967, no. 63
87 RAUPOLZ			Garbsch 1967, no. 64
88 WORINGEN			Garbsch 1967, no. 65
89 DICKENREIS	13 × 13m, walls 1.8m thick		Garbsch 1967, no. 66
90 MEMMINGEN	12.8 × 12.8m, walls 1.8m thick		Garbsch 1967, no. 67
91 SENNHOF	10–11 × 10–11m, walls?		Garbsch 1967, no. 68
92 BELLENBERG			Garbsch 1967, no. 69
93 FINNINGEN	11 × 11m, walls 1.3m thick		Garbsch 1967, no. 70
94 STRASS	Walls 1m thick		Garbsch 1967, no. 71
95 WAIZENRIED			Garbsch 1967, no. 72
96 FÖRGHOF	12.6 × 12.6m, walls 1.5–1.6m thick	Valentinianic?	Garbsch 1967, no. 73
97 BAISWEIL	12.6 × 12.8m, walls 1.5m thick	Coins date to late C4, but late C3 period	Garbsch 1967, no. 74; Oblenroth *BRGK* 29 (1939), 122f
98 SCHLINGEN			Garbsch 1967, no. 75
99 LAUTERBACH			Garbsch 1967, no. 76
100 MÜHLHART	10 × 10m, walls 1.18m thick		Garbsch 1967, no. 77
101 KREUT	10 × 10m, walls 0.8–1m thick		Garbsch 1967, no. 78
102 WEICHERING	10 × 10m		Garbsch 1967, no. 79
103 STÜRMENKOPF			von Uslar 1964, 318; *US* 32 (1968), 17
104 PORTIFLÜH			*JSGU* 44 (1954–5), 120
105 MANDACHER EGG			*JSGU* 22 (1930), 91
106 BAD GÖGGING			Garbsch 1967, no. 80
107 EINING			Garbsch 1967, no. 81

APPENDIX 2

Site	Description	Dating	References
108 EICHBERG			Garbsch 1967, no. 82
109 FRAUENBERG	Rectangular, 41 × 15m, Tower at N end, flanking entrance	Glass sherd, dated 350–400 found in wall mortar	K. Spindler, *Die Archaeologie des Frauenberges* (1981), 102f
110 UNTERSAAL	17.2 × 17.2, walls 1.95–2m thick	Pottery of uniform late C4 date	Garbsch 1967, no. 83 (see fig. 65, p. 164)
111 ABBACH			Garbsch 1967, no. 84
112 OBERNDORF			Garbsch 1967, no. 85
113 REGENSBURG			Garbsch 1967, no. 86
114 UNTERIRADING			Garbsch 1967, no. 87
115 STADTWALD			Garbsch 1967, no. 88
116 OBERRANNA	12.5 × 17m, walls 1.5m thick	Round tower defends s wall, ?2-period occupation	*PAR* 1960, 26f; *RFS* 12, 589
117 WILHERING	Nearly square 9.9 × 9.95m	Late Roman finds	*RFS* 12, 589
118 AU	? square	Late Roman	*RFS* 12, 590
119 YBBS		Site of discovery of *CIL* XIII, 5670, dating construction of *burgus* to 370	
120 SPIELBERG	Round tower? 15m diam.	Late Roman	*RFS* 12, 591
121 BACHARNSDORF	Rectangular 12.2 × 12.4m	Late Roman	*RFS* 12, 591
122 ROSSATZ	12.4m square	Late Roman	*RFS* 12, 591
123 MARIA PONSEE	6m square	C2–C3 date?	*RFS* 12, 592
124 GREIFENSTEIN			*RFS* 12, 59
125 WIEN-OBERDOBLING		Tile stamps of URSICINUS LEG X, LEG XIV and others	*RLiO*, 1949, 175
126 WIEN-GUMPENDORF			*RLiO* 1949, 173
127 HOFLEIM			*RLiO* 1900, 52
128			*RLiO* 1900, 52
129			*RLiO* 1949, 22–3
130 ST MARGARETHEN	9 × 15.7m, walls 0.6m thick	Internal rooms. Occupation in late C3–C4	*OJh* 1948, Bb 283f
131 SAUERBRUNN			Graf, 1936, 72
132 BEZENYE	Possible tower: tile and pottery finds		Fitz (ed) 1976, 13
133 MARIAKÁLMOK	Rectangular walled enclosure 6 × 3.5m: landing station?		Fitz (ed) 1976, 17
134 MOSONMAGYAR-OVAR	Possible tower?		Fitz (ed) 1976, 17
135 HORVATKIMLE	Traces of watchtower		Fitz (ed) 1976, 17
136 OTTEVENY	Stamped tiles of Leg. XIII. Possible tower		Fitz (ed) 1976, 20
137 KUNSZIGET	On Danube bank, rectangular enclosure 8m long: landing station?		Fitz (ed) 1976, 21

APPENDIX 2

Site	Description	Dating	References
138 GYÖR-ESZTERGETO	Tower? 10m square	Samian dating to C2–C3	Fitz (ed) 1976, 25
139 GYOR-SZENTIVAN	Possible tower		Fitz (ed) 1976, 25
140 GONYU	Tower? 10m square, building debris. Tile stamp of Leg. XIIII		Fitz (ed) 1976, 25
141 ACS-PAPISTA	Large watchtower on small promontory into Danube. Building debris		Barkoczi, *Brigetio*, p. 6
142 KOMAROM-MILCH	Tower 9.55m square, central pillar walls 1m thick, entrance to s		Fitz (ed) 1976, 31
143 KOMAROM FERIENHAUS-GYURKI	Tower recorded in 1942: no trace remains		Fitz (ed) 1976, 31
144 KOMAROM	Possible watchtower		Fitz (ed) 1976, 31 no. 4
145 KOMAROM	Tower? 16 × 14m, building debris		Fitz (ed) 1976, 31 no. 6
146 IZA	Tower	Possibly C2 in date	*TIR* L 34
147 SZONY KURUCDOMB	Tower 7.7m diam., part excavated	Valentinianic	Fitz (ed) 1976, 37; *AErt* 47 (1934) 139f
148 ALMASFUZITO	Unexcavated tower		Fitz (ed) 1976, 37
149 PUSZTADOMB	Rhomboid tower 16 × 17m	Late Roman	Fitz (ed) 1976, 41
150 ALMASFUZITO-BULCSIKBRUCKE	Unexcavated tower		Fitz (ed) 1976, 41
151 DUNAALMAS	Possible tower under Reformed Church		Fitz (ed) 1976, 41
152 NESZMELY	Tower 7.5m square, entrance to s surrounded by ditch	Coins from Constantine-Valentinian; late Roman tile-stamp	Fitz (ed) 1976, 41; *AErt* 89 (1962), 142f
153 NESZMELY KATALIN-BERG	Remains of round building with triple surrounding ditch	Coins of Constantius II and Valens	Fitz (ed) 1976, 41
154 LABATLAN-PISZKE	Partly destroyed tower	Late Roman tile-stamp	*AErt* 87 (1960) 209
155 NYERGESUJFALU	Rectangular tower, now destroyed	C2 date, but also late Roman	Fitz (ed) 1976, 45
156 ESZTERGOM ZSIDOD	Rectangular tower 8.65 × 8.7m, walls 1.1m thick, ditch surrounds it	Late Roman tile-stamps	*AErt* 87 (1960), 207f
157 ESZTERGOM SZENTKIRALY	Under remains of medieval abbey		Fitz (ed) 1976, 49
158 ESZTERGOM SZENTGYORGY-MESO I	Rectangular tower 7.18 × 7.22m, walls 1m thick, surrounded by ditch	Late Roman tile-stamps. Valentinianic	Soproni 1978, 21f
159 ESZTERGOM SZENTGYORGY-MESO II	Part destroyed, on river bank: rectangular 8.25m × ?	Late Roman stamped tiles. Valentinianic	Soproni 1978, 24
160 ESZTERGOM SZENTGYORGY-MESO III	Part destroyed, on river bank: rectangular 7.9 × ?	Late Roman tile-stamps. Valentinianic	Soproni 1978, 24–5

Site	Description	Dating	References
161 ESZTERGOM SZENTGYORGY- MESO IV	On flood plain of Danube almost completely destroyed	Late Roman tile-stamps	Soproni 1978, 25
162 ESZTERGOM DEDA I	Tower, c 10 × 10m, destroyed		Soproni 1978, 25
163 ESZTERGOM DEDA II	Rectangular tower, building debris	Late Roman pottery, tile-stamp. Valentinianic??	Soproni 1978, 25
164 ESZTERGOM BUBANAT-VOLGY I	Largely destroyed, rectangular tower	Late Roman pottery. Valentinianic?	Soproni 1978, 26
165 ESZTERGOM BUBANAT-VOLGY II	Rectangular tower, 7.14 × 7.18m, walls 1m thick		Soproni 1978, 26
166 HELEMBA	Tower site? 10 × 10m: building debris		Soproni 1978, 77
167 PILISMAROT BASAHARC I	Tower site? 10 × 10m, building debris	Late Roman coins and pottery	Soproni 1978, 30
168 PILISMAROT BASAHARC II	Largely destroyed by Danube. Walls 1.05–1.1m thick	Late Roman tile stamps. Valentinianic	Soproni 1978, 30–1
169 PILISMAROT BASAHARC III	Partly destroyed; tower 10 × 10m square		Soproni 1978, 31
170 SZOB	Large tower, 17.6 × 9.5m, spur walls 2.8m thick, down to Danube	Late Roman tiles. Diocletianic?	Soproni 1978, 77–8
171 PILISMAROT BASAHARC IV	Rectangular watchtower, 10 × 10m	Late Roman tiles. Valentinianic	Soproni 1978, 31
172 PILISMAROT BASAHARC V	Rectangular? tower, building debris		Soproni 1978, 32
173 PILISMAROT DUNA	Spread of tile and building rubble 10 × 10m		Soproni 1978, 32
174 PILISMAROT DUNA MELLEKE II	Rectangular tower, 8 × 8.07m, walls 1m thick	Late Roman tiles. Valentinianic	Soproni 1978, 32–3
175 PILISMAROT SCHIFFSTATION	Rectangular tower 8.59 × 8.62m, walls 0.96m thick, surrounded by ditch	Late Roman tiles. Valentinianic	Soproni 1978, 33–6
176 PILISMAROT MALOMPATAK	Small fortlet, tower with central pillar 12.35 × 12.25m. Surrounding wall and ditch. On Danube side of tower, a small 4-roomed house	Coins from Constantius II. Valentinianic. Late Roman tile-stamps. Valentinianic	Soproni 1978, 36–46 (see fig. 74, p. 187)
177 DOMOS KOVESPATAK	Rectangular tower 8.87 × 8.95m, surrounded by ditch	Valentinianic coins and C4 pottery	Soproni 1978, 49–50
178 DOMOS HARBOUR	Tower 10 × 10m square	Late Roman tile-stamps	Soproni 1978, 50
179 VISEGRAD GIZELLA	Traces of 10 × 10m tower		Soproni 1978, 51

APPENDIX 2

Site	Description	Dating	References
180 VISEGRAD QUARRY	Rectangular tower 8.9 × 8.9m, central pillar. Oven in corner	Building inscription of Leg. I. Martia dated 372. Valentinianic coins	Soproni 1978, 51–5
181 VISEGRAD REV UTCA	Foundations of tower 11 × 11m	Valentinianic?	Soproni 1978, 55
182 VISEGRAD SZENTGYORGY-PUSZTA I	Tower 9.35 × 9.15m	Late Roman tiles	Soproni 1978, 60
183 VISEGRAD SZENTGYORGY-PUSZTA II	Rectangular, possibly a courtyard	Late Roman metalwork. Valentinianic?	Soproni 1978, 60–1
184 KISOROSZI CHAPEL	?Watchtower, building debris	Late Roman tile-stamps	Soproni 1978, 72
185 KISOROSZI PUSZTATEM-PLOM	Damaged tower of unknown dimensions	Fourth-century pottery finds	Soproni 1978, 73–4
186 KISOROSZI PASZTORKERT	Possible bridgehead fort 9.48 × ?m	Late Roman tile-stamps	Soproni 1978, 73–4
187 VEROCE	Bridgehead fort, 23 × 18m, walls 2.8m thick. Enclosing walls 2m thick, towers at angles	Late Roman tile-stamps	Soproni 1978, 78 (see fig. 74, p. 187)
188 DUNABOGDANY KOSZEGTO	Partly destroyed tower 14.06 × 13.06m, surrounded by wall	Late Roman tile-stamps	Soproni 1978, 61
189 TAHITOTFALU SZENTPETER	Traces of tower: building debris	Fourth-century pottery	Soproni 1978, 72
190 TAHITOTFALU BALHAVAR	Bridgehead: rectangular tower with spur walls, 19.4 × ?m. Spur walls end in rectangular towers	Diocletianic?-Constantine II	Soproni 1978, 74–5
191 TAHITOTFALU NYULASI PATAK	Traces of rectangular tower 10 × 10m		Soproni 1978, 62–3
192 LEANYFALU	Tower 16.12 × 16.25m in rect. courtyard. 4 central supports	Late Roman coins, tile stamps. Re-used inscript.	Soproni 1978, 63–4 (see fig. 74, p. 187)
193 SZENTENDRE HUNKA DOMB	Small rectangular fort 30 × 40m; further enclosure outside	Late Roman tile-stamps. Built in early C3, used in Valentinianic period	Soproni 1978, 66–7 (see fig. 74, p. 187)
194 SZIGET-MONOSTOR GOD FERRY	Unexcavated tower		Soproni 1978, 75
195 SZIGET-MUNOSTOR HORANY	Bridgehead: tower 16 × 22m, with spur walls. Towers 4 × 4m	Late Roman tile-stamps	Soproni 1978, 75–6 (see fig. 74, p. 187)
196 DUNAKESZI	Bridgehead?	Many late Roman tile-stamps	Soproni 1978, 79

Site	Description	Dating	References
197 SZENTENDRE DERA-PATAK	Bridgehead, tower 20 × 20m with L-plan wing walls, and 2 small towers 7.5m square, on Danube bank	Late Roman tiles. Diocletianic?	Soproni 1978, 71–2 (see fig. 74, p. 187)
198 SZIGET-MONOSTOR FACANOS-KERT	Not excavated?		Fitz (ed) 1976, 79
199 BUDAKALASZ LUPPA CSARDA	Tower 16.3 × 15.5m: 4 support pillars, courtyard 39m square	Late Roman tile-stamps	Fitz (ed) 1976, 79
200 BUDAKALASZ BARAT PATAK	Unexcavated watchtower		Fitz (ed) 1976, 79
201 SZIGET-MONOSTOR SZIGETCSUCS	Round tower 4–6m diam., walls 1.7m thick	Late Roman?	Fitz (ed) 1976, 75
202 BUDAPEST, UJPEST WATERWORKS	Not excavated		Fitz (ed) 1976, 121
203 BUDAPEST, BEKASMEGYAR	Tower 8.1m square	Late Roman tile-stamps	Fitz (ed) 1976, 81
204 BUDAPEST, UJPEST PALOTA PATAK	Reported remains of watchtower		Fitz (ed) 1976, 121
205 BUDAPEST, BEKASMEGYAR	Tower 7 × 6.9m, walls 1.1m thick	Late Roman tile-stamps	Fitz (ed) 1976, 81
206 BUDAPEST, GAS FACTORY	Tower 7m square, with outer wall 14m square	Late Roman: area around used as cemetery in C4	Fitz (ed) 1976, 81
207 BUDAPEST, LAJOS UTCA	No details		Fitz (ed) 1976, 87
208 BUDAPEST, MARGARET IS.	Round Tower		Fitz (ed) 1976, 87
209 BUDAPEST, MARGARET IS.	Now destroyed on site of medieval monastery. Detailed description remains		Fitz (ed) 1976, 87
210 BUDAPEST, PARLIAMENT	Building debris, altars		Fitz (ed) 1976, 121
211 BUDAPEST, LANCHID UTCA	Buildings and debris	Late Roman stamped tiles	Fitz (ed) 1976, 88
212 BUDAPEST, LANCHID	Roman finds under medieval remains – watchtower?		Fitz (ed) 1976, 121
213 BUDAPEST, TABAN	Tower 10 × 15m	Late Roman stamped tiles	Fitz (ed) 1976, 89
214 BUDAPEST, GELLERT HOTEL	Late Roman buildings and debris above earlier remains	Stamped tiles	Fitz (ed) 1976, 89
215 ERCSI, EOTVOS MON.	Traces of a tower		Fitz (ed) 1976, 97
216 IVANCSA	Tower 4m square		Fitz (ed) 1976, 97
217 LORER	Foundations of tower found opposite Vetus Salina		Fitz (ed) 1976, 100

Site	Description	Dating	References
218 RACALMAS-KULCS	Watchtower	Late Roman. Valentinianic	Fitz (ed) 1976, 100
219 220 221 DUNAUJVAROS 222 223 224	6 watchtowers		Fitz (ed) 1976, 105
225 BOLCSKE	Round tower, 6.5m diam.		Fitz (ed) 1976, 109
226 LEANYVAR	Watchtower indicated from surface finds		Fitz (ed) 1976, 109
227 DUNASZENTGYORGY	Watchtower indicated from surface finds		Fitz (ed) 1976, 110
228 ZADER-IMSOS	Bridge head fort remains visible at low water, area 100 × 55m	Late Roman stamped tiles	Fitz (ed) 1976, 125
229 DUNAFALVA (=CONTRA FLORENTIAM)	Bridgehead fort on Danube bank: rectangular tower and enclosing walls, 85 × 59m	Reused C3 material	Fitz (ed) 1976, 125 (see fig. 74, p. 187)
230 BAR	Stray finds – watchtower?		Fitz (ed) 1976, 115
231 MOHACS	Finds suggest watchtower		Fitz (ed) 1976, 115
232 HATVAN	Tower 9.4 × 10.02m in rect. courtyard, 14.87 × 15.7m. Two-roomed building 58m away. Surrounded by ditch	Late Roman tile-stamps	Soproni 1978, 81–6

Appendix 3 HILL-TOP FORTIFICATIONS BUILT OR USED IN THE LATE ROMAN PERIOD

The numbers refer to the location maps, figs 87 and 88

Site	Size/shape	Defences/dating	References
1 AHRWEILER	Hill-top?	Possible traces of ring-wall	Kleeman 1971, 79
2 ALTRIER Ginstegestell	Promontory fort	Bank and ditch, dry stone wall in part	Schindler & Koch 1976, 17–19
3 ALTWIES Kaaschtel		Roman coins of C3 and C4	Schindler & Koch 1976, 19
4 ANGLEUR			Brulet *Caesarodunum* 1978, p. 12
5 ANNWEILER Der Trifels			F Sprater, *Der Trifels bei Annweiler* (1953), 11f
6 BAILEUX	Promontory	Coin hoard of Postumus	Graff, 1963, 128
7 BEN AHIN	Promontory	Coins of C3, laetic cemetery nearby	Graff, 1963, 148, *ABelg* 104
8 BERTRIX Château des Fées	Promontory site, 0.25ha	Early wooden/earth defences replaced in stone. Towers at entrance	A Matthys & G Hossey, *ABelg* 146 (1973)
9 BONNERT Katzenknapp	Promontory	One side protected by bank and ditch possible tower. Roman finds	Graff, 1963, 149
10 BODANGE	Hill-top	Roman? origin, medieval fort	Mertens, *ABelg* 76
11 BUZENOL	Walled promontory	Stone walls 1.2m thick cut off inner *refugium*. Square towers and wall added to former prehistoric bank. Re-used Roman sculpture and some late C2 pottery in rampart-bank	J Mertens, *ABelg* 16 and 42
12 CAMP DE LA BURE			*BPV* 90–1 (1964–5) 61, 92 (1966) 5
13 CARIGNAN Montilleul			*ABelg* 146 (1973)
14 CHESLAIN D'ORTHO	Walled ridge-top site, 1.5ha, triangular	Drystone enclosing wall 2.8m thick. 3 phases: (a) earth bank, wooden towers (b) stone wall (c) large stone towers added. Finds of Argonne pottery	Mertens, *ABelg* 129 (1971)
15 CINEY			Brulet *Caesarodunum* 1978, p. 12

Site	Size/shape	Defences/dating	References
16 CLAIREFONTAINE	Spur defended by bank and ditch 1.5ha, triangular	Banks and ditches re-employed and recut in Carolingian times. Roman? watchtower. C4 Roman pottery	Mertens, *ABelg* 49 (1960)
17 CONSDORF Buurgkapp	Promontory 3.75ha	Broad earthen wall across E side. Finds of Roman material	Schindler & Koch 1976, 29f
18 CORA	Rough oval jointed to rest of plateau at neck	Hallstatt defences surmounted by wall 2m thick, with projecting towers 8m diam. and 23–25m apart. Coin finds of late C3–C4	Von Uslar, 1964, no. 75; Grenier 1931, 430
19 COUVIN La Butte aux Roches			Brulet *ABelg* 160, p. 8 (map)
20 DONNERSBERG		Large walled site. Scatter of late Roman coins	Sprater, 1929, II, 49; *Mainz F* 12, 102f
21 DOURBES La Roche à Lomme	Naturally defended	Defences with 2 walls and ditches on east. Two periods in C4	Brulet, *ABelg* 160 (1974)
22 DRACHENFELS	Hill-top	Double bank round whole hill-top, naturally well defended	Von Uslar, 1964, no. 85; Sprater 1929, I, 72
23 ECHTERNACH	Hill-top, circular, 1.5ha	Single mortared wall 1.4m thick, regular blocks, 4 gates with some evidence of 2 periods. Inturned entrance to S. Re-used material and some Roman pottery	Meyers, *Hémecht* 1964, 176f
24 ÉPRAVE	Roughly triangular, 0.37ha	2 ditches and encircling wall on less steep south side. Wall to W and N occupied early C4, given up early C5	Graff 1963, 131–2; Bogaers & Rüger 1974, 237
25 ÉPRAVE II			Brulet *Caesarodunum* 1978, p. 12
26 ESCOMBRES La Forteresse			Brulet *ABelg* 160, p. 8 (map)
27 FAINS	Hill-top, rectangular, c 2ha	Stone wall embedded in earth rampart	Grenier 1931, 253
28 FALAEN			Brulet, *Caesarodunum* 1978, p. 12
29 FRATIN		Military cemtery (*laeti*) nearby	Mertens *ABelg* 76, 197–8
30 FURFOOZ	Small plateau, 0.8ha	Lines of defences on plateau side; 2 earthworks, wall, rampart banks with at least one tower. *Laeti* cemetery; finds late C2–late C4	Nenquin, *Furfooz*: Graff 1963, 13; Brulet *La Fortification ... de Furfooz* (1978)
31 GEROLSTEIN Dietzentley	Pear-shaped hill-top, 240 × 90m	Prehistoric ring-wall, late Roman use possible. Coins of Magnentius	*Mainz F* 33, 318f

Site	Size/shape	Defences/dating	References
32 GEROUVILLE		Records of semicircular tower, possibly Roman	Mertens, *ABelg* 76, 200–1
33 GOUGNIES	Hill-top, flattened plateau, 1ha	Thought to be on site of former *oppidum*. Finds of all periods from site	Graff 1963, 132–3
34 GREVENMACHER	Large site, 8.45ha	Semicircular wall, 37 late C3–C4 coins	Schindler & Koch, 1976, 35
35 GRÖNIG Mommerich	Plateau site, 0.12ha	Wall, 48m long, cuts off part of plateau of former hill fort. Roman material found	Schindler 1968, 105f
36 HAMBUCH Burgberg	Long narrow site on ridge	Late Roman finds, coins, glass, and pottery. Fragmentary remains of mortared wall	von Uslar 1964, no. 143; *BJ* 153 (1953) 136
37 HAN SUR LESSE			Brulet *Caesarodunum* 1978, p. 12
38 HEFFINGEN Albuurg	Roughly circular hill-top	Encircling wall of Roman period, combined with barrier walls. Roman finds	Schindler & Koch, 1976, 38
39 HÉRAPEL	Promontory site, *c*3.5ha	Circuit wall, with projecting towers facing plateau side. Roman finds from C1 onwards, but coin series Tetricus-Gratian	von Uslar 1964, 160; *RE* Supp. III, 1121; Toussaint *La Lorraine à l'époque Gallo-Romain*, 173f
40 HONTHEIM Entersburg	Irregular, small hill-top plateau	No known defences, but 42 coins from Aurelian-Magnentius, pottery, glass and metalwork	KJ Gilles, *TZ* 37 (1974) 99f
41 HOTTON			Brulet *Caesarodunum* 1978, p. 12
42 HOUFFALIZE	On rocky hill-top, in river bend	Earth banks, rectangular tower, late C3 coins	Brulet *Caesarodunum* 1978, p. 12
43 HOVELANGE	Narrow hill-top	Defended area cut off by triple or quadruple ditches and one bank. Roman finds from C1–C4, and square tower	Schindler & Koch, 1976, 39
44 INSUL Bergberg	Promontory site	Prehistoric stone ring-wall, with large stone wall and ditch across plateau neck. Late Roman pottery from site	von Uslar 1964, 198; *BJ* 153 (1953), 138; Kleeman, 1971, 91
45 ITZBACH	Promontory site	2 parallel blocking banks, outer one enclosing *c* 7ha, inner one *c* 1ha	Schindler 1968, 19–20
46 ITZIG Torbelsfiels	Promontory	Double ditch and bank across plateau. Roman tile from bank	Schindler & Koch 1976, 42
47 JAMOIGNE		Site of laetic cemetery, possible hill fortification	Mertens *ABelg* 76

Site	Size/shape	Defences/dating	References
48 JEMELLE	Promontory site, 1.5ha	Blocking wall, and defensive circuit with round and rectangular towers. Small building (baths?) inside. Late Roman date	Bogaers & Rüger 1974, 240
49 JUNGLINSTER	Long promontory	Short blocking wall. Coin hoard of 275–6	Schindler & Koch 1976, 43
50 KASTEL	Large trapezoidal promontory	Neck of promontory defended by large blocking bank 12m wide: wall in centre of bank. Roman finds of C1–C3 date, and re-used material in bank	von Uslar 1964, 210; *TZ* 14 (1939) 231; *TZ* 24–6 (1956–8) 319; *Mainz F* 5, 179f
51 KATZENBURG	Irregular oval	Double ditch surrounds 3 sides, and wall with round towers on 4th. Coins C4 date, late Roman finds from ditches	von Uslar, 1964, 211; *Germania* 1921, 25f
52 KATZENKÖPFCHEN	Promontory fort	Stone wall and earth bank 15m wide cuts off neck. Ditch 10m wide, 3–4m deep. No dating evidence	Steinhausen, 1932, 129
53 KEMPFELD	Ringwall	Possibly Roman, but earlier uses damaged in medieval period	*BJ* 153 (1953) 138
54 KINDSBACH Grössenberg	Triangular, 1.5ha	Encircling wall, 1m thick. Ditch to N and S, but steep drop elsewhere. 200 coins, from Trajan to Magnentius	*Germania* 1961, 225f; A Fehr, *Vor-und Fruhgeschichte der Kreise Kaiserslautern* (1972) 53 and 103
55 KOPSTAL Muesbierg	Small promontory	Bank 6m wide, entrance to N. No finds	Schindler & Koch 1976, 44–5
56 KORDEL Burgberg	Ringwall	Dry stone wall encircles hill, ring ditch with extensions, Roman re-used prehistoric site	*TZ* 16–17 (1941–2) 220
57 KOTTELBACH	0.3ha enclosed	Ring wall, Roman finds of tile and pottery	*BJ* 137 (1937) 242
58 KREIMBACH	1ha enclosed	Ring wall: gates to E and W, 3 towers on curtain wall. Wooden buildings against outer walls? Coins from 260–360, large amounts of ironwork found	Sprater 1929 I, 67f
59 LANAYE	Large plateau, 17.5ha	Curved wall cuts off plateau neck	Graff, 1963, 135
60 LAROCHETTE	Promontory, 1ha	Bank and triple ditch as blocking wall. Roman?	Schindler & Koch 1976, 48
61 LÉBOUS	Irregular site	Ring wall, probably prehistoric, but late Roman thickening. Late Roman coins	*Germania* 41 (1963) 241
62 LEIKÖPPCHEN	Triangular	Blocking wall of large stones cuts off plateau. Nr Palzkyll villa. Roman finds	von Uslar 1964, 236

Site	Size/shape	Defences/dating	References
63 LIMBACH	Oval, 1.5ha	Wall of re-used material: late Roman pottery from site	Schindler, 1968, 17; *Mainz F* 5, 190–4
64 KOMPRET	Promontory	Coins of Postumus found	Graff 1963, 135–8
65 LORENZTWEILER	Promontory and ditch	Natural defences strengthened with bank and ditch. Bank contains Roman tile	Schindler & Koch 1976, 51
66 LUSTIN			Brulet, *ABelg* 160, p. 8 (map)
67 MACQUENOISE	Plateau site	7 coin hoards found, 3 date to *c*270	Graff 1963, 158
68 MARIALSKOPF	Oval	Drystone ring-wall and ditch. Roman? use of prehistoric site	*BJ* 153 (1953) 128
69 MATAGNE-LA-GRANDE Bois des Noels		Rectangular structure with bastion. Late Roman wall?	Brulet, *ABelg* 160 (1974) 8
70 MONTCLAIR	Plateau 700 × 150m	Prehistoric hill-fort. Late Roman reuse?	Schindler, 1968
71 MUNSHAUSEN Kukigt	0.3ha	Blocking bank, 18m wide and 45m long. No dating, but late Roman?	Schindler & Koch 1976, 56
72 NAMUR	Large *oppidum*, reduced to 40ha	Wall with reused material round circuit	Bogaers & Rüger, 1974, 231
73 NEEF Petersberg		Identified as temple site, but finds more typical of refuge. Constantinian coins	*TZ* 37 (1974) 114, fn 30; Gilles, in *Heimat zwischen Eifel und Hunsruck* 22 (1974) 3f
74 NEUSTADT Hambacher Schloss			Sprater 1929, I, map
75 NISMES Sainte Anne			Brulet, *ABelg* 160 (1974) map p. 8
76 NISMES La Roche Trouée		Wall and ditch, re-used material	Brulet *ABelg* 160 (1974) map p. 8
77 OBERKIRCHEN Weisselberg	Terraced hill-top	Drystone wall and bank to N and S. Late Roman pottery and coin of Magnentius	*TZ* 13 (1938) 255; Schindler 1968, 21
78 OBERSTEUFFENBACH Heidenburg	Originally ring-wall, now quarried	Wall 1.5–1.8m thick, with ditch. Casemate buildings	Sprater 1929, I, 67f; *MHVP* 68 (1970) 112
79 ODILIENBURG	Irregular, large contour fort, 140ha	Wall 1.8m thick encircles hill-top. Small gates with towers. Coins of C1–C3. Radio carbon date for wooden piles of C3	*Germania* 1958, 113; *A Korr* 3 (1973) 459
80 OTZENHAUSEN			Schindler, 1968
81 PÖLCH Burgberg	Small plateau	No defences known. Late Roman pottery and Laetic equipment. Cemetery at base of hill	von Uslar 1964, 275; *BJ* 148 (1948) 439

Site	Size/shape	Defences/dating	References
82 PRY			Brulet, *Caesarodunum* 1978, p. 12
83 PÜNDERICH Marienburg	High-level plateau 120m long above Mosel	Built over by medieval castle. Much late Roman pottery from site	*TZ* 37 (1974) 114f
84 PÜTZ	Triangular promontory	Ditch across plateau neck. C4 finds	*Archäologische Funde der Kreise Bergheim* (1974) 92–3
85 REIFFERSCHEID Alte Burg	Part quarried, small hill-top	Circuit wall of basalt stone. Entrance to SW possible ditch and bank on this side. Roman and medieval pottery finds	Kleeman, 1971, 102
86 RHEINBACH Tomburg			*Römische Kunststatten* (1973) 10f
87 RHEINECK Reutersley	Promontory	Double banks on S, single bank on N. Roman pottery and coin found	Kleeman, 1971, 108
88 RISSENTHAL Alte Schloss	Narrow plateau, 4ha	Bank and outer fortifications	Schindler 1968, 18–9
89 RÖMERBURG		Re-used stone from Tarquimpol nearby	*ASHAL* 1892 pt ii, 116f
90 ROTKOPF	Promontory	Thick wall cuts off plateau. Roman finds of nails	*TZ* 14 (1939) 230; *BJ* 153 (1953) no. 25
91 ROUVEROY	Irregular rectangular promontory, 350 × 240m	Earth bank round site: Roman finds drystone revetment	Graff 1963, 126–7
92 ST INGBERT Grosse Stiefel	Long narrow hill, 300 × 500m	Encircling bank, with blocking wall 0.75m high. Roman finds from the site	Sprater 1929 II, map; *Mainz F* 5, 118f
93 SARS LA BUISSIÈRE	Promontory	Cemetery nearby, Roman coins from site	Graff 1963, 142–3
94 SAXON	Former *oppidum*	Solid wall, follows hill crest all round. No dating evidence	Toussaint, *Répertoire Arch éologique du Merthe et Moselle* 182f
95 SCHEIDSKOPF	Ring-wall, quarried away	C19 disruption suggests main wall in front of lesser bank	*BJ* 49 (1870), 185; 124 (1917), 201f; 125 (1919), 17; 126 (1921), p. 2
96 SCHLOSSBURG			*ASHAL* 1892, 116f
97 SINSIN	Oval hill-top, roughly circular	Stone wall encircles hill, C4 occupation	Graff, 1963
98 SOMMERAIN	1ha	Earth wall and drystone foundations, semicircular towers on E and S, and tower at S gate. Mid- and late Roman sherds in earth wall	Bogaers & Rüger 1974, 245; Brulet, *Revue des Archéologues et Historiens de l'Universite de Louvain* vii (1974)
99 STEINBACH			Sprater 1929 II, map
100 STEINEBURGER LEY			Schindler, *Mainz F* 33, 330f

Site	Size/shape	Defences/dating	References
101 TEMPELBERG	Promontory	Ditch cuts off the plateau. Possibly Roman	*BJ* 153 (1953) no. 39; *TZ* 3 (1928) 16
102 THÉOU	Protected valley	Approaches cut off by wall and gate	Site of discovery of *CIL* XII 1524; *FOR* I no. 78
103 THOLEY	Promontory, 130 × 80m	Triple ditch and bank cuts off the plateau neck Roman sculpture on site	von Uslar 1964, 294; Schindler 1968, 21; *Mainz F* 5, 202
104 THON			Brulet, *ABelg* 160 (1974) p. 8
105 UDANGE		Possible Roman occupation	*ABelg* 76 (1964)
106 VIRTON	1.5ha	Defensive wall 1.6m thick encircles hill. Re-used material in wall. Late Roman pottery	*ABelg* 206 (1978) 82; 213 (1979) 112f; 223 (1980) 67f
107 VOGENÉE			Brulet *Caesarodunum* 1978 p. 12
108 VÖLKLINGEN	Elevated promontory, 100 × 70m	Surrounded by bank and ditch gate, with 2 semicircular towers with linking building. Near laetic cemetery	*RFS* 7, 49; *AKorr* 3 (1973) 231
109 WALDFISCHBACH Heidelsburg	Triangular plateau, 1.5ha	W side defended by wall and gate. Re-used material 2 periods of occupation. Hoard of ironwork and *saltuarius* inscription. C4 finds	Sprater 1929 I, 67f; *RFS* 7, 114
110 WALSDORF		Defended site? Tower 8.5m square. Re-used material found	*TZ* 14 (1939) 246
111 WEILER	Promontory	Earth and stone blocking wall: interior face of wall mortared. Roman pottery	von Uslar 1964, 338; *BJ* 142 (1942) 247
112 WELLESWEILER Maykesselkopf	Plateau, 2ha	Double ditch blocks approaches. Late C3 coin hoard found	Schindler, 1968, 20
113 WILLIERS	Roughly rectangular, 4ha	Wall 3m wide surrounds site: particularly well defended on approach side. Coins and pottery of early-late C4 date	*ABelg* 74, 221
114 WITTLICH	Promontory	Drystone wall cuts off plateau	Steinhausen 1932, 77–8
115 ZELL Alteburg	Promontory, 0.25ha	Encircled by drystone wall, rectangular building within. Coins of late C3–early C4	*AKorr* 3 (1973) 67f
116 ARNBERG		Constantinian coins from site	Jantsch 1947
117 BÜRKLI	Triangular, steep sides	Earthen bank, with stone gate. Tower at highest point: some internal buildings. No dating evidence	32 *JSGU* (1940), 147; *Festschrift E Vogt*, 271; *JSGU* 59 (1976), 203f.
118 CASTELS			Garbsch 1970, map

Site	Size/shape	Defences/dating	References
119 CHASTLI-BURG	Promontory site, triangular c 1ha	Bank 90m long with drystone wall 3m thick. Buildings inside defended area	*Festschrift Laur-Belart* 354
120 FRASTAFEDERS	Promontory	Wall, cutting off plateau neck	Jantsch 1947, 185
121 GROSSENKASTEL			*Historia* 1956
122 HEIDENBURG	Plateau, 180 × 40m	Surrounding wall with casemate buildings poorly built. Tower in NE, several building periods	Jantsch, 1947, 172f.
123 HORWA		8m square tower, defended by earthwork at lower level	von Uslar, 1964, 178; Jantsch, 1947, 182
124 KANINCHENINSEL	Island in Bielersee	Late Roman finds, Probus-Constantine coins	*JSGU* 40 (1949), 130–1
125 KOBLACH			Garbsch, 1970, map
126 LÜTZENGÜTLE			*JbHVFL*, 37 (1937) 85f.
127 MANTIKEL			Garbsch, 1970, map
128 MUCIASTER	0.15ha	Tower 11m diam., wall round lower hill	Jantsch, 1947, 179
129 RENGGEN		Finds of Roman pottery	
130 SCHAAN	Rough oval, 70 × 35m	Rough wall round hill-top. 2 buildings inside, 2-period construction. Tower 6m square. Coins suggest 2 periods – late C3, and mid-C4	*JbHVFL* 64 (1964) 7f
131 SCHLOSSLE	Low hill, 0.3ha	2 defended areas, hill to S, with wall round plateau	Jantsch, 1947, 180
132 STELLFEDER		Wall surrounding hill. Some C4 pottery	Jantsch, 1947, 183
133 STÜRMENKOPF		Towers and closing wall. Wall round top of hill, very steep. Aurelianic coin, C4 pottery	von Uslar, 1964, 318; *US* 32 (1968) 17f
134 TIEFENCASTEL			Garbsch 1970, map
135 WITTNAUER HORN	Triangular promontory, 0.5ha	2-period blocking wall, 1.50m wide on prehistoric bank. Tower at each end, and gate to N. Coins suggest 2 periods c 270 and early C4	Bersu, *Das Wittnauer Horn* (1945)
136 ALTENSTADT	Low hill, roughly oval	Late Roman cemetery nearby	Keller, 1971, 156
137 'ENDIDAE'	1.5ha approx	2 areas of hill-top; lower part defended by wall 3.5m wide and 'Andernach' gate. Highest point of hill has wall 1.10m thick	von Uslar 1964, 97a; *Germania* 1926, 150
138 FRAUENBERG	Triangular promontory in bend of Danube, 7.5ha	Prehistoric fortification used as blocking wall: gate passage with rectangular tower at S end	K. Spindler and others, *Die Archäologie des Frauenberges* (1981), 102f

Site	Size/shape	Defences/dating	References
139 GEORGENBERG		Wall 1.4m thick round part of hill, used from late Roman times to early medieval	von Uslar 1964, 118; *JOEAI* 1956 Bb 123
140 GRÜNWALD	Triangular promontory, *c* 1ha	Ditch systems (only inner one Roman?) cut off low-lying site. Wall in bank – small *vicus* defence? Late Roman finds from site	Reinecke, *Kleine Schriften*
141 KUCHL	Plateau *c* 6ha	Late Roman wall found at various places round hilltop. ¾-round towers?	*PAR* 1963 and 1967
142 LORENZBERG	Oval, *c* 4ha, low hill	Wall completely encircles site; one simple entrance, square towers, casemate buildings. Various period from late C3 to late C4	Werner *Der Lorenzberg*
143 MOOSBERG	Roughly oval, low hill	Wall 0.7–0.9m thick, square and semi-circular bastions. 2 gates, one Andernach type, interior free-standing and casemate buildings. 2 periods *c* 260 and *c* 340	Garbsch *Der Moosberg*
144 STAFFERSBERG		Site near late Roman cemetery	Keller, 1971
145 VALLEY			Keller, 1971, 158–9
146 WIDDERSBURG			Keller, 1971, Abb 4
147 AIDOVSKI GRADEC	Hill-top, contour fort, long polygonal	Walls all round hill; tower at SE corner. Coins from Augustus-Valens	Šašel and Petw, 1971, 87
148 CASTEL DEI PAGANI	Hill-top		Puschi 1902, map
149 CASTUA	Hill-top		Puschi 1902, map
150 DRASCHITZ			*TIR* Tergeste, p 35
151 DUEL	Hill-top, plateau 240 × 110m, *c* 2.5ha	Walls 0.9m thick round hill; 2 periods of construction? Buildings lean against walls dated *c* 400	Von Uslar 1964, 87; *JOEAI* 25 Bh 189
152 FALKENBURG	Hill-top	Fort with towers defended on S, E and N by slope. Bank and ditch to W	Jantsch 1938 no 53; *Car*, 1934, 15f
153 GLODNITZ	Hill-top/eperon 3 sides sheer, 300 by 100m, 3ha	Wall to N making attachers expose their left flank: NE corner rectangular tower	Jantsch 1938 no. 34
154 GOLO	Hill-top trapezoidal, 2.5ha		Puschi 1902 map
155 GRAFENBRUNN	Hill-top	Single wall 350m long with tower	Puschi, 1902 map
156 GRAFENBRUNN	Hill-top triangular	2 sides walled, but sheer drop on side	Puschi 1902 map

Site	Size/shape	Defences/dating	References
157 GRAZERKÖGEL	Hill-top	Fortifications known, and also church. Roman cemetery connected	Von Uslar 1964, 125; Jantsch 1938, no. 5; Meyer, Die Gurina 1885
159 HEMMABERG	Promontory, roughly oval, 370 × 170m at largest	Wall cuts off long s side, 1.2–1.8m thick. 2 gates, 75m apart, and church. Re-used material in church	Von Uslar 1964, 129
160 HOISCHÜGEL	Hill-top, roughly rectangular plateau	Walls adapted very closely to hill shape; formerly church and temple site. Dated to early C3	Von Uslar 1964, 172; *Car* 1950
161 KARNBURG	Hill-top	Former prehistoric site defended by wall	Von Uslar 1964, 209; *Car* 1939, 265
162 LAMPRECHTS-KÖGEL	Hill-top, roughly circular, 1ha	Surrounded by wall with 2 gates and interior buildings against wall	Jantsch, 1937 no. 55; *Car* 1931 11 & 1934, 14
163 LAVANT	Hill-top, circular, top of very steep hill, 2.5ha	Wall surrounds hill, 2.1–2.3m thick. Gate in centre of N side protected by rectangular towers. Former temple site, later church	*JOEAI* 1950Bb 31f
164 LIMBERG	Hill-top		*GMDS* 1939, 147f
165 MAUTHERN	Hill-top	One known tower, rectangular, 7.75 × 9.0m	Jantsch 1938, p 325
166 NASIRZ	Hill-top		Puschi 1902, map
167 PRIMANO	Hill-top		Puschi 1902, map
168 RIFNIK			*ArchKarte Jugoslaven* Rogatec 52; *A Vest* 21–2 (1970–1), 127f
169 SEMBIJE	Hill-top, large circuit of 1000m	Wall round whole hill-top, with double wall to E. Possibly also large towers	Puschi 1902 map
170 SILER TABOR	Hill-top, large contour fort	Strong towers, wall and ditch	Puschi 1902 map
171 SILER TABOR	Hill-top		Puschi 1902 map
172 SILER TABOR	Hill-top		Puschi 1902 map
173 TERNOVO	Hill-top		Puschi 1902 map
174 TINJE			*ArchKarteJug* Rogatec 65
175 TREBINEC VRH			*GMDS* 1939
176 TREFFLING	Hill-top, sheer hill, 121 × 94m, c1ha	Tower placed at top of hill; whole site defended by wall	Jantsch 1938, no. 13
177 TESCHELTNIG-KÖGEL	Hill-top	Strong wall cuts off whole hill, re-used material in it. Gates and towers known. Some interior buildings including hypocaust. C2–C4 pottery. Re-used material	Jantsch 1938, no. 20

Site	Size/shape	Defences/dating	References
178 ULRICHTSBERG	Hill-top	Small part of defences on w side of hill. Early church with mosaics	Jantsch 1938, no. 52
179 VELIKE MALENCE	Hill-top, rough oval, 440 × 300m, *c* 8ha	Wall 2.1–1.65m wide all round. Towers and gates. Re-used material and 2 periods of defences. Church set centrally. *Tpq* of 253 from re-used inscriptions C3–4	*GMDS* 10 (1929) 11f; Werner & Ewig (ed) 1979
180 WEISSENSTEIN	Hill-top	Mortared walls and tile fragments	Jantsch 1938, no. 17
181 ZAGRAD	Hill-top		Puschi 1902 map
182 ZALOG		Walls 2m thick, small fort (on hill?) of *burgus* type	*GMDS* 20, 147

Bibliography

ANCIENT WRITERS AND COLLECTIONS OF SOURCES

AM	Ammianus Marcellinus, Gardthausen, Teubner Leipzig 1874–5
CIG	*Corpus Inscriptionum Graecarum*
CIL	*Corpus Inscriptionum Latinarum*
CJ	*Codex Justinianus* ed Krüger (Berlin 1877)
CSEL	*Corpus Scriptorum Ecclesiasticorum Latinorum* (1866–
CT	*Codex Theodosianus* ed T. Mommsen and P. Meyer (Berlin 1905)
ILS	*Inscriptiones Latinae Selectae*
ND	*Notitia Dignitatum* O. Seeck (Berlin 1876)
NGall	*Notitia Galliarum*, published in O. Seeck, *Notitia Dignitatum* (Berlin 1876)
Pan Lat	*Panegyrici Latini*, ed R.A.B Mynors (Oxford 1969)
SHA	*Scriptores Historiae Augustae*, ed R. Magie (Loeb)

MODERN WORKS CITED IN THE TEXT

Alfoldy, 1974	G. Alfoldy, *Noricum* (London 1974)
Anthes 1917	E. Anthes, 'Spätrömische Kastelle und feste Städte in Rhein – und Donaugebiet', *BRGK* 10 1917, 86ff
BAR	*British Archaeological Reports*
Balil 1960	A. Balil, 'La Defensa de Hispania en el Baio Imperio', *Zephyrus* xi (1960) 179f
Balil 1970	A. Balil, 'La Defensa de Hispania en el Baio Imperio', *Legio VII Gemina* (León, 1970) 269f
Bartholomew and Goodburn, 1976	P. Bartholomew and R. Goodburn (eds.), *Aspects of the Notitia Dignitatum*, *BAR* S 15
Bechert, 1971	T. Bechert, 'Römische Lagertöre und ihre Bauinschriften', *BJ* 171 (1971) 201f
Belart, 1966	R. L. Belart, *Führer durch Augusta Raurica* (Basel 1966)
Blanchet, 1907	A. Blanchet, *Les Enceintes Romaines de la Gaule* (Paris 1907)
Blazquez, 1968	J. M. Blazquez, 'La Crisis del siglo III en Hispania y Mauretania Tingitana', *Hispania* xxviii (1968) 5f
Boethius and Ward Perkins, 1971	A. Boethius and J. B. Ward Perkins, *Etruscan and Roman Architecture* (London 1971)
Bogaers and Rüger 1974	J. E. Bogaers and C. B. Rüger, *Der Niedergermanische Limes*, Kunst und Altertum am Rhein no 50, 1974
Breeze and Dobson, 1976	D. Breeze and B. Dobson, *Hadrian's Wall* (London 1976)
Brünnow and von Domaszewski, 1904–9	R. E. Brünnow and A von Domaszewski, *Die Provincia Arabia* (Strasbourg 1904–9)
Butler (ed) 1971	R. M. Butler (ed) *Soldier and Civilian in Roman Yorkshire* (Leicester 1971)
Degrassi, 1954	A. Degrassi, *Il confine nord-orientale dell' Italia* (Bern 1954)
Démougeot, 1969	E. Démougeot, *La formation de l'Europe et les Invasions Barbares* (Paris 1969)
FGH	*Die Fragmente der Griechischen Historiker* ed F. Jacoby (Berlin 1923–30)

Fitz (ed) 1976	J. Fitz (ed) *Der Römische Limes in Ungarn* (1976)
FOR	*Forma Orbis Romani – Carte Archéologique de la Gaule Romaine*
	I Alpes Maritimes
	VIII Gard
	X Hérault
	XII L'Aude
	XIII Indre et Loire
Frere, 1974	S.S. Frere, *Britannia* (London 1974)
Garbsch 1967	J. Garbsch, 'Die burgi von Meckatz und Untersaal', *BVbl* 32 (1967) 51f
Garbsch, 1970	J. Garbsch, *Das Donau-Iller-Rhein Limes* Museum of Aalen publications 1970
Graf, 1936	A. Graf, *Übersicht der Antiken Geographie von Pannonien* (1936)
Graff, 1963	Y Graff, 'Oppida et Castella du pays des Belges', *Celticum* viii, 1963
Grenier, 1931	A. Grenier, *Manuel d'Archéologie Gallo-Romaine* I (1931)
Grenier, 1935	A. Grenier, *Manuel d'Archéologie Gallo-Romaine* II (1935)
Grenier, 1960	A. Grenier, *Manuel d'Archéologie Gallo-Romaine* III (1960)
Hagen, 1931	J. Hagen, *Römerstrasse der Rheinprovinz* (1931)
Jantsch, 1938	F. Jantsch, 'Die spätantike und Langobardische Burgen in Kärnten', *MAGW* 68 (1938) 331
Jantsch, 1947	F. Jantsch, 'Die spätantiken Befestigungen in Vorarlberg', *MAGW* 73–4 (1947) 168f
Johnson, 1976	S. Johnson, *The Roman Forts of the Saxon Shore* (London 1976)
Johnston, 1977	D. E. Johnston (ed), *The Saxon Shore* CBA Research Report 18 (1977)
Jones, 1964	A. H. M. Jones, *The Later Roman Empire* (Oxford 1964)
Kähler, 1942	H. Kähler, 'Die römischen Torburgen der frühen Kaiserzeit', *JDAI* (1942) 1–104
Keller, 1971	E. Keller, *Die spätrömische Grabfunde in Südbayern* (Munich 1971)
Kellner, 1971	H. J. Kellner, *Die Römer in Bayern* (Munich 1971)
Kleeman, 1971	O. Kleeman, *Die Vor- und Frühgeschichte des Kreises Ahrweiler* (1971)
Koethe, 1942	H. Koethe, 'Zur Geschichte Galliens zur Mitte des dritten Jahrhunderts', *BRGK* 32 (1942) 199f
LF	*Limesforschungen* 1– (Berlin, 1959)
MGH AA	*Monumenta Germaniae Historica* Auctores Antiquissimi 15 vols (Berlin 1877–1919)
Macmullen, 1963	R. Macmullen, *Soldier and Civilian in the Later Roman Empire* (Cambridge Mass. 1963)
Mainz F	*Führer zu vor- und frühgeschichtliche Denkmälern* (RGZMainz publications, since 1964)
Mayer (ed) 1956	Th. Mayer, *Studien zu den Anfangen des europäischen Städtewesens*, Vorträge und Forschungen 4 (1956)
Mocsy 1974	A. Mocsy, *Pannonia and Moesia Inferior* (London 1974)
Petrikovits, 1938	H von Petrikovits, 'Reichs- Macht- und Volkstumgrenze am linken Niederrhein' *Festschrift Oxe* (1938) 221f
Petrikovits, 1960	H. von Petrikovits, *Das römische Rheinland* Beihefte BJ 8 (1960)
Petrikovits, 1971	H. von Petrikovits, 'Fortifications in the North-Western Roman Empire from the third to the fifth centuries AD', *JRS* lxi (1971) p 178–218
Poidebard, 1934	R. Poidebard, *Le trace de Rome dans le desert de Syrie* (1943)
Poidebard and Mouterde, 1945	R. Poidebard and P. Mouterde, *Le Limes de Chalcis* (1945)
Ponsich, 1970	M. Ponsich, *Recherches Archéologiques à Tanger et dans sa region* (Paris 1970)
Puschi, 1902	A. Puschi, 'I valli romani delle Alpe Guilie' *Archeografo Triestino* 24, Supp. 1902
RCHM	Royal Commission on Historical Monuments
RE	Paulys *Realencyclopädie der Altertumswissenschaft*
RFS I	E. Birley (ed.) *The Congress of Roman Frontier Studies* 1949 (Durham 1952)

RFS 2	E. Swoboda (ed) *Carnuntina*, Vorträge beim internationalen Kongress der Altertumsforscher (Carnuntum 1955) in Römische Forschungen in Niederösterreich 3 (Graz-Köln 1956)
RFS 3	*Limes Studien* Vorträge des 3 internationalen Limes Kongresses in Rheinfelden-Basel, 1957 Schriften des Instituts für Ur- und Frühgeschiche der Schweiz (Basel 1959)
RFS 4	Limes-Konferenz Nitra (1961)
RFS 6	*Studien zu den Militärgrenzen Roms,* Vorträge des 6 internationalen Limes Kongresses in Süddeutschland. Beihefte der *B. J.* Bd 16 (Köln 1967)
RFS 7	S. Applebaum and M. Gichon (eds) *Roman Frontier Studies* 1967 (Tel Aviv 1971)
RFS 8	M. G. Jarrett (ed), *Roman Frontier Studies* (Cardiff 1971)
RFS 9	D. M. Pippidi (ed) *Actes du IXe Congrès International d'études sur les frontières Romaines* (Mamaia 1972). (Bucharest/Köln-Wien 1974)
RFS 10	H. G. Horn (ed) *Studien zu den Militärgrenzen Roms* II (Bonn/Köln 1977)
RFS 11	*Limes* Akten des XI Internationalen Limes Kongresses (Budapest 1978)
RFS 12	W. S. Hanson and L. J. F. Keppie (eds) *Roman Frontier Studies* 12 (1979) *BAR* S 71 (1980)
Richmond, 1930	I. A. Richmond, *The City wall of Imperial Rome* (Oxford 1930)
Richmond, 1931	I. A. Richmond, 'Five town walls in Hispania Citerior', *JRS* xxi (1931)
RLiO	*Römische Limes in Österreich* XIX F. Pascher – *Römische Siedlungen und Strasse in Limesgebiet zwischen Enns und Leitha* (1949) XXI R. Noll – *Römische Siedlungen und Strasse in Limesgebiet zwischen Inn und Enns* (1958) XXVI H. Stiglitz, *Das Römische Donaukastell Zwentendorf in Niederösterreich* (1975) XXV R. Eckhardt – *Das römische Kastell Schlögen in Niederösterrich* (1969)
Rodwell and Rowley 1975	W. Rodwell and T. Rowley (eds), *The Small Towns of Roman Britain BAR* 15 (1975)
Šašel and Petru, 1971	J. Šašel and P. Petru, *Claustra Alpium Juliarum I, Fontes,* Limes u Jugoslaviju II (1971)
Schindler, 1968	R. Schindler, *Studien zur vorgeschichtlichen Siedlungs – und Befestigungswesen des Saarlandes* (1968)
Schindler and Koch, 1976	R. Schindler and F. Koch, *Vor- und frühgeschichtliche Burgwälle des Grössherzogtums Luxembourg* (1976)
Schönberger, 1969	H. Schönberger, 'The Roman Frontier in Germany: an archaeological survey' *JRS* lix (1969) 144–197
Schültze, 1909	H. Schültze, 'Die Römischen Stadttore' *BJ* 118 (1909) 280–352
Soproni, 1978	S. Soproni, *Der spätrömische Limes zwischen Esztergom und Szentendre* (Budapest, 1978)
Sprater I, II, (1929)	F. Sprater, *Die Pfalz unter den Römern* I and II (Speyer 1929)
Stähelin, 1948	F. Stähelin, *Die Schweiz in Römischer Zeit* (1948)
Stehlin, 1957	K. Stehlin and V von Gonzenbach, *Die spätrömische Wachttürme am Rhein von Basel bis zum Bodensee* I (1957)
Taracena, 1948	B. Taracena, 'Las fortificaciones y la poblacion de la Espana Romana', *IV Congreso Arqueologico del Sudeste Espanol* (1948) 421f
Thompson, 1965	E. A. Thompson, *The Early Germans* (Oxford 1965)
TIR	*Tabula Imperii Romani* L33 Tergeste L34 Aquincum
Todd, 1978	M. Todd, *The Walls of Rome* (London 1978)
von Uslar, 1964	R. von Uslar, *Studien zu frühgeschichtlichen Befestigungen* Beihefte der *Bonner Jahrbücher* no 11, 1964

VCH	Victoria County History
Wacher, 1974	J. Wacher, *The Towns of Roman Britain* (1974)
Werner and Ewig, 1979	J. Werner and E. Ewig, *Von der Spätantike zum frühen Mittelalter*, Vorträge und Forschungen XXV (1979)
Wightman, 1967	E. M. Wightman, *Roman Trier and the Treveri* (London 1967)
Wilkes, 1967	J. J. Wilkes, *Dalmatia* (London 1967)

PERIODICALS CITED AND ABBREVIATIONS USED

AAWW	*Anzeiger der Österreichischen Akademie der Wissenschaften in Wien*
AArchHung	*Acta Archaeologica Academiae Scientiarum Hungaricae*
AArchSlov	*Acta Archaeologica. Arheološki Vestnik*
ABelg	*Archaeologia Belgica*
AC	*L'Antiquité Classique*
ACamb	*Archaeologia Cambrensis*
ACant	*Archaeologia Cantiana*
AEA	*Archivo Español de Arqueologia*
AEp	*Année Epigraphique*
AErt	*Archaeologiai Ertesitö*
AIEC	*Annuari de Institut d'Estudis Catalans*
AJ	*The Archaeological Journal*
AKorr	*Archäologisches Korrespondenzblatt*
APreg	*Arheoloski Pregled*
ARoz	*Archeologicke Rozhledy*
ASA	*Anzeiger für Schweizerische Altertumskunde*
ASAN	*Annales de la Société Archéologique de Namur*
ASHAL	*Annuaire de la Société d'Histoire et d'Archéologie de la Lorraine*
Alba Regia	*Alba Regia*
All Gesch	*Allgäver Geschichtsblättes*
Ann Bret	*Annales de Bretagne*
Antike Welt	*Antike Welt*
AntJ	*The Antiquaries Journal*
Antiquity	*Antiquity*
Archaeologia	*Archaeologia or Miscellaneous Tracts relating to Antiquity*
Archaeology	*Archaeology*
Arrabona	*Arrabona*
BAM	*Bulletin d'Archéologie marocaine*
BCTH	*Bulletin Archéologique du Comité des Travaux Historiques*
BJ	*Bonner Jahrbücher*
BM	*Bulletin monumental*
BPV	*Bulletin de la Société Philogique Vosgienne*
BRGK	*Bericht der Römisch-Germanischen Kommission des Deutschen Archäologischen Instituts*
BROB	*Berichten van de Rijksdienst voor het oudheidkundig Bodemonderzoek*
BSAB	*Bulletin et mémoires de la Société archéologique de Bordeaux*
BSAGers	*Bulletin de la Société Archéologique de Gers*
BSANorm	*Bulletin de la Société des Antiquaires de Normandie*
BSAP	*Bulletins trimestriels de la Société des Antiquaires de Picardie*
BSAS	*Bulletin de la Société Archéologique de Soissons*
BSDASD	*Bulletin de la Société Départmentale de l'Archéologie et Statistique du Drôme*
BSSY	*Bulletin de la Société des Sciences . . . . de l'Yonne*
BTSAT	*Bulletin Trimestriel de la Société Archéologique de Touraine*
BVbl	*Bayerische Vorgeschichtsblätter*
BadFund	*Badische Fundberichte*
Britannia	*Britannia*
BudReg	*Budapest Regiségei*
BullNantes	*Bulletin de Nantes (de la Société Archéologique et Historique de Nantes)*
CAAH	*Cahiers alsaciens d'Archéologie, d'Art et d'Histoire*

CAF	Congrès Archéologique de la France
CAHC	Cuadernos de Arqueologia e Historia de la Ciudad
CAN	Congreso Arqueologica Nacional
CTEERoma	Cuadernos de Trabajos de la Escuela Espanola ... en Roma
Caesarodunum	Caesarodunum
Car	Carinthia I
Celticum	Celticum
EHR	English Historical Review
Emerita	Emerita
FA	Fasti Archaeologici
FMGTV	Forschungen und Mitteilungen zur Geschichte Tirols und Vorarlbergs
FO	Fundberichte aus Österreich
FHess	Fundberichte aus Hessen
FolArch	Folia Archaeologica
GMDS	Glasnik Muzejskega drustva za Slovenijo
Gallia	Gallia
Genava	Genava
Germania	Germania
Gymnasium	Gymnasium
HA	Helvetia Archaeologica
HSPh	Harvard Studies in Classical Philology
Hémecht	Hémecht
JBB	Jahreshefte Bayerische Bodendenkmalpflege
JDAI	Jahrbuch des Deutschen Archäologischen Instituts
JGPV	Jahrbuch der Gesellschaft Pro Vindonissa
JHVFL	Jahrbuch des Historischen Vereins für das Fürstentum Liechtenstein
JOEAI	Jahreshefte des Österreichischen Archäologischen Instituts
JRGZ	Jahrbuch des römisch-germanischen Zentralmuseums (Mainz)
JRS	Journal of Roman Studies
JSG	Jahrbuch für Solothurnische Geschichte
JSGU	Jahrbuch der Schweizerischen Gesellschaft für Ürgeschichte
LincsHA	Lincolnshire Historical and Archaeological Association Bulletin
MCACO	Mémoires de la Commission des Antiquités ... de la Côte d'or
MDAI(R)	Mitteilungen des Deutschen Archäologischen Instituts (Rom Abt)
MHVP	Mitteilungen der Historischen Verein für Pfalz
MSABayeux	Mémoires de la Société d'Agriculture ... de Bayeux
MSAIlle-Vil	Mémoires de la Société Archéologique ... d'Ille et Vilaine
MSHABret	Mémoires de la Société d'Histoire et d'Archéologie de Bretagne
MSHACS	Mémoires de la Société d'Histoire de d'Archéologie de Chalon-sur-Saône
MZ	Mainzer Zeitschrift
NA	Nassauische Annalen
NAH	Notiziario Arqueologico Hispaniarum
NC	Numismatic Chronicle
NHeim	Nassauische Heimatblätter
NSA	Notizie degli Scavi di Antichita
NSJFS	North Staffordshire Journal of Field Studies
NorfArch	Norfolk Archaeology
OJh	Österreichische Jahreshefte
OMRL	Oudheidkundige Mededelingen uit het Rijksmuseum van Oudheiden te Leiden
PAR	Pro Austria Romana
PBSR	Papers of the British School at Rome
PfMus	Pfalzisches Museum
Proc Arch Inst	Proceedings of the Archeological Institute
Pyrenae	Pyrenae
RA	Revue Archéologique
RACF	Revue archéologique du Centre (de la France)
RAnjou	Revue d'Anjou
REA	Revue des Études Anciennes

REL	*Revue des Études Latines*
RdN	*Revue du Nord*
RS&A	*Revue de Saintonge et Aunis*
RSHAM	*Revue de la Société Historique et Archéologique de Maine*
RVM	*Rad Vojvodzanskih Muzeja*
RevEE	*Revista de Estudios Extremeños*
SAC	*Sussex Archaeological Collections*
SJ	*Saalburg Jahrbuch*
Sept	*Septentrion*
Syria	*Syria*
TAADN	*Transactions of the Architectural and Historical Society of Durham and Northumberland*
TWFC	*Transactions of the Woolhope Field Club*
TZ	*Trierer Zeitschrift*
Tamuda	*Tamuda*
US	*Urschweiz*
Vastro Spomenikov	*Vastro Spomenikov*
VHAD	*Vjestnik za Arheologiju i Historiju Dalmatinsku*
WAM	*Wiltshire Archaeological Magazine*
YAJ	*Yorkshire Archaeological Journal*
ZArch	*Zeitschrift für Archäologie*
Zephyrus	*Zephyrus*

References

1 Roman Imperial Fortifications (pages 9–30)

1 Boethius and Ward-Perkins, 1971, 58f; F. R. Winter, *Greek Fortifications* (1971), 326f
2 Cicero, *De Lege Agraria* 2, 73
3 On Aosta, see I. A. Richmond, (ed P. Salway), *Roman Archaeology and Art* (1969), 249f
4 On Augustan gateways, Schültze, 1909 and Kähler 1942
5 I. A. Richmond, *PBSR* x, 12–22; Boethius and Ward Perkins 1971, 203–4
6 Schültze and Steuernagel, *BJ* 98 (1895), 1f; Anthes 1917, 89f; Petrikovits 1960, 85f; U. Sussenbach, *Die Stadmauer des römischen Köln* (1981)
7 Stähelin, 1948, 604f; Schwarz, *Die Kaiserstadt Aventicum* (1964)
8 Wacher, 1974, 137f (on Gloucester); 120f (Lincoln); 104f (Colchester)
9 Aquincum, Mocsy, 1974, 161; Xanten: Bogaers and Rüger 1974, 106; Tongres: Bogaers and Rüger 1974, 214f
10 Ladenburg: *Germania* XXX (1952), 67. Heddernheim: *Germania* xxii (1938), 161f
11 Garcia y Bellido, *Archivos del Instituto de Estudios Africanos* VIII (1955), 31f
12 R. Laur-Belart, *Führer durch Augusta Raurica*, 30f
13 *JOEAI* 1935, Bb 1f: 1953 Bb, 93f
14 Wacher, 1974, 75f
15 R. F. Hoddinott, *Bulgaria in Antiquity* (1975), 169, 187, 339
16 K. Dietz and others, *Regensburg zur Römerzeit* (1979), 194f
17 E. M. Wightman, *Roman Trier and the Treveri* (1970), 92f; E. Gose, *Die Porta Nigra in Trier* (1969), 59f
18 Bechert, 1971, 260f
19 Romanelli *Topografia e Archeologia dell' Africa Romana*, (1970), 46–8; G. C. Ricard, *Castellum Dimmidi* (1974)
20 Bu Ngem and Gheria-el-Garbia: R. G. Goodchild, *PBSR* xxii (1954), 59. Theilenhofen: Schönberger, 1969, map B no 143
21 Brünnow and van Domaszewski, 1904–9. Poidebard, 1934; Poiderbard and Mounterd, 1945
22 M. Gichon, *RFS* 6, 175f
23 For fuller descriptions of both sites, see Brünnow and Domaszewski 1904–9: II, 24f, and I, 433f
24 Brünnow and Domaszewski 1904–9, III, 187f
25 R. G. Goodchild, *PBSR* xxii (1954), 59
26 J. Wagner, in S. Mitchell (ed) *Armies and Frontiers of Anatolia*, *BAR* forthcoming. I am grateful to Mr J. G. Crow for this reference
27 Brünnow and Domaszewski 1904–9, II, 49f
28 MAR College, *Parthian Art* (London 1977), 30f

2 The New Design in Late Roman Defences (pages 31–54)

1 Vitruvius, *De Architectura* 1, 4
2 Vegetius, *De re militari*, iv, 1f
3 Dilke, *Introduction to the Agrimensores* (1971), 66–70
4 Vegetius, i, 21f
5 Bogaers and Rüger, 1974, 33f
6 AM 17, 9, 1; 18, 2, 3
7 Eg AM 17, 1, 11
8 B. W. Cunliffe, *Excavations at Portchester* I (1975), 19f
9 S. Johnson, in A. P. Detsicas (ed) *Essays in Memory of Stuart Rigold* (1981) 23f
10 J. Wilpert, 'Die Malereien der Grabkammer des Trebius Justus' in Dölger, *Konstantin der Grösse und seine Zeit* (Freiburg 1913) 276f
11 S. Johnson, *loc, cit.* (n 9) above
12 Ausonius, *Ordo Nobilium Urbium* XVIII
13 P. Corder, 'The Reorganisation of the defences of Romano-British towns in the fourth century', *AJ* cxii (1955), 20f
14 R. M. Butler, in *Soldier and Civilian in Roman Yorkshire* (1971) 97f
15 Vegetius, iv, 6
16 Richmond, 1930, 76f and fig 14
17 Forrer, 'Zür Bedahung der römischen Torburgen', *Germania* 1918, 73f

18 See, eg, Bluma L. Trell, 'Architectura Numismatica Orientalis', *NC* 1970, 29f
19 Grenier, 1931, 538. Forrer, '*Strasbourg, Argentorate*', 223, fig 133
20 Petrikovits, 1971, 201. G. Stein, *SJ* (1961), 8f; (Andernach) and *SJ* 23 (1966) 106 (Boppard)
21 Richmond 1930, 91f; Todd 1978, 35f
22 Ausonius *Ep* XXV, 124f, also XIX, 144. Paulinus of Pella, *Eucharist*, 46–7
23 *CIL* XIII, 6256, with a more mutilated but parallel text *CIL* XIII 5249, from Oberwinterthur
24 Gersbach, in *Provincialia* Festschrift R. L. Belart (Basel 1968), 551f; L. Beyer, *JSGU* 59 (1976) 206f and E. Gersbach, in *Helvetia Antiqua* (1966) 271f
25 Wightman, 1969, 117f
26 C. M. Daniels, *RFS* 12, 173f
27 J. J. Wilkes, in *Britain and Rome* (1966) 114f
28 G. Bersu, *Die spätrömische Befestigung 'Bürgle' bei Gundrzmmigen* (1964), 45
29 J. Garbsch *All. Gesch* 73 (1973) 43f
30 Stehlin 1957, nos 11–12
31 R. Fellmann, *Antike Welt* 2 (1979) 47f

3 Contemporary Records (pages 55–66)

1 *CIG* 3747–8. H. Karnapp, *Die Stadtmauer von Isnik*
2 M. Price and B. L. Trell, *Coins and their Cities* (London 1977), *passim*
3 XII Panegyrici Latini, ed R. A. B. Mynors (Oxford 1964)
4 *CSEL* XIX, XXVII, ed S. Brandt and G. Laubmann
5 *Ammianus Marcellinus*, ed Gardthausen (Leipzig 1874–5): English translation and parallel text in Loeb series by J. C. Rolfe
6 *Sextus Aurelius Victor de Caesaribus*, ed Fr. Pichlmayr (Leipzig 1970)
7 Eutropius, *Breviarium ab Urbe condita*, ed H. Droysen *MGH* AA 2 (Berlin 1879)
8 Zosimus, *Historia Nova*, ed L. Mendelssohn (Leipzig 1887). For English verion see J. J. Buchanan and H. T. Davis *Zosimus Historia Nova* (1967)
9 Zonaras, *Epitome Historiarum*, ed L. Dindorf (Leipzig 1868–75)
10 The bibliography on the Historia Augusta is extensive. The theory here outlined is argued persuasively by R. Syme, *The Historia Augusta* and *ibid. Ammianus and the Historia Augusta* (1968). For the text, *Scriptores Historiae Augustae*, ed E. Hohl (Teubner 1971). English version and parallel text by D. Magie, in Loeb series. See also the introduction (by A. R. Birley) to *Lives of the later Caesars*, (Harmondsworth 1976)

11 Ausonius, ed K. Schenkl, *MGH* AA 5, 2 (Berlin 1883); for English version and parallel text see Loeb ed by H. G. Evelyn White
12 Sidonius Apollinaris, *Poems and Letters* ed C. Luetjohann, *MGH* AA (Berlin 1887): for English version and parallel text, see Loeb edn, ed W. B. Anderson
13 Avienus *Carmina* (ed A. Holder) (1965)
14 Julian, *Letters* (Loeb edn. ed.   )
15 For example, Themistius, *Orationes* (in Greek, no English translation) ed G. Dindorf (Leipzig 1965)
16 Vegetius, *De Re Militari*, ed K. Lang (Leipzig 1967)
17 K. Miller, *Die Peuteringische Tafel* (Stuttgart, 1962)
18 O. Cuntz, *Itineraria Romana* i (Leipzig, 1929)
19 O. Seeck, *Notitia Dignitatum* (Frankfurt 1962)
20 *Codex Theodosianus*, ed T. Mommsen and P. Meyer (1905): translation and notes by C. Pharr, 1952
21 *Codex Justinianus*, ed P. Kruger (Berlin 1963), *Corpus Iuris civilis* 2
22 For the Chronicler of 354, see *MGH* AA ix (Chronica Minora)
23 *MGH* AA ix (Chronica Minora 1), 230
24 Malalas, XII, Richmond 1930, p 29
25 *CIL* III, 734
26 *CIL* V, 3329
27 *CIL* III, 3653
28 *AM* 17, 1, 13
29 *AEp* 1942–3, 81, from Aqua Viva, on the Syrian frontier, which was built in 303
30 *CT* XV, 1, 13
31 R. Macmullen, 'Roman Imperial Building in the Provinces', *HSPh* 64 (1959), 207f, and especially p 219 and n 95
32 Malalas, XII, 409; Procopius, i XIX, 34–5; Procopius *De Aed* II, VIII, 7
33 *AM* 16, 11, 11 : 17, 9, 1. Julian, *Letter to the Athenians*, 297A
34 *AM* 28, 2, 1; 29, 6, 2
35 *AM* 30, 3, 1
36 *AM* 28, 2, 2–4
37 Themistius *Or* X, 136a–138b. The quotation actually comes from 137b
38 *ILS* 773, in Arabia, built by the Equites VIII Dalmatae; *ILS* 774 at Ips by auxiliary troops; *ILS* 775, built at Gran by legionaries
39 *CIL* VIII 22774. Similar tituli are also found elsewhere in Africa – *CIL* VIII 9725, 19328, 21531
40 Macmullen, 1963, 12–3; *CT* VII, 21, 1; VII, 15, 1
41 Anonymus, *De Rebus Bellicis* XX
42 *ND* Occ XXV, XXX, XXXI
43 *Gai Institutiones* II: Seston, in *Mélanges Piganiol*, III 1489f
44 *Digest* 43, 6, 3 ; 1, 8, 9. See also *Digest* 1, 1, 6

45 *CIL* XII, 2229
46 Cities which changed names: Orléans (Cenabum, became Civitas Aurelianorum, *NGall* IV, 7) Tac (Gorsium, changed at some stage to Herculea) Coutances (Cosentia? changes to Constantia, *ND* Occ XXXVII, 20 and *NGall* II, 8)
47 *CIG* 3747–8
48 Macmullen, *HSPh* (cited n 31), 207, and especially the table of works done by troops on p 218
49 SHA, *Probus* 9
50 Julian, *Convivium* 314b
51 Libanius, *Discours* (ed. Harmand), 109; Eumenius, *PanLat* IX (IV) 4, and 18, 4. Lactantius, *De Morte Persecutorum*, 7, 8–9
52 AM XVIII, 2, 6
53 SHA *Gallienus*, 13, 6
54 Liebenam, *Stadtverwaltung* 141f
55 Jones, 1964, 1013–1014, and fn *ad loc.* 62–5
56 Diocletian's Price Edict, section VII, where a professor of architecture earns 100 denarii for each pupil per month, lower than a teacher of Greek or Latin (200 denarii) or a sophist or orator (250 denarii).
57 *CT* XIII, 4, 1 (AD 334), see also XIII 4, 3 (AD 344), XIII 4, 2 (AD 337)
58 *Pan Lat* IX (IV), 4, 3; *Pan Lat* VIII (V) 21, 2
59 Macmullen, 1963, 33f and n 40 *ad loc.*
60 *CT* XV, 1, 36
61 *CT* IV, 12, 5. A similar arrangement was made for Autun after the Bagaudic raids, *Pan Lat* V (VIII) 11f, where the city was allowed a remittance of one fifth of its normal contribution for rebuilding.
62 *CT* XV 1, 18 and V 14, 35. Jones, 1964, 1304, n 58
63 *CJ* XI, 42
64 AM 29, 6, 11
65 *CT* XV, 1 passim. AM 30, 5, 14
66 Other cities are also said to be in a bad state of repair – Autun AM 16, 2, 1, Avenches 15, 12, 12, Carnuntum, 30, 5, 2
67 *CT* XV, 1, 16, 17 and 37
68 *CT* XV, 1, 3; 11; 15; 21; 27; 28; 29
69 *CT* XV, 1, 4; 22; 38; 39; 46
70 *CT* XV, 1, 41. This is also implied by XV, 2, 9 and X, 3, 5. For repairs made from civic funds, *CT* XV, 1, 18; 32; 33 and V, 14, 35. Jones, 1964, 1301f
71 Richmond, 1930, p 29, Malalas *chron* XII, 299. On collegia in general, see *RE* V, 1375. Jones 1964, 708 and nn 49–50 *ad loc.*
72 J. W. Liebeschutz, *Antioch* (Oxford 1972), Libanius *Or* 46, 21: *Or* 50, passim: Jones 1964, 1198 n 125
73 On exemptions from public works (thereby implying that there was some kind of compulsion) *CT* XV, 1, 5; 1, 7; 1, 33; and instances of a general levy *CT* XV, 1, 49; XV 1, 23; XI 17, 4, and *Digest* 50, 4, 4.
74 H. A. Thompson, 'Athenian Twilight, AD 267–600', *JRS* xlix (1959), 64 n 20: *Inscriptiones Graecae* 11² 5199, 5200
75 See H. Pflaum, *Syria* 1952, 307–9
76 Jones, 1964, 734 and n 51 *ad loc*
77 His order is: Rome, Constantinople, Carthage, Antioch, Alexandria, Trier, Milan, Capua, Arles, Seville, Cordova, Tarragona, Braga, Athens, Catane, Syracuse, Toulouse, Narbonne and Bordeaux

4 The Pressures on Rome: Barbarian Invasions and Tactics (pages 67–81)

1 E.g. Koethe, 1942, 199f; Kellner 1971
2 Alfoldy, 1974, 155–6, Kellner 1971, 71f
3 On Regensburg: see Kellner, 1971, 75
4 Alfoldy, 1974, 155f
5 Rodwell and Rowley, 1975, 94
6 See Johnson 1976, 17f; Cunliffe, in Johnston 1977, 5–6
7 Breeze and Dobson, 1976, 134f
8 G.y Bellido, in *Archivios dell'Instituto de Estudios Africanos* VII no 33 (1955), 31f. Thouvenot, *Essai sur la province Romaine de Bétique* (1940) 379f
9 See, for example on the Pannonian frontier, J. Fitz *Alba Regia* 4–5 (1963–4), 75f
10 Orosius VII, 22; SHA *Marcus Aurelius* 11, 7
11 A. H. M. Jones, 'Inflation under the Roman Empire' in *EHR* 5 (1953), 293f
12 Herodian, VI 9, 7
13 Herodian, VII, 2, 6–8
14 SHA *Gordian*, 34, 4
15 *CIL* XIII 4131
16 Sites claimed to have been damaged or destroyed in this raid are Dambach, Pförring, Welzheim, Eining and Pfunz. The raid is documented by Herodian VI, 7 and 8. The coin hoards are documented in Kellner, 1971, p 140
17 Klein and Klumbach, *Die römische Schatzfund von Straubing* (1951)
18 Alfoldy, 1974, 169; Mocsy 1974, 202f
19 *RIC* IV, 2, nos 2–4, and 18–23 for coins bearing the title *Germanicus* and celebrating a *Victoria Germanica*.
20 Kellner, 1971, 143f
21 Demougeot 1969, 484f
22 Aur. Vict, *Caes* 32, 5
23 Aur. Vict. *Caes* 33, 1; Zosimus 1, 30, and 37
24 *AEp* 1935, 140
25 Demougeot, 1969, 493f
26 Koethe, 1942, 199f
27 Aur. Vict. *Caes* 33, 3
28 There is a large literature on these Spanish

invasions. Two of the main articles were contributed by A Balil 'Las invasiones Germanicas en Hispania durante le segunda mitad del siglo III' *CTEERoma* ix (1957), 95f, and 'Hispania en los anos 260 a 300 despues de J.C.' *Emerita* xxvii (1959) 269f. See also Blazquez, 1968
29 Tarradell 'La crisis del siglo III en Marruecos', *Tamuda* III (1955) 75f
30 *REL* 1938, 266–8
31 Demougeot 1969, 442f. Mocsy 1974, 206f
32 Kellner 1971, 149
33 *JOEAI* VI (1903) Bbl 107f; *FolArch* VI (1954), 73
34 Wightman, 1967, 98; *CIL* XIII 3679
35 SHA *XXXTyr*, V, 4
36 R. Roeren *JRGZ* vii (1960), 214f; Weidemann *JRGZ* xix (1972), 99
37 Grenier, 1931, 447–9
38 Demougeot, 1969, 393f
39 On Dexippus, see F. Millar, in *JRS* lix (1969), 27f
40 On the course of these events, see Demougeot, 1969, 409–16
41 Demougeot, 1969, 422
42 Millar, *loc. cit.* (n 39)
43 H. A. Thompson, *JRS* xlix (1959), 63f
44 Thompson, 1965, 134f
45 Demougeot, 1969, 425f
46 SHA *Aurelian* 39, 7
47 *Pan Lat* VIII (V), 10, 2
48 Belart, 1966, 12
49 *CIL* XIII 5203
50 Velkov. *Archéologie (Sofia)* X (1958), 124f
51 Demougeot, 1969, 521f
52 Gregory of Tours, *Hist. Franc.* III, 19
53 SHA *Probus* 13, 5–7
54 Julian, *Convivium*, 314b. Other sources are Zosimus, 1, 67, Orosius VII, 24
55 *FGH* 101, fragment 2, para 5
56 Balil, *Emerita* (quoted in n 28), 281–2; Koethe, 1942, 199f
57 Johnson, 1976, 16f
58 Eutropius IX, 21; Aurelius Victor *Caes* XXXVII, 20
59 Jones, 1964, 625f
60 Thompson, 1965, 140–1
61 Alfoldi, in *RFS* 1, p 1f
62 AM 28, 5, 1
63 AM 28, 5, 7
64 E.g. the Goths in 260, Thompson 1965, 144
65 see, for example, D. A. White, *Litus Saxonicum* (1961), 39–40
66 Tacitus, *Annales* 1, 57, 1; 12, 29, 2
67 *FGH* 101 Frag. 2f
68 Ibid.; AM 31, 5, 16; Zosimus I, 29, 2
69 See above p 73, and n 44
70 See above p 76, and n 55
71 Dexippus' accounts of the sieges at Marcianopolis (frag. 25) and Side (frag. 29), end, like that at Philippopolis, with the barbarians giving up in frustration
72 AM 16, 2, 6–7
73 AM 17, 6
74 AM 30, 3, 3
75 Thompson, 1965, 131f
76 P. V. Marsden, *Greek and Roman Artillery, Historical Development* (1969), 139f
77 AM 23, 4; see Marsden (*op. cit.* n 76), p 192f
78 Schramm, *Die antiken Geschütze der Saalburg* (1918), 29f; Marsden (*op. cit.* n 76) 86f
79 Richmond, 1930, 90
80 *ND* Occ IX, 33 at Autun, and IX, 38 at Trier, probably providing equipment for specialist troops like the *milites ballistarii* at Boppard (*ND* Occ XLI, 23)
81 Vegetius *De Re Militari* 2, 11
82 AM 19, 5, 2
83 *ND* Occ VII, 97
84 Zosimus 1, 70
85 AM 18, 9, 1
86 AM 17, 1, 12
87 For example, AM 18, 7, 6; 19, 1, 7; 19, 5, 6; 20, 7, 2; 31, 15, 12
88 J. Garbsch, in *Realeneyclopedie der Germanischen Altertumskunde*, s.v. Bewaffnung
89 For this view surviving into the fifth century, see Sidonius Apollinaris. *Letters* VIII, 6, 13–15
90 See R. Tomlin, in J. Casey (ed) *The End of Roman Britain BAR* 71 (1979), 253
91 *FGH* 101, fig 28; see Millar *loc. cit.* (n 39) 27f
92 *AEp* 1928, 38
93 *REL*, 1938, 266–8
94 See Tomlin, *loc. cit.*, (n 90) 262
95 AM 20, 11, 5

5 Town and City Fortifications (pages 82–135)

1 On the terminology and nomenclature, see F. Vittinghoff in Mayer (ed) 1956, 11f, and A. L. F. Rivet, in Bartholomew and Goodburn 1976, 134–5
2 Published by O. Seeck, *Notitia Dignitatum* (Berlin 1876), 261f
3 J. Harries, *JRS* lxviii (1978), 26f
4 A. L. F. Rivet, in Bartholomew and Goodburn, 1976, 119f
5 A. Audin, *Essai sur la topographie de Lugdunum* (1956), Ch. II; *Caesarodunum* Supp 28 (1978), 243f
6 De Fontenay and De Charmasse, *Autun* (1889), 1–48; Blanchet 1907, 14f; Grenier 1931, 337f; Kähler 1942, 590f; Duval *BSAF*, 1950–1, 81–7
7 *Pan Lat* IX (IV)
8 *Pan Lat* VIII (V), 21, 2
9 AM XVI, 2, 1

10 *ND* Occ IX, 33-4
11 Blanchet, 1907, 21f
12 Eutropius *Breviarium*, i, 9, 23
13 *MSHACS* ii (1850), 5f; xxxvi (1960-1), 54f; H. Armand Colliat, *Le Chalonnais Gallo-Romain* (1937), 48f; *Gallia* xix (1961), 447f
14 *ND* Occ XL, 111, the base of the Praefectus Classis Rhodanensis – see also *Pan Lat* VII (VI), 18, 26
15 AM XV, 11, 11
16 Jeanton, *Le Maconnais Gallo-Romain* (1926), 4f; *Gallia* xix (1961), 452
17 *ND* Occ IX, 32
18 *Acta Sanctorum*, I, 155
19 *Historia Francorum* III, 19
20 *MCACO* xii (1940-6), 310f; xxiv (1954-8), 115f; *Gallia* xii (1954), 477 xiv (1956), 288f; xviii (1960), 336f
21 H. Thévenot, *Le Beaunois Gallo-Romain* (1971)
22 *BCTH* (1920), 155f
23 Morel, *REA* 26 (1924), 68f; *Gallia* xv (1957), 128f
24 Blanchet, 1907, 33f; *Gallia* xxiii (1965), 272f
25 H. de Caumont, *Statistique Monumentale de Calvados* (1851), 450f; Blanchet, 1907, 36f; Lièvre, *MSABayeux* i (1891), 67f; v (1900), 85
26 L. Coutil, *Archéologie Gauloise, Gallo-Romain, Franque et Carolingien du département de l'Eure*, iv (1921), 23f; Blanchet, 1907, 37f; *Gallia* v (1947), 450
27 *BSANorm* 53 (1955-6), 169f; *Caesarodunum* Supp 28, (1978) 188f
28 De Vesly, *Le Castrum de Juliobona* (1915); Lantier, *RA* xxi (1913), 189f; Grenier, 1960, 895; *Gallia* xx (1962), 427
29 *BTSAT* xxvi (1937), 495f; xxvii (1938), 175f; xxviii (1939), 235f; *REA* 50 (1948), 313f; *FOR* XIII, 75f; *Gallia* xxxii (1974), 317; xxxiv (1976), 321; xxxvi (1978), 279
30 R. Charles, *RSHAM* ix (1881), 107f; 250f; x (1882), 325f; Triger *RSHAM* liv (1926), 267f; Ligier *La Cénomanie Romaine* (1904); de Linière, *RSHAM* lxxxvii (1931), 3f; *Gallia* ix (1951). 97f; xii (1954), 172f; xv (1957), 202f – Butler *JRS* xlviii (1958), 33f; Biarne, *Caesarodunum* Supp 28 (1 78), 145f
31 *MSAIlle-Vil* xx (1889), 71f; *MSHABret* xxx (1950), 73f; *Ann Bret* lxv (1958), 97f; *Gallia* xviii (1959), 343f; xxi (1963), 423f; xxvii (1969), 241f
32 *CAF* 1862, 38f; 1871, 28f; *RAnjou* vii (1870-1), 375f; lxiii (1911), 401f; lxviii (1914), 5f; *Gallia* xiii (1955), 164; *AnnBret* lxxi (1964), 85f M. Prévost, *Angers Gallo-Romain* (1978), 133f
33 Blanchet, 1907, 56f; Durville *Fouilles de l'Evêché de Nantes* (*Bull Nantes* Supplement) 1913; *Bull Nantes* lxxii (1932), 259f; lxxiii (1933), 289f; lxxvii (1937), 35f; lxxxvii (1947), 20f; xc (1951), 36f; *Gallia* xiii (1955), 157; xvii (1959), 350f

34 Blanchet, 1907, 60f; De la Martinière, *Ann Bret* (Mélanges Loth) 1927, 104f
35 Langouet, in Johnston 1977, 38f; *Dossiers du Centre Régionale Archéologique d'Alet*, 4 (1976), 72f
36 Laurain, *Les Ruines Gallo-Romaines de Jublains* (1928), 22f; Oelmann, *RFS* I, 80f; *Gallia* xii (1954), 171; *Caesarodunum* Supp 28 (1978), 334f
37 Sanquer, in Johnston, 1977, 45f
38 Galliou, *RFS* 12, 400f
39 *BM* 1854, 379f; 1906, 161f; Laval, *La Station Romaine de Rubricaire* (1913)
40 *BM* 1856, 308f; C. R. Smith, *Collectanea Antiqua*, iv (1857), 8f; *Gallia* xiii (1955), 162; *FOR* XIII, 64; *Gallia* xxxii (1974), 316
41 S. Johnson, *Britannia* iv (1973), 210f
42 *BCTH* i (1843), 331f; (1915), 26f; *Gallia* vii (1944), 112; ix (1951), 62; xii (1954), 144; xv (1957), 165; xviii (1959), 282f; xix (1961), 301; xxv (1967), 196; xxvii (1969), 231; xxix (1971), 225f
43 Jullian, *REA* v (1903), 35f; Matherat, *BSAF* 1943, 196f; *Gallia* v (1947), 438f; vi (1948), 453f; viii (1950), 220f; ix (1951), 83; xii (1954), 146; xv (1957), 165f; xxxiii (1975), 305; Roblin *REA* 67 (1965), 368f
44 Blanchet 1907, 103f; *BSAS* vii (1853), 199f; *BSAS* (3rd. ser) vii (1897), vif; *Gallia* xxvi (1968), 189f
45 *BSAP* xliii (1949), 20f; 37f; 226f; 352f; xliv (1951-2), 146f; xlv (1955), 16f; xlvii (1957), 224f *RdN* xl (1958), 467f; *Gallia* vii (1949), 103f; ix (1951), 72f; xii (1954), 129f; xiii (1955), 141f; xv (1957), 159; xvii (1959), 260f; xxxiii (1975), 312; J. L. Massy, *Amiens Gallo-Romain* (1979), 81f
46 Heurgon, *Hommages à Bidez et à Cumont* (*Latomus* II, 1948), 127f; *REA* xlvi (1944), 299f; l (1948), 101f; *RA* i (1958), 158f; ii (1959), 40f; *RdN* 42 (1960), 364f; *Gallia* xxvii (1969), 226f; xxix (1971), 229f; xxxiii (1975), 281f; xxxv (1977), 289f; xxxvii (1979), 285f; Seillier, in Johnston, 1977, 35f; *Sept* 8 (1978), 18f
47 *CAF* 135 (1977), 74f; *Gallia* xxxi (1973), 413; xxxiii (1975), 409f; xxxv (1977), 408f; xxxvii (1979), 424
48 Blanchet, 1907, 108f; *Gallia* ix (1951), 83f; xxxv (1977), 308
49 *Gallia* ii (1944), 159f; v (1947), 301f; viii (1950), 55f; xii (1954), 136f; xiii (1955), 143f; xiv (1956), 120f; xv (1957), 152f; xvii (1959), 247f; xx (1962), 88f; xxiii (1965), 291f; xxxiii (1975) 269f; xxxv (1977), 279f; *RdN* 44 (1962) 391 : 357f; 46 (1964), 183f; 47 (1965), 577f; 48 (1966), 557
50 *Gallia* xix (1961), 159f
51 Blanchet, 1907, 61f; *BCTH* (1903), 222f; *BSSY* xci (1937), 221f; *Gallia* xvi (1958), 313f; xviii (1960), 362f; *Caesarodunum* Supp 28 (1978), 214f; A. Huré, *Le Senonais Gallo-Romain*, (repr. 1978), 205f

52 Chartres: Blanchet, 1907, 67f; *REA* xv (1913), 60f; Troyes: Blanchet, 1907, 71f; *Gallia* xxii (1964), 296f; Melun: Blanchet, 1907, 83f
53 De Pachtère, *Paris à l'Époque Gallo-Romain* (1912); *REA* 53 (1951), 302f; *Gallia* xxv (1967), 205f; xxxv (1977), 321
54 Julian *Misopogon* 340 D
55 Blanchet, 1907, 68f; *BSSY* 58 (1904), 50f; *Gallia* xii (1954), 510
56 Blanchet, 1907, 73f; Debal, *Caesarodunum* Supp 28 (1978), 160f; *Gallia* xxxiv (1976), 325
57 Blanchet, 1907, 82f
58 See Ch. 6, pp 142–3
59 See Ch. 6, pp 158
60 *FOR* XII, 86f
61 M. Clavel, *Béziers et son Territoire dans l'Antiquité* (1970)
62 M. Labrousse, *Toulouse Antique* (1968)
63 *FOR* VIII, s.v. Nimes; Grenier 1935, 314; Grenier 1960, 150f; *Gallia* xxxvi (1978) 454
64 Blanchet, 1907, 144f; *Celticum* vi (1963), 307f
65 L. Constans, *Arles Antique* (1921), 97f and 217f; *JRS* xvi (1926) 174f; *Gallia* i (1943), 279f; vi (1948), 210; xii (1954), 429
66 *Genava* ii (1924), 109f; vii (1929), 135f; *REA* (1925), 125; *HA* i (1970), 71f; *AKorr* v (1975), 209f
67 Blanchet, 1907, 148f; *Gallia* xxii (1964), 519f
68 *FOR* X (1957), 44f; *BSDASD* (1933–4)
69 *Gallia* xiv (1956), 322f; xvii (1959), 293f; xix (1961), 311f; xxi (1963), 376f; xxiii (1965), 242f; *RACF* ii (1964), 303f; *Celticum* vi (1963), 297f; *Gallia* xxxi (1974), 307; xxxv (1977) 115f
70 R. Étienne, *Bordeaux Antique* (1962), 203f
71 Ausonius, *Ordo Nobilium Urbium* XX, lines 140f
72 *RS&A* ii, (1953), 33f; ii, 2 (1954), 83f; *Gallia* xiii (1955), 169f; xvii (1959), 478; xxiii (1965), 359; xxv (1967), 251; xxxv (1977), 379; L. Maurin, *Saintes Antique* (1978), 325f
73 Durand, *Fouilles de Vésone 1912* (1920); *Gallia* 11 (1944), 245f
74 Blanchet, 1907, 164f
75 Grenier 1931, 490; Blanchet 1907, 176f
76 P. Bistandeau, *Caesarodunum* Supp. 28 (1978), 382f
77 *BSAGers* lxii (1961), 293f
78 *BM* xxii (1856, 572f; C. R. Smith, *Collectanea Antiqua* v (1861), 226f; *REA* iii (1901), 211f; Blanchet 1907, 186f; *Gallia* xv (1959), 404f
79 Blanchet, 1907, 194
80 M. Lizop, *Les Convenae et les Consoranni*, (1931), 82f; *Gallia* xv (1957), 261f; xxiv (1966), 422f; *Revue de Comminges* lxxviii (1965), 83f
81 Blanchet, 1907, 195f; Lizop, *op. cit.* (n 80), 102f; *Gallia* xv (1957), 261f
82 *CAF* 1888, 396f; *REA* (1905), 147f; *CAF* 1939, 511f; *Gallia* xxxv (1977), 472

83 R. Cros-Mayreiville, *Les Monuments de Carcassonne* (1906); *REA* 24 (1922), 313f; Poux, *La Cité de Carcassonne*, (1922), 117f; *FOR* XII (1956), 166f; *Gallia* xx (1962), 613; xxxiii (1975), 495
84 *REA* iii (1901)
85 SHA, *XXX Tyranni* XXIII, 13, 6
86 W. Seston, in *Mélanges Piganiol* III (1966), 1489f; *Digest* 43, 6, 3; 50, 1, 6
87 SHA *Probus* 13, 8
88 Julian *Convivium*, 314b
89 Julian *Misopogon* 340D
90 Richmond, 1930; Todd 1978
91 *CIL* V, 3329; Richmond, *PBSR* xiii (1935), 69f
92 *NSA* 1900, 465–6; *NSA* 1938, 328f; Pan Lat IV (X), 21
93 Korosec, *Report on Archaeological Excavations on the Castle Hill Ptuj in 1946*: *RE* XXI, 1167f; *VHAD* lvi-lix (1954–7), 156f
94 *BRGK* XV (1924), 205f: *AArchSlov* v (1953), 109f; *RE* Supp XII, 139f
95 *Sirmium I and II* (1971–6)
96 Augsburg: *Germania* 32 (1954), 76f; Radnoti, *JBB* 2 (1961), 62f; Kempten: W. Schleiermacher, *Cambodunum* (1972)
97 *AErt* 89 (1962), 66–7; 94 (1967), 137f; 95 (1968), 131f
98 AM XXX, 5, 14f
99 *Bud Reg* xxiv (1976), 149
100 K. Sagi, in *Magyarorszag Regeszeti Topografiaja* I (1966), 81f
101 *AArchHung* ii (1952), 215f; *AErt* 87 (1960), 204, 234; Radnoti, in *Laureae Aquincenses* II, 91f
102 *AErt* 86 (1959), 177f
103 *RE* VIIA, 82f; *AArchHung* 14 (1966), 99f; *AErt* 99 (1972), 258
104 Soproni, 1978, 138–42
105 In general on the Spanish late Roman walls see Taracena 1948, Balil 1960 and Balil 1970.
106 Lacarra, *Estudios de Edad Media de la Corona de Aragon* i (1945), 193f
107 XI *CAN* (1970), 627f
108 Richmond 1931
109 Garcia y Bellido, *La Edad Antigua* (1968), 39f; Richmond 1931, 90–1; *NAH* ii (1933), 143f; v (1962), 152f
110 A. Balil, *Las Murallas Romanas de Barcelona* (1961); *CAHC* xiii (1969) 5f
111 Taracena 1948, 438; *Carta Arqueologica de España* (Soria), 125f
112 *AEA* 41 (1968), 121f; Mélida, *Monumentos de España* (1925), 45f
113 *AEA* 16 (1941), 257f; 21 (1948), 283f
114 Mélida, *op. cit.* (n 112), 45–6; *RevEE* xii (1956), 263f; *REA* (1961) 331f – Dias Martos, *Las Murallas de Coria* (1956)
115 *AIEC* vii (1927–31), 69f; *AEA* xv (1942), 114f

116 Nieto, *El Oppidum de Iruña* (1958); Balil 1960, 189f
117 Richmond 1931, 90f; Garcia y Bellido, in *Legio VII Gemina* (León, 1970), 569f
118 Richmond 1931, 86f; Balil 1960, 193f; Vilas, *Las Murallas Romanas de Lugo* (1972)
119 Balil, 1960, 188; V *CAN* (1957), 253f
120 Ausonius, *Ordo Nobilium Urbium* XIV
121 *ND* Occ XLII, 25–32
122 Richmond 1931, 93–4
123 *ND* Occ XLII, 19
124 Blazquez, in *VI Congreso Internacional de la Mineria* (León, 1970) 145–6
125 See Ch. 4, p 70f
126 Richmond 1931, 98–9
127 See Ch. 4, p 76f
128 Wacher 1974, 75; Frere 1974, 285f
129 *Norf Arch* xxx (1947–52), 146; *JRS* li (1961), 131
130 *JRS* xxxix (1949), 111; xliii (1953), 127; xlv (1955), 143; xlvi (1956), 144; li (1961), 191; *Arch Cant* lxi (1948), 1f
131 J. S. Wacher, *Excavations at Brough on Humber* (1969)
132 *AJ* cxix (1962), 114f
133 *AJ* cvii (1950), 94f; *Britannia* i (1970), 296; ii (1971), 278
134 *TWFC* 35 (1956), 138f; 36 (1958), 100f; 37 (1962), 149f; 38 (1966), 192f
135 *NSJFS* 2 (1962), 37f
136 M. Todd, in Rodwell and Rowley 1975, 215f
137 *RCHM Cambridge* vol. I, p lix; *Britannia* i (1970), 290
138 H. J. M. Green, in Rodwell and Rowley 1975, 204f
139 *AntJ* xl (1960), 175f; J. B. Whitwell, *Roman Lincolnshire* (1970), 69f
140 *Lincs HA* 2 (1967), 36; 4 (1969), 101; Whitwell *op. cit.* (n 139), 73f
141 M. Cotton, and I. Gathercole, *Excavations at Clausentum*; *AntJ* xxvii (1947), 151f; *JRS* xlii (1962), 100
142 *JRS* xliii (1953), Pl XIII; xlvii (1957), 222–3; xlix (1959), 131; xl (1960), 223–5; *WAM* 45 (1930–2), 197; 57 (1957–8), 233; 58 (1958–63), 245; 60 (1965), 137
143 *VCH Essex* III, 72f
144 Webster, in *RFS* 7, 38f
145 *RCHM York*, vol I (1962), 49
146 J. S. Wacher, in Butler (ed.) 1971, 167f
147 P. Salway, *The Frontier People of Roman Britain* (1965), 45f
149 R. M. Butler, in Butler (ed) 1971, 97f
148 Frere 1974, 290f
150 *Britannia* iii (1972), 302
151 G. Clarke, *Pre-Roman and Roman Winchester II, The Cemetery at Lankhills* (1979), 377f
152 *ND* Occ IX, 46f. For the possible identification of a spatharia (sword factory) at Amiens, see *Gallia* xxxv (1977), 314
153 *ND* Occ XLII, 33f

6 The Frontiers I: The Rhine (pages 136–168)

1 J. C. Mann, in *Aufstieg und Niedergang*, Pt. 2 vol 1, pp 508–33
2 *ND* Occ, I, 47; XXVI, XLI: Seeck suggested that the Dux Germaniae Secundae had dropped out and that there was a gap in the text
3 Wightman, 1967, 98
4 *CIL* XIII, 11975–6
5 *MZ* 48–9 (1953–4) 70f; *MZ* 58 (1963) 27f; *JRGZ* 15 (1968) 146f; *BJ* 172 (1972), 212f; 178 (1978) 29f
6 *LF* 4 (1962), 82; *FHess* 2 (1962) 158f; 4 (1964) 224f; *BJ* 172 (1972) 237f
7 Schumacher, *MZ* 1 (1906) 25
8 Petrikovits, 1960, 83
9 See Ch. 5 n 49
10 See Ch. 5 n 50
11 R. Brulet & J. Mertens *ABelg* 163 (1974)
12 Bogaers & Rüger 1974, 220f
13 Bogaers & Rüger 1974, 217f
14 *AC* 17 (1948) 119f
15 Hagen 1931, 139f
16 Holwerda *Nederlands Vroegste Geschiednis* (1925) 219f
17 Holwerda *Nederlands Vroegste Geschiednis* (1925) 221f
18 *Rheinische Ausgrabungen* 3 (1968) 1
19 Grenier 1931, 447f
20 Grenier 1931, 449f
21 Grenier 1931, 452–3
22 *BJ* 98 (1895) 1f; Bogaers & Rüger 1974, 160f
23 Bogaers & Rüger 1974, 106f
24 Bogaers & Rüger 1974, 214f
25 Wightman 1967, 92f
26 Sprater 1929 I, 23f
27 R. Forrer *Strasbourg-Argentorate* (1927); *CAAH* 130 (1949), 257f; 133 (1953), 73f; *CAAH* 2 (1958), 27f; 3 (1959), 39f; 13 (1969) 73f; *Gallia* 26 (1968) 423; 28 (1970) 317
28 Anthes, 1907, 108f; *MainzF* 13 p 16f
29 See n 25
30 R. Toussaint, *Metz à l'Époque Gallo-Romain* (1948), 165f; *Germania* 27 (1943–4) 79f
31 R. Toussaint, *Répertoire Archéologique de Meurthe et Moselle* (1947) 116; *Gallia* 6 (1948) 233f; 7 (1949) 88f
32 *Gallia* 26 (1968) 401
33 Y. Dollinger Leonard, in Mayer (ed) 1956 208f
34 Julian *Ep* 26
35 *RE* VIII A2, 1965f; *CAF* 118 (1960), 18f;

Levat, in *Mélanges Piganiol* (1966), 1037f

36 Stähelin 1948, 91f, 613f; *Ur und Frühgeschichtliche Archäologie der Schweiz* V (1975), 33f

37 Stähelin 1948, 205f, 604f

38 Belart 1966, 29f

39 Stähelin, 1948, 617f

40 F. Dijkstra and H. Ketelaar, *Brittenburg* (1965)

41 Bogaers & Rüger, 1974, 72

42 Bogaers & Rüger, 1974, 58f

43 Bogaers & Rüger, 1974, 74f

44 Bogaers & Rüger, 1974, 76f; *OMRL* 2 (1921) 6f; *Numaga* 16 (1969) 1f

45 Bogaers & Rüger 1974, 100

46 Bogaers & Rüger, 1974, 84f; *RFS* 7, 71f

47 Bogaers & Rüger, 1974, 96f; *BJ* 142 (1942) 325f, under 'Schneppenbaum'

48 Bogaers & Rüger, 1974, 101f

49 Bogaers & Rüger, 1974, 128f

50 Rüger, *BJ* 179 (1979), 499f

51 Bogaers & Rüger, 1974, 135f

52 Bogaers & Rüger, 1974, 139f; *BJ* 161 (1961) 475f; Petrikovits, 1960, 81f

53 Haberey, *BJ* 157 (1957), 295f; Bogaers & Rüger 1974, 147

54 Bogaers & Rüger, 1974, 163f

55 *CIL* XIII, 8502

56 *RFS* 12, 531f

57 Bogaers & Rüger, 1974, 208f

58 *ND* Occ I 47, V 141: Dux Mogontiacensis – Occ XLI

59 *BJ* 107 (1901) 1f; Anthes 1917, 96; *Germania* 39 (1961) 208f; *SJ* 19 (1961) 8f

60 Anthes 1917, 99f; *BJ* 142 (1937) 60f; *RFS* 12, 574f

61 Anthes, 1917, 100f; *BJ* 146 (1941) 323f; *SJ* 23 (1966) 106f; *RFS* 12, 567f

62 Anthes, 1917, 105f; Behrens, in *Katalog west- und süddeutscher Altertumssammlungen* IV (Bingen) 1920, 49f

63 Anthes, 1917, 107f; *NHeim* 43 (1953) 21f; *FHess* 4 (1964) 224; *BJ* 172 (1972), 232f

64 Anthes, 1917, 115f; *MZ* 62 (1967) 182; *MainzF* 12, 162f

65 *BJ* 122 (1922), 137f; *Germania* 13 (1929), 177f; *Germania* 38 (1960) 393f; 49 *BRGK* (1968); *MZ* 24–5 (1929–30) 71f; 27 (1934) 43f

66 Sprater 1929 I, 47f

67 *Gallia* 11 (1953), 150f; 12 (1954) 475f; 14 (1956) 300f; 22 (1964) 363f; 24 (1966) 316f

68 *PfMus* 45 (1928) 3f; G. Bersu, in G. Rodenwald *Neue Deutsche Ausgrabungen* (1930), 170f

69 See, for details and references, Appendix 2, nos 13, 14, 15, and 17, respectively (p 270–1)

70 Anthes, 1917, 104f; *TZ* 10 (1935) 1f; 24–26 (1956–8) 533f

71 Anthes, 1917, 103f; *TZ* 11 (1936) Beiheft 50f

72 Anthes, 1917, 101f; *TZ* 24–26 (1956–8) 502f

73 *Germania* 9 (1925) 68f

74 Anthes 1917, 121f; Forrer, *L'Alsace Romaine* (1935), 83f; *Gallia* 24 (1966), 219f

75 *ABelg* 103 (1967)

76 Ausonius *Mosella* 11

77 These include, in addition to those already mentioned, Maastricht (Bogaers & Rüger 1974, 186), Jülich, (Bogaers & Rüger 1974, 170), Scarponne (Grenier 1931, 119f), Pachten (*TZ* 11 (1936) Beiheft 107f; *Germania* 41 (1963), 28f); Saarebourg (17 *BRGK* (1927), 140), but see also *ASHAL* 60 (1960) 45, 52f; and 65 (1965), 6f, and Tarquimpol (*ASHAL* 4 (1892), 153f; 7 (1895), 182f)

78 Anthes 1917, 124f, Forrer, *L'Alsace Romaine* (1935), 74f, *CAAH* 8 (1964) 44f, *Gallia* 26 (1968), 434

79 *RFS* 11, 123f

80 *Germania* 24 (1940), 37f, *AKorr* 6 (1976), 309f

81 R. Fellmann, *Basel in römischer Zeit* (1955), 44f; *BZG* 60 (1960), 24f; 62 (1962), 17f

82 See Appendix 2, p 271 no 21; *AKorr* 4 (1974), 161f

83 Belart 1966, 165f; *JSGU* 57 (1972), 183f

84 *Bad Fund* 3 (1933), 105f; Anthes 1917, 131f

85 Anthes 1917, 132f; *JSGU* 50 (1963), 88f; *HA* 1 (1970), 50–3

86 Anthes 1917, 134f; Stähelin 1948, 272f; *JSGU* 31 (1939), 104, *HA* 6 (1975), 38f

87 *CIL* XIII 6526

88 *CIL* XIII 5249

89 *RFS* 3, 34f

90 Stähelin, 1948, 276, 597; *JSGU* 34 (1943), 58f; *US* 28 (1964), 1f

91 Stehlin, 1957

92 Garbsch, 1970; Garbsch 1967, 51f

93 *CIL* XIII, 11537

94 *CIL* XIII, 11538

95 Stähelin 1948, 273f; 633f; *JSGU* 50 (1963), 87f; *US* 32 (1968), 14f

96 Stähelin 1948, 307f; *JSG* 13 (1940), 143f; 36 (1963), 251f; 37 (1964), 306f; 38 (1965), 268f, *JSGU* 31 (1939), 1f

97 Stähelin 1948, 620f; *JSGU* 17 (1965), 74f; *REA* 25 (1923), 57f; *RE* XVII, 2502f

98 Stähelin, 1948, 631f; *JGPV* 1966, 12f

99 Anthes, 1917, 136f; *ASA* (1924), 212f; *JSGU* 28 (1932), 78f; 32 (1940–1), 150f, 36 (1945), 60f

100 Stähelin 1948, 276f, 595; *JSGU* 20 (1928), 82f; 23 (1931), 76

101 E. Vogt, *Der Lindenhof in Zürich* (1948)

102 Anthes, 1917, 137f; Stähelin 1948, 274f; Meyer, *Das Kastell Irgenhausen* (1954)

103 Anthes 1917, 140; Stähelin 1948, 275f; *JHVFL* 57 (1957), 231f; 59 (1959), 227f; *JSGU* 49 (1962), 29f

104 Pan Lat IX (IV), 18, 2, 2

105 AM XVI, 3, 1
106 AM XVI, 2–4, 11–12, XVII, 1–2, 8–10
107 AM XVIII, 2, 4
108 AM XVII, 9, 1
109 AM XVIII, 2, 3
110 CT VI, 35, 8
111 See Ch. 7, p 193–4
112 R. Pirling, *Das römisch-fränkische Gräberfeld von Krefeld-Gellep* I (1966); II (1974); III (1980).

7 The Frontiers II: The Danube (pages 169–195)

1 Garbsch, 1970
2 Anthes, 1917, 143: *PAR* 19 (1969), 13f
3 W. Schleiermacher *Cambodunum-Kempten* (1972)
4 Garbsch, 1970, 12f; *Germania* 49 (1971), 137f, *All Gesch* 73 (1973), 43f
5 *BVbl* 26 (1961), 60f, *Germania* 41 (1963), 128f
6 *ND Occ* XXXV 19
7 Anthes 1917, 144f; *BVbl* 20 (1954), 119f; Kellner and Wagner *Das spätrömische Kellmünz* (1957)
8 Petrikovits 1971, 211
9 Garbsch 1967, 51f
10 *ND Occ* XXXV
11 Keller, *AKorr* 1 (1971), 177
12 Eugippius *Vita Sancti Severini* 15, 1
13 G. Bersu, *Die spätrömische Befestigung 'Bürgle' bei Gundremmingen* (1964); *BJ* 165 (1965), 493
14 Anthes, 1917, 146f; *BRGK* 7 (1914), 19f; Wagner, *Die Römer in Bayern* (1928), 43f; Kellner, 1971, 62–4
15 R. Christlein, in Werner and Ewig (1979), 91f
16 Anthes 1917, 153; Reinecke *Germania* 3 (1919), 57f; *SJ* 15 (1956) 77f; *Neue Ausgrabungen in Deutschland* (1958), 312
17 *RLIÖ* 21 (1958), 28; *BVbl* 24 (1959) 132, *LF* (1962), 132f
18 K. Dietz (ed.) *Regensburg zur Römerzeit* (1979) 125f
19 See Ch. 10, pp 239–40
20 Anthes 1917, 154
21 *JOEAI* 44 (1959) Bb 5f; *Germania* 43 (1965) 427
22 See n 17
23 In general, see Ubl, in *RFS* 12, 587f
24 *CIL* III 5670 = *ILS* 774
25 *PAR* 8 (1958) 3f; 9 (1959) 7f; 10 (1960) 11f; Eckhardt, *RLIÖ* 25 (1969)
26 *ND Occ* XXXIV, 39 and 43
27 Noll, *RLIÖ* 21 (1958) 46f; *RFS* 12, 590
28 15 *BRGK* (1923–4) 166f; *FÖ* 10 (1971) 87; *Römisches Österreich* 5–6 (1977–8) 109f
29 *RE* XVII, 1, 1000
30 *RFS* 12, 591

31 *RFS* 12, 591
32 H. Stiglitz, *RLIÖ* 26 (1975)
33 *RFS* 12, 592; *RFS* 10, 251f
34 *RFS* 12, 593
35 R. Neumann, *Vindobona* (1972) 54f; *RFS* 12, 594
36 *RFS* 12, 594
37 A recent, useful summary is Fitz (ed) 1976
38 Fitz (ed) 1976, 19; *AErt* 92 (1965) 235f; *Arrabona* 8 (1966) 67f
39 Fitz (ed) 1976, 23; *RE* Supp X 91–8; *Arrabona* 9 (1967) 27
40 Fitz (ed) 1976, 27; *AErt* 99 (1972) 232f
41 *AArchHung* 2 (1952) 201; 13 (1965) 215f; Barkoczi, *Brigetio* (1951); Fitz (ed) 1976, 33–5
42 Dobias, *Il limes Romano nelle terre della Reppublica Cecoslovacca* (1938; Swoboda, *Rimsky tabor u Leanyvaru* (1957); *A Roz* 10 (1958) 548; *Slov Arch* 10 (1962) 422f
43 Fitz (ed) 1976, 38–9
44 Fitz (ed) 1976, 42–3
45 Mocsy, *Tokod* (1980)
46 *AErt* 83 (1956) 98, 194; 87 (1960) 207f; 92 (1965), 243; Fitz (ed) 1976, 48–9
47 *RFS* 3, 137–8; Fitz (ed) 1976, 52–3; Soproni 1978, 26–7
48 *RFS* 3, 134f; Fitz (ed) 1976, 58–9; Soproni 1978, 46–8
49 Soproni, 1978, 55–9
50 Fitz (ed) 1976, 68–9
51 Fitz (ed) 1976, 77f; *AErt* 69 (1942) 276f
52 *Bud Reg* XXIV (1976) 149; Fitz (ed) 1976, 82–5
53 Fitz (ed) 1976, 121
54 Fitz (ed) 1976, 123
55 *RE* Supp XI, s.v. Campona; *AErt* 85 (1958) 85; 91 (1964) 257; Fitz (ed) 1976, 93
56 *RE* Supp IX 398f; 635; *AErt* 96 (1964), 257; *AErt* 97 (1965) 236
57 *AArchHung* 4 (1954) 129f
58 Fitz (ed) 1976, 101f; *AErt* 105 (1973) 385f; *RFS* 12, 681f
59 See Fitz (ed) 1976, 107–117
60 *RE* XV, 2, 1568; *RE* Supp IX, 642
61 *APreg* (1963) 109f
62 *Limes u Jugoslaviju* I (1961) 89: *RVM* 11 (1962) 123f. In general on the area in *Pannonia II* see Graf 1936, 111f, and *Limes u Jugoslaviju* I (1961) passim: for further references see *TIR* L34 (1968) under site names
63 See also Ch. 10, pp 240
64 Jantsch, 1938
65 On Capidava, see Condurachi, in *RFS* 3. See also G. Florescu, R. Florescu and P. Diaconu, *Capidava* I (1958)
66 On Drobeta; Barcacila, '*Drubeta, azi Turnu Severin*' (1932); Tudor *Oltenia Romana* 1968, p 449f and also Florescu, *RFS* 6 144f, esp. p 147–8

and plan p 148
67 ND Occ XXXII, 41; XXXIII, 48
68 Mocsy 1974, 271f
69 Soproni, 'Limes Sarmatiae' AErt 96 (1969) 43f
70 Soproni RFS 9. 197f
71 Mitt. der Prähistorische Kommission der Akademie der Wissenschaften, in Wien II, 6 (1930) 439f
72 Pascher, RLiÖ XIX (1942) 145f
73 Kolnik, RFS 4 27f
74 RFS 11, 181f
75 Mocsy, Fol Arch X (1958) 85f
76 Porphyrius, carm, VI, 14; Zosimus II, 21
77 AM 16, 10, 20. Zosimus III, 1
78 CIL III 10596
79 CIL III 5670
80 AM 27, 4, 6
81 AM 14, 2, 3
82 AM 30, 5, 17
83 AM 30, 5, 2
84 AM 29, 6, 2. Watchtowers with stamped tiles of Valentinianic date have been found beyond the Danube at Wagram am Wagram, Etsdorf and Zeilberg – RLiÖ XIX 1949, 33, 66, 159
85 Wagner 'Das Ende der Römerherrschaft in Rätien', Mocsy 1974, 339f. Alfoldi, Untergang der Römerherrschaft in Pannonien I, 1924
86 Kellner, 1971, 182f; Kent RFS 2 85f
87 Eugippius, Das Leben des Heiligen Severin ed. Noll (1963)
88 Vetters, Gymnasium 76 (1969) 481f; ibid AAWW 106 (1969) 75

8 The Frontiers III: The North Sea (pages 196–214)

1 CIL XII, 616. In general on the Classis Britannica see D. Atkinson, 'Classis Britannica', in Historical Essays in Honour of James Tait, (Manchester, 1933), p 1–11 and 'The British Fleet', in B. W. Cunliffe, Fifth Report on the Excavations of the Roman Fort at Richborough, Kent. (Soc. of Ant., Res. Rep. 23), p 225f
2 Johnson 1976, 15f
3 J. L. Warner, Proc. Arch. Inst. (Norwich, 1851), 9f; J. K. St. Joseph, Ant.J. 16 (1936) 444f; D. A. Edwards and C. J. S. Green in Johnston 1977, 21f
4 On Reculver: R. Jessup, Antiquity X (1936), 179f; F. H. Thompson ACant lxvi (1953); B. Philp, ACant lxxiii (1960), 96f; B. Philp, The Roman Fort at Reculver (1969)
5 I. A. Richmond, Ant.J lxi (1961) 224–8; J. C. Mann, in Johnston 1977, 15
6 See, in general, J. Mertens, in Johnston 1977, 51–2; Johnson 1976
7 See n 26 for fuller references to Oudenburg
8 Aardenburg: D. de Weerd, BROB 18 (1968), 237f; J. A. Trimpe Burger, BROB 23 (1973) 135f
9 J. Trimpe Burger, 'Zeeland in Romeinse Tijd' in Deae Nehalenniae (1971), and BROB 23 (1973) 135f
10 See n 27 for fuller references to Brittenburg
11 J-Y Gosselin and others, Sept 6 (1976) 5f; Sept 8 (1978) 18f
12 B. Philp, The Classis Britannica Fort at Dover (1981)
13 Eutropius, ix, 21, Sextus Aurelius Victor, De Caesaribus xxxix, 20
14 In general for these sources, see N. Shiel, Carausius and Allectus, BAR 40, 1977, or Johnson, 1976, chapter 2, p 23f
15 Burgh Castle: C. F. C. Hawkes and J. Morris, AJ cvi (1949), 68f; J. Morris PSIA xxiv (1948), 102f; JRS li (1961), 183; lii (1962), 178; S. Johnson RFS 12, 325f
16 Walton Castle: C. Fox, VCH Suffolk, I (1911), 287, 305
17 Bradwell, RCHM Essex IV (1923), 13f; VCH Essex III (1963), 52f
18 Richborough: J. P. Bushe-Fox, Excavations at the Roman Fort of Richborough (Soc. of Ant. Research Reports) I (1926); II (1928); III (1932); IV (1949); V (ed. B. Cunliffe, 1968); S. Johnson, Britannia i (1970) 240f
19 Dover, R. E. M. Wheeler and C. J. Amos AJ lxxxvi (1929), 47f; S. E. Rigold, AJ cxxvi (1969), 78f; B. Philp, Roman Dover (1973); Britannia ii (1971), 286, iii (1972), 351; iv (1973), 322; v (1974), 459; vi (1975), 283; vii (1976), 376; viii (1977), 424; ix (1978), 471; xi (1980), 401
20 Lympne: C. Roach Smith, The Antiquities of Reculver, Richborough and Lympne (1850); B. W. Cunliffe, Britannia xi (1980) 227f
21 Pevensey: VCH, Sussex, III (1935), 5f; L. Salzman SAC 51 (1907), 99f; ibid 52 (1908), 83f. J. P. Bushe-Fox, JRS xxii (1932), 60f
22 Portchester, B. W. Cunliffe, Excavations at Portchester Castle I (Soc. of Ant. Res. Report), 1975
23 Lancaster: Britannia, ii (1971), 254; v (1974), 418; vi (1975), 239
24 Piercebridge: Keeney TAADN 7 (1936), 235f; ibid 9 (1939–41), 43f, 127f; ibid 10 (1950), 285f. Elslack: T. May YAJ 21 (1911), 113f; Newton Kyme: JRS xlvii (1957), 209
25 Cardiff: Archaeologia 57 (1899), 335f; ACamb 63 (1908), 29f; ibid 68 (1913), 159f; ibid 69 (1914), 407f; AntJ ii (1932), 361f
26 Oudenburg: J. Mertens, Helinium ii (1962), 51f; J. Mertens and L. Van Impe, Het laat romeins gravfeld van Oudenburg, ABelg 206 (1978), 73f
27 H. Dijkstra and F. C. Ketelaar, Brittenburg (1965); Mertens, in Johnston 1977, 51. G. Rickman, Roman Granaries (1971), 268f
28 AM xviii, 2, 3; Libanius Or. XVIII, 83
29 Boulogne, see n 11

30 Alet: L. Langouet, in Johnston, 1977, 38f
31 See p 94
32 ND Occ XXXVII, and XXXVIII
33 Described as 'militaris' by Ausonius, Ep. IV, 16
34 R. Sanquer, in Johnston, 1977, 45f
35 On these sites in more detail, see Johnson, 1976, 88f
36 See p 114–5
37 See n 13–14
38 E.g. D. A. White, *Litus Saxonicum* (Madison Wisconsin, 1961)
39 ND Occ XXVII
40 ND Occ V, 126–132, which despite textual problems includes the *Comes Litoris Saxonici* among a list of other *Comites*, two of whom at least held frontier troops under their command
41 J. S. Wacher, *Excavations at Brough on Humber* 1969
42 P. A. Rahtz, *Ant.J* xl (1960), 175, M. Todd, *The Coritani* (London, 1973), 42f
43 C. F. C. Hawkes, *AJ* cii (1964), 22f
44 P. W. Gathercole and M. A. Cotton, *Excavations at Clausentum* (1952)
45 B. W. Cunliffe, in Johnston 1977, 2f
46 J. P. Bushe-Fox, *JRS* xxii (1932), 60f
47 In more detail, see S. Johnson, in Bartholomew and Goodburn 1976, 81f

9 The Defence of Italy and Spain (pages 215–225)

1 On Grenoble and Die, see Ch. 5, nn 67 and 68
2 Roberti, in '*Provincialia, Festschrift Laur Belart*' (Basel 1968), 386f
3 Borghi, 'Il castrum di S. Stefano di Lecco', in *Oblatio Raccolta di Studi di Antichita ed Arte in Onore di Artistide Calderini*, Soc. Arch. Commense 1971, 232 and f29. The fortification at Lecco is very similar to those on the Rhine frontier and also to the towers at Castelseprio. On Coccaglia, see Roberti, in *Storia di Brescia* I (1961), 319
4 Zecchinelli, 'Fortificazioni Romani sul Lario' – *Archeologia e Storia nella Lombardi pedemontana occidentale*, (Como 1969), 157f
5 Hartmann, *JOEAI* 2 (1899), Bb 1ff; Bullough, *PBSR* X (1955), 148f. This Byzantine limes includes a series of watchtowers at Glemona, Gradišča and a large number of similar sites in the foothills of the southern side of the Alps near Friuli.
6 Stähelin, 1948, 275f; *JHVFL* 57 (1957), 231f; *JHVFL* 59 (1959), 227f; *JSGU* 49 (1962), 29f
7 *JOEAI* 44 (1959) Bb 5f
8 Anthes 1907, 154; *FMGTV* 10 (1913), 177f; *RE* Va, 727f
9 *ASA* 1903–4, 137f; *ASA* 1923, 78f; Poeschel, *Kunstdenkmäler des Kantons Graübunden* 7 (1948)
10 Jantsch, 1937, 325f
11 See Appendix 3 for details
12 *AEp* 1893, 88 (*ILS* 8977), erected soon after 170
13 'Claustra', see *Thesaurus Linguae Latinae* III, 1319f; Šašel and Petru 1971, 17f
14 Šašel and Petru 1971, nos 1–5
15 For example the section at Jelenje, Šašel and Petru 1971, no 2
16 On these sites in general, see the individual site surveys in Šašel and Petru 1971
17 *RFS* 4, 39f; Šašel and Petru 1971, 93f; T. Ulbert, *Ad Pirum* (Munich, 1981)
18 Šašel and Petru, 1971, 76f
19 Schmid, 15 *BRGK* (1923–4),183–9
20 *RFS* 8, 178; Šašel and Petru 1971, 98f
21 *RFS* 6, 122f; *Arch Pregled* 4 (1962), 224f; *Vastro Spomenikov* 9 (1962–4), 146f
22 *RFS* 6, 124f; *Vastro Spomenikov* 9 (1962–4), 192f
23 On Aquileia, see Chapter 5 n 98. See also Herodian VIII 2, 4–5
24 Šašel and Petru, 1971, no 6, Loski Potok. See also *GMDS* 1937, 17f, 132f and 1939, 118f
25 Šašel and Petru, 1971, no 11. Puschi *AT*, 144
26 Puschi 1902, passim; See also Puschi, in *Atti e memorie della Societa Istriana di archeologia e storia patria* 17 (1901), 376–401
27 Šašel and Petru, 1971, no 33
28 Šašel and Petru, 1971, no 26 and no 15 respectively
29 See n 5 above
30 Jantsch 1938
31 On the Hoischügel, see *Car* I 1951
32 P. Petru and T. Ulbert, *Vranje bei Sevnica* (1975). See also T. Ulbert, in Werner and Ewig 1979, 141f
33 *AEp* 1934, 230; Šašel and Petru 1971, 23
34 Herodian VIII, 1, 6
35 Julian, *Letters* III, 17, 20–5
36 Julian, *Letters* III, 18, 1–4; AM 21, 12, 21
37 AM 29, 6, 1
38 Philostorgius, XI, 2
39 Schmidt, *Geschichte der Deutschen Stamme, Die Ostgermanen* (1941), 437, 443
40 *CJ* I, 31, 4; I, 46, 4
41 Prosper, *Chron* 1367, on Attilla, in 452; Cassiodorus, *Varia*, II, 19, on Theodoric in 489
42 Wilkes, 1967, 419–20, Mocsy, 1974, ch 10, 339f
43 Blazquez, *RFS* 9, 485f; *RFS* 12, 345f
44 ND Occ XLII, 25–32. See Ch. 5, p 129–130 and n 109–19
45 ND Occ VII, 118–134
46 J. M. Lacarra, *Textos Navarros del Codice de Rota*, (1945). A. H. M. Jones, *X Congrès d'Études Byzantines*, (Istanbul 1957), 223f

47 See n 43
48 *ND* Occ XLII
49 Carcopino, *Le Maroc Antique*, 231f
50 Jones, *JRS* xliv (1954), 21–9
51 *ND* Occ XXVI
52 *Tamuda* i (1953), 59f; iii (1955), 75f; *BAM* 6 (1966), 394
53 *AEA* 22 (1949), 86f; *Zephyrus* v (1954),
54 *BAM* 6 (1966), 418
55 *BAM* 6 (1966), 374
56 *I Congreso Arqueologico del Marruecos Español*, 331f
57 *Tamuda* i (1953), 302f; *BAM* 5 (1964), 281- Ponsich, 1970, 352
59 See, e.g. Goodchild *JRS* xxxix (1949), 53 and xl (1950), 30f
60 Aurelius Victor *DeCaes*, 33, 3
61 J. Carcopino, *Le Maroc Antique*, (Paris 1943), 244; Tarradell, *Zephyrus* v (1953), 129f
62 M. Ponsich and M. Tarradell, *Garum et industries antiques de salaison dans la Mediteranée Occidentale* (Paris, 1965), Blazquez, *Hispania* xxviii (1968), 16–17
63 Ponsich, 1970, 401–2

10 Local and Rural Protection: Hill-top Defences (pages 226–244)

1 General works on hill-top defences are von Uslar 1963 and 1964, where the material for the Rhineland provinces is catalogued and discussed. All periods of defensive earthwork are there dealt with. For Belgium, Graff 1963 provides perhaps a rather uncritical list of late Roman hill-top sites: more precision is given in J. Mertens 'Le Luxembourg meridional au Bas Empire' *ABelg* 76 (1964). The basic list of sites given by von Uslar and Graff are supplemented by more recent material from other sources on individual sites: the footnotes to this chapter deal in the main only with the sites mentioned by name in the chapter. Short bibliographies of other sites are given in an appendix (Appendix 3)
2 In general, on this area, see M. Hartmann, in *Ur- und Frühgeschichtliche Archäologie der Schweiz* Band V (1977), 21f
3 In general on this area see Alfoldy 1974, 214f
4 In general on this area, see Puschi 1902: the defences in this area are discussed in Ch. 9, p. 218
5 See, for example, the complete excavation of the Altburg nr Bundenbach, in *Trierer Grabung und Forschungen* X (1977)
6 An introduction to this area is afforded by Wightman 1967, 176f, and by her paper 'Some aspects of the late Roman defensive system in Gaul', in *RFS* 7, 46f

7 J. Mertens and H. Rémy, *Le Cheslain d'Ortho*, *ABelg* 129 (1971)
8 A. Matthijs and G. Hossey, *Le Château des Fées à Bertrix*, *ABelg* 146 (1973)
9 R. Brulet, *La Roche à Lomme à Dourbes*, *ABelg* 160 (1974)
10 J. Mertens and R. Brulet, *Le 'Vieux Château' de Sommerain à Mont* (Acta Archaeologica Lovanensia 9, 1974).
11 Strasbourg: R. Forrer *Strasbourg-Argentorate* (1927), p 123 pl xviii; Saxon Shore forts: Johnson 1976, 57, 141
12 Bogaers and Rüger 1974, no 75 (p 240)
13 Graff, 1963, 126–7
14 Bogaers and Rüger, 1974 no 78 (p 247)
15 For bibliographical details on these sites, see Appendix 3
16 J. Mertens, *ABelg* 74 (1963), 220f
17 E. Wightman, in *RFS* 7, 49
18 Sites frequently quoted in this context are: Leikoppchen, near the villa of Palzkyll, Eprave and Jemelle near the villa estate of Nassogne, and a cluster of sites near Polch and Mayen, including the Katzenberg. Polch, Hambuch, Weiler and Kottelbach
19 See gazetteer in Appendix 3
20 J. Mertens, *Le refuge antique de Montauban sous Buzenol*, *ABelg* 16 (1953), and J. Mertens 'Sculptures Romaines de Buzenol, *ABelg* 42 (1958)
21 E. Wightman, *RFS* 7, 49: Weisgerber *AKorr* 3 (1973), 231
22 F. Sprater, 67f
23 *Ibid*.
24 Kaiser, *Germania* 39 (1961), 225f: G. Fehr, *Vor-und Frühgeschichtliche Beseidlung der Kreise Kaiserslautern* (1972) 53f; 103f
25 E. Wightman, *RFS* 7, 49
26 J. Nenquin, *La Nécropole de Furfooz* (Bruges 1953): R. Brulet, *La Fortification de Hauterecenne à Furfooz* (Pubs. de l'Art et d'Archéologie de L'Université Catholique de Louvain XIII, 1978)
27 W. Haberey, *BJ* 148 (1948), 439f
28 Bogaers and Rüger, 1974, no 74, p237; J. Mertens and H. Rémy *ABelg* 144 (1973) 72f
29 Graff, 1963, 148
30 See Brulet (*op. cit.* n 9), 48–9. In general on the laetic cemeteries in this area, see H. Roosens, 'Laeti, Foederati und andere spätrömische Bevölkerungsniederschlage im belgischen Raum', *ABelg* 104 (1968), 92f; A. Dasnoy, in *ASAN* liii (1966) 226f
31 *ND* Occ XLII 33–44; a general discussion of this and other literary evidence on laetic settlement in the area, see R. Gunther, 'Laeti, foederati und Gentilen im nord und nord-ost Gallien', *ZArch* v (1971), 39f
32 *ND* Occ XLII, 64–70
33 R. Brulet, *op. cit.* n 9, 49

34 J. Mertens, in *RFS* 12, 447f
35 R. Roeren, 'Zur Archäologie und Geschichte Südwestdeutschlands in 3–5 Jht'. *JRGZ* 7 (1960), 214f
36 Glauberg: J. Werner, 'Zu den alamannischen Burgen', *Speculum Historiale* (Festschrift J. Sporl, 1965)
37 G. Fehring, Frühmittelälterliche Wehranlagen in Sud-West-deutschland, *Chateau Gaillard studies* V (1972) 37f
38 Meyer, *Hémecht* 10 (1964), 179f
39 A. Rieth *Germania* xxxvi (1958), 113; T. Biller, *AKorr* 3 (1973) 459f
40 On this area see n 2
41 A. Gerster *US* 32 (1968) 17f
42 J. Garbsch, 1970 p 16 and map, Kellner 1971, 163f
43 G. Bersu *Das Wittnauer Horn* (Monographien zur Ur-und Frühgeschichte der Schweiz, 4) 1945
44 E. Gersbach, in *Provincialia* (Festschrift R. L. Belart) 1958, 551
45 L. Berger *JSGU* 59 (1976) 206f: K-J. Gilles, *Germania* liv (1976) 440f
46 J. Werner *Der Lorenzberg bei Epfach vol II* (1969)
47 J. Garbsch *Der Moosberg bei Murnau* 1966
48 Wilkes, 1967, 419
49 For example at Sisteron (see below, n 76) or le Camp de la Bure, in *BSPV* 90–1 (1964–5), 61, 92 (1966), 5
50 Alfoldy, 1974, 214f
51 Teurnia: R. Egger, *Teurnia*, (1970)
52 Karnburg: Von Uslar, 1964, no 209; *Car* I (1939), 265: (1948), 198
53 Grazerkogel: Von Uslar, 1964, no 125; Jantsch, 1938, no. 5
54 Lavant *JOEAI* 38 (1950) Bb, 31f
55 Duel *JOEAI* 25 (1929) Bb 189f
56 Hemmaberg, Von Uslar 1964, no 129; W. Schmid *BRGK* xv (1923/4), 236
57 Hoischügel, Von Uslar 1964, no 172; *Car* I, 1950, 178; W. Schmid *BRGK* xv (1923/4), 237
58 See Ch. 9, p 218f; Alfoldy, 1974, 219–20
59 Schaan: H-J Kellner, in *JHVFL* 64 (1965) 5f
60 Gurina: Meyer *Die Gurina* (1885)
61 Vranje: P. Petru and T. Ulbert *Vranje bei Sevnica* (1975); T. Ulbert, in Werner & Ewing 1979, 141f
62 *AEA* xv (1942), 21f; Taracena *CASE*, 436
63 Palol, *Boletin de la Seminario de Estudios de la Universidad de Valladolid* XXXII 1966, 33f
64 *Carta Arqueologica de España, Salamanca*: Yecla, 121–8, Lumbrales, 70–87, and Fuenteguinaldo, 63. On Lumbrales (Las Merchanas), see also Maluquer, in *Pyrenae* 4 (1968), 101f
65 Garcia Guinea, IX *CAN* (Valladolid) 1965, 33–4

66 Moreno, *Catalogo Monumental Salamanca* (Valencia, 1967), 57 and pl. 12. This is quadrilateral; 28.83 by 26.80m, and is clearly a type of block house. There is no published plan.
67 An exhaustive article, giving details of all such sites so far known is given by Palol, *Boletin del Seminario de Estudios de Arte y Arqueologia* XXXII (1966), 5f. Particularly important are his maps I and II, and the inventory/bibliographies which accompany them on 23–34. Hill-top sites known to have associated cemeteries are Suellacabras and Lumbrales.
68 An article giving an overall survey of the relevance of these Laetic sites in late Roman Spain see Blazquez, *RFS* 12, 345f
69 P. Rahtz and P. J. Fowler, 'Somerset AD 400–700' in *Archaeology and the Landscape* (ed P. J. Fowler) 1972, 187f
70 M. G. Welch, *Highdown and its Saxon Cemetery* (1976).
71 P. J. Fowler, Hill-forts AD 400–700, in M. Jesson and D. Hill (eds) *The Iron Age and its Hill Forts* (1971), 203
72 H. von Petrikovits, in *Reallexicon der Germanischen Altertumskunde* IV, 197f s.v. Fluchtbefestigung
73 J. Werner, *Germania* 34 (1956), 243f
74 N. Walke, *PAR* 1963, 29f, G. Pohl & H. Stiglitz, *PAR* 17 (1967), 14
75 See, for example, R. Macmullen, 1963, 115f, 138–9, 146–7
76 *CIL* XII, 1524 (=*ILS* 1274)
77 E. G. *CIL* VIII, 22774; see Macmullen 1963, 149f
78 Sidonius Apollinaris *Epistles*, V, 14, 1
79 Ibid., *Carmina*, XXII, 101f
80 *CIL* V, 5418
81 Cassiodorus, *Variae*, 5, 9; R. Heuberger, *Rätien im Altertum und frühen Mittelälter* (1932), 257
82 R. Paribeni, 'Le Dimore dei Potentiores nel Basso Impero', *MDAI(R)* 55(1940), 131f
83 Wightman, 1967, 168–9
84 E. Dyggve & H. Vetters, *Mogorjelo* (1966)
85 A. Mocsy, 1974, 305f
86 Boethius & Ward-Perkins, 1971, 524f
87 Ausonius, *Mosella*, 9
88 H. Vetters, 'Zum Episcopus in Castellis', *AAWW* 106 (1969), 75f
89 *NGall*. I, 6, 7; IX, 6–9; XV, 9
90 Venantius Fortunatus, *Carmina*, 3, 12
91 See, in general, P. D. C. Brown, *Britannia* 2 (1971), 225f
92 See n 54
93 H. Vetters, *Tutatio, Die Georgenberg bei Michelsdorf*, *RLiO* 28 (1976)
94 See n 61

11 Late Roman Frontier Policy (pages 245–261)

1 AM 16, 11, 4
2 AM 16, 2, 6–7; Thompson 1965, 130f
3 Themistius *Oration* X 188b-c
4 AM 28, 5–9
5 E.g. *CIL* III, 6151, from Transmarisca
6 *ILS* 770
7 *CIL* III, 3653 = *ILS* 775
8 AM 27, 5, 7
9 R. Noll, *Eugippius, Das Leben des heiligen Severin*, (Berlin 1963) (Schriften und Quellen der Alten Welt, 11)
10 Baileux and Lompret, see Appendix 3
11 Hoddinott, *Bulgaria in Antiquity* (London 1975), 215f
12 *JOEAI* 1935 Bb 153–9
13 Kalinka, *Antike Denkmäler aus Bulgarien* (Wien 1906), 66, no 70 and 71
14 *CIL* III, 7450
15 SHA *XXX Tyr* xxiii, 13, 6
16 Gregory of Tours, III, 19
17 SHA *Aurelian*, xxvi, 10, 2
18 SHA *Aurelian* 21; Julian, Convivium 314b; SHA *Probus* 13, 8; 14, 5–7; Zosimus I, 68
19 Sidonius Apollinaris, Letters, VIII, 6, 14
20 *CIG* 4649 (dated to 278?)
21 Eumenius, Pan Lat IV (IX) 18, 4; Eunapius, frag. 5; Zosimus II, 34, 2
22 Malalas, *Chron* XII, 308
23 *CIL* XIII, 5256
24 *CIL* XIII, 5249
25 *CIL* XII, 2229
26 *ND* Occ XXXII (Pannonia II) 39 Ad Herculem; 42 Castra Herculis; Occ XXXIII (Valeria) 32, 46 (Both Ad Herculem); 61 'Iovia'; Occ XXXIV (Pannonia I) 20 Ad Herculem
27 *CIL* III 6151
28 *AEp* 1936, 256 no 10, dating to 302
29 *CIL* III 14450
30 Procopius II, viii, 7
31 AM 23, 5
32 *ILS* 638
33 Procopius I xix, 34–5
34 *CIL* III 14380: see Poidebard 1934, Pl XLV
35 Poidebard, 1934, Pl XXVII–III; Brünnow and von Domeszewski 1904–9, II, Pl XLII
36 *ILS* 617
37 Zosimus II, 34
38 *CIL* XIII, 8502
39 *ILS* 8938 = *CIL* III 13734
40 *ILS* 724
41 *ILS* 740
42 On Constans' visit to Britain, AM 27, 8, 4
43 AM 16, 11, 11
44 AM 17, 9, 1
45 AM 18, 2, 3
46 AM 17, 1, 11
47 AM 18, 7, 6
48 AM 18, 9
49 AM 22, 7, 7
50 AM 28, 2, 1
51 AM 30, 3, 1
52 AM 28, 3, 2
53 AM 29, 6, 2
54 *CIL* XIII 11537–8; *ILS* 774; Soproni 1978, pl 57; *ILS* 775
55 *ILS* 762
56 See nn 23, 24
57 Florescu, *RFS* 6, 144f

Index

Reference numbers in *italic* refer to illustrations

Aardenburg 197, *198*
Abrittus 73
Adamklissi 191, 255
Adraha 65
Ad Flexum 180
Ad Herculem see Pilismarot
Ad Mures 180
Ad Status 180, 191
Ad Tricensimum 218
Aemilianus 70
Africa 61, 62, 68, 72, 80, 255
Agen 106
Ager *126*
Agrimensores, corpus of 40
Aguntum (Lienz) 20, 240
Aidovski Gradec 218
Aidovščina 216, 218, *219*
Aire 108
Aix les Bains 15
Albing 68
Alderney 212, *213*
Alemanni 20, 69, 70, 72, 74, 84, 150, 167, 193, 257
Aleth 87, 93, 94, 209
Alicante 131
Allectus 199
Alpes Maritimae 104, 215
Alsóheténypuszta *122*, 123
Altenburg-Brugg 164, *165*, 166, 167
Altkaltar 146
Altrip 33, *151*, 154–5, 167, 259
Alzey 49, 50, 52, 53, *149*, 150, 153–4, *154*, 155, 158, 164, 167, 168, 235
Amida 79, 81, 257
Amiens *96*, 97, 113, 114
Ammianus Marcellinus 57, 61, 78–9, 81, 84, 121–3, 161, 167, 221, 246, 257
Ampurias 130
Ancaster 132, 133
Anderidos see Pevensey
Andernach 43, *48*, 49, 150–2, *151*, 167
Angers 33, 92, *92*, 117
Anse 47, *85*, 86, 249
Antinonupolis 257
Antonine Itinerary 58
Antioch 65
Aosta 13, *14*, 15, 16, *17*, 23, 45, 119
Aquileia 16, *42*, 121, 215, 216, 218
Aquincum 19, 20, *122*, 123, 185–6, 187, 192
Aquitanica I 76, 104
Aquitanica II 106, 113
Arabia 24, 27
  frontier 224, 225
Arbon *162*, 163, 166, 190
Arcadius 44, 150, 260
Aristotle 32

Arles 15, 35, 104, *105*
Arlon 155, *157*, 166
Armorica 199
arms factories 79
Arrabona (Györ) 180
Arras 95, 234
Arras Medallion 42, *43*
Asperden *140*, 145
Astorga 125, *126*, 130
Asturias 221
Athens 22, 65, 73, 80
Au 175
Auch 108, *110*
Augsburg 121, 169, 195
Augst (Augusta Raurica) 20, 143
Augustus 11, 13, 30, 84
Aurelian 70, 74, 113, 118, 119, 130, 166, 211, 218, 249, 259, 261
'Aurelius' 84
Ausonius 45, 58, 65–6, 106, 130, 152, 166, 243
Austria 51
Autun 13, *14*, 15, *17*, 35, 37, 47, 63, 82, 84, *85*, 86, 116
Auxerre 101, *103*, 113
Avenches 16, *19*, 20, 23, 40, *42*, 74, 101, 116, 143
Avienus, poems of 58
Avranches 86, 94, 209
Azaum 182

Bacharnsdorf 175
Badalona 22, 130, 131
Bad Kreutznach 49, 53, *149*, 152–3, 155, 158, 167, 168, 235
Baelo 20, 22
Baileux 231
Balbinus 69
ballistae 44, 79
barbari 72, 224
  dealings with 78, 246–7
Barbarian invasions 67–81
  early C3 67–72
  effectiveness of fortifications 78–80, 134–5
  Gothic raids 73–4
  late C3 raids in west 74–6
  motives for raiding 76–8
  resistance to raids 80–1
Barcelona 13, 15, 32, 37, 45, *46*, 125, *127*, 129, 130
Basel 101, 143, 158–162, *160*, 166, 257
Basques 221
Bassianae *19*, 121
Batavis, see Passau
Bavai (Bagacum Nerviorum) 33, 35, *96*, 98–9, 116, 138, 249
Bavaria see Raetia
Bayeux 87, 87, 234
Bayonne *110*, 111–2, 130
Bazas 108
Beaune *85*, 86, 249
Beauvais 37, 95, *96*, 97, 98, 113, 114
Belgica 199
  hilltop sites in 226–236
Belgica I 101, 116, 143, 253
Belgica II, 33, 95, 98, 113, 253
Belisarius 119
Ben Ahin 234
Beroe 22
Bertrix 228, *229*, 230, 231
Besançon 101, 143
Béziers 104, 116
Bigorre 244
Bingen 152
Bitburg 69, 155, *156*, 166
Bitterne 132, *133*, 212
Blaye 106, 209
Boas 108
Boiodurum see Passau Innstadt
Bonn 148–50
Bonnert 231
Bononia (Banostor) 188, fort opposite 192
Boppard 43, 53, *149*, 152, 167
Bordeaux 45, 76, 106, *107*, 117
Bostra 251
Boulogne *96*, 97–8, 196, 198–9, 208–9, 211, 213
Bourges 45, *103*, 104
Bradwell 35, 201, 202, *203*
Braga 130
Braives 138, *147*
Brancaster 68, 196, *198*, 201, 210
Bregenz 54, 169
Breisach 49, 50, 53, 158, *160*, 167, 259
Brest 93–4, 209
Brigetio (Oszöny) 180, 182, 194, 255
Britain 18, 20, 22, 24, 35, 38, 51, 68, 69, 70, 74, 94, 117, 131–134, 196–214, 241, 251, 257
Brittenburg 145, *147*, 198, 208
Brixia 121, 215
Brough-on-Humber 131, 212
Buciumi 25
Bulgaria 22
Bu Ngem 24, *25*, *26*
Burg-bei-Stein-am-Rhein 47, 50, *162*, 163, 166, 253, 259
Burgh Castle 35, 44, 201–2, *203*, 207, 210, 211
Burgheim 173, 190
Burghöfe (Summuntorium?) 173

INDEX

Bürgle (Pinianae) 53, *170*, 173, 190
Burgo de Osma 125, 131
burgus 146, 173, 178–9, 182, 185, 246–7
Burkli *238*
Buzenol *228*, 232

Caceres 125–6, *127*
Caer Gybi 213
Caerwent 133
Caesar, Julius 11, 13
Caistor 132, *205*, 212
Caistor St. Edmunds (Venta Icenorum) 35, 131, 133, 210
Cambrai 95
Cambridge *133*
Campona *186*, 187, 188, 191, 193
Cantabrians 221
Canterbury 131, 210
Capidava 191
Carcassonne 37, 38, 43, 104, *110*, 112, 116
Cardiff 48, *49*, *203*, 207–8
Carhaix 116
Carnuntum (Bad-Deutsch Altenburg) 25, 52, 68, 179, 180, 194, 195
Carausius 76, 199, 211, 213, 251
Casei 26
Cassiliacum 171–2
Castell Collen 25, *26*
Castellum see Kastel
Castellum Onagrinum 188, 192
Castelseprio 215
castra 101, 116, 143, 164, 195, 243–4
Castra ad Herculem (Pilismarot) 46, 59, *183*, 185
Castra Constantia (Ulcisia Castra) 51, 185, 253
Castra Herculis 145, 167
Catterick 132, *133*
Celamantia (Leányvár) 182, *183*, 255
Celje (Celeia) 121
centenaria 255
Cercusium 253
Cerevic 188
Chalon-sur-Saône 82, 84, *85*, 244
Châlons-sur-Marne 95
Chancy *160*
Channel coast defences,
    Carausius 119–200
    Continental coasts 208, 211
    Late third century 201–7
    Litus Saxonicum (see also Saxon Shore forts) 211–214
    Third century 196–9
Charsovo 61
Chartres 99–101
Cheslain d'Ortho 227–8, *228*
Chester 37
Chesterholm 51
Chur 193, 215
churches 189, 243–4
Cifer-Pac 192
Cirencester 22
Cirpi 185
City wall building
    dating 55–9, 113–116
    first century 13–18
    gates 22–3, 44–50
    late Roman, in Britain 131–134
        in Eastern provinces 121–124
        in Gaul 82–117
        in Italy 117–121
        in Rhineland 142–5
        in Spain 124–131

second and third centuries 20–24
siting and construction 31–38, 116–7
towers 38, 40
Classis Britannica 76, 196, 199, 204
Classis Germanica 76
Claudius 16, 142
Claudius II 63, 74, 106
Claustra Alpium Juliarum 189, 216–221
Coccaglia 215
coin hoards 67–8, 72, 76, 88, 131, 133, 245
Colchester 20, *23*, 35
Como 15, *17*, 32, 36, 119
Concordia 218
Condeixha-a-Velha *46*, *127*, 128
Conimbriga 131
Constans 185, 257
Constantine 63, 101, 119, 123, 133, 143, 146, 153, 155, 161, 164, 166, 171, 172, 175, 180, 185, 191–2, 193, 218,
    Frontier policy 255–7, 259
Constantinople 60, 65
Constantius 32, 84, 251, 253
Constantius II 57, 64, 79, 123, 178, 185, 193, 220, 221, 253, 257, 259
Consularia Constantinopolitana 59
Contraquincum *186*, 189, 192
Contra Florentiam 187
Corbridge 132
Coria *46*, *127*, 128
Corinth 73
Corseul 116
Cortanovci 188
Coutances 86, 94, 209
Coz Yaudet, le 94
Crumerum 182
Cuijk 53, 146, *147*, 166
Cyrene 80

Dacia (Rumania) 24, 69, 72, 74
Daganija (Da-ganiya) *28*, 54
Dalmatia 121, 221, 240, 243
Danube 68, 70, 73, 245, 246, 247, 253, 255
    frontiers along 169–195
Danube-Iller-Rhine frontier 169–172
Daubian provinces 10, 46, 51, 58, 61
Daphne 255
dating of defences
    archaeological 55–6
    historical records 56–9
    typological 56, 280
Dax 33, 76, 108–9, *110*, *111*, 114, 130
De Rebus Bellicis 62
Decius 73
defences, construction of, 59–66
    civilian assistance 62–6
    cost of constructing 59–60, 62–4
    cost of materials 64
    duration of construction 66
    imperial patronage 61–3
    labour force 60–5
    sign of pressure 245–7
    siting 60–1
defended villas 242–4
Deutz 46, 47, *47*, 53, 148, *149*, 166, 255
Dexippus 73, 76, 80
Devil's Dyke, in Hungary 192, 253
Deyr-el-Kahf 255
Die 45, 104, *105*, 215
Digne 215
Dijon 84, *85*, 86, 74, 113, 249
Dimmidi, Castellum 24, *26*
Diocletian, 50, 61, 63, 77, 115, 121, 123, 130, 143, 161, 166, 167, 171, 172, 188, 189, 190, 192, 193, 199, 218, 220, 243, 249
    frontier policy 253–5, 259, 260
Diocletian's Edict 63
Dionysias 54
Dorchester on Thames 132
Dover 196, 204, *205*, 209, 210, 211, 213
Dourbes 228, *229*, 230, 234
Draschitz 220
Drobeta (Turnu Severin) 260
Druten 145
Duel 240, *241*
Dura Europus 30
Duero (river) 221
Durostorum 253
Dyrrachium 73

Eauze 108
Echternach 235
Egypt 61, 63
Eisenberg *141*
Eining *170*, 173–4
El Benian 223–4
Elche 131
El-Dmer 28
El-Leggun 27, ,*28*, 54
El-Kastal 27, *28*
Elslack 207
Embrun 215
Emona see Ljubljana
Endidae (Castelfeder) *241*
Engers *141*, 155, 193, 259
Éprave *229*, 234
Ermelo 32
Eugippius, Life of St. Severinus 173, 174, 178, 247
Eumenius 63, 84
Euphrates 24, 27, 253
Eutropius 57, 58, 76
Evreux *87*, 88

Falkenburg 220
Famars 98, *98*, 99, 138, 249
Favianis (see also Mautern) 247
Fenékpuszta see Valcum
fort defences (see also defences)
    dating of 55–9, 166–168, 190–5 209–11
    first century 18–20
    gates 24, 46–50
    in East 27–30
    late Roman in
        Channel coasts 196–214
        Danube frontier 169–195
        Julian Alps 215–221
        Rhine frontier 136–168
        Tingitana 223–5
    second and third centuries 24
    siting of, and construction of 31–38
    towers 38–43, 190–2, 253
    typology of 166
forts, internal buildings 50–54, 207
Forum Julii see Fréjus
Franks 69, 72, 76, 94, 193, 199, 213, 251
Frastafeders 215
Fréjus (Forum Julii) 13, *14*, 15, 16, *18*, 35, 38, 40
Friglas 223
Friuli 215, 218
Froitzheim 138
frontiers
    Danube 169–195
    General Survey 249–261

INDEX

Julian Alps 215–221
North Sea 196–214
Rhine 136–168
Tingitana 223–5
Furfooz 229, 233–4, 242
Füssen 189, 242

Galerius 119
Gallic Empire 70, 72, 74, 121, 249
Gallienus 57, 63, 65, 70, 72, 73, 74, 249
Gamzigrad 243
Gandori 224
Gap 215
Garbsch 54
gates
  first century 13–15
  late Roman 44–50
  late Roman in hilltop forts 231–3
  second century 22–4
Gaul 10, 15, 35, 43, 44, 45, 61, 63, 69, 70, 72, 74, 76, 94, 130, 135
  Coastal defences in 209–10
  hilltop sites 249
  town defences in 82–117, 249–51
Gelbe Burg 235
Gellep (Gelduba) 148, 168
Gemersheim 154
Geneva 104, *105*, 113, 116, 242
Georgenberg 244
Germania I 101, 142
  frontier in 150–158
  hilltop sites in 226–236
Germania II 101, 253
  frontier forts in 142–9, 167
Germans 234
  besiege towns 76, 78–9
  pirates in English Channel 68
  raids in Africa 224
  weapons 79–80
Germany 15, 18, 60, 69, 72, 74
Gerona 46, *46*, *126*, 128, 131
Gerulata (Rusovce) 179–80
Gheria el Garbia 25, *26*
Glauberg 235
Gloucester 20
Godmanchester 132, *133*, 133
Goldberg 170, 171, 172, 189, 190, 193, 249
Goldsborough 213
Gordian III 69
Goths 69, 73–4, 78–9, 80, 247
Goudsberg 138, *140*
Graiae et Poeninae 104
Grand 143
Grannona 201, 209, 213
Gratian 178
Grazerkögel 240
Grcarevec 220
Great Chesterford 132, *133*
Greece 73–4
Gregory of Tours 74, 84, 86, 113, 249
Grenoble 63, 104, *105*, 215, 253
Grünwald 238
Günzburg (Guntia) 173
Gurina 240

Hadrian 9, 20
  Hadrianic forts 24
Hadrians Wall 51, 52
Han Aneybe 255
Hatra 30
Hatvan 192
Haus Bürgel *140*, 148, 166
Heddernheim 20

Heerlen 138
Heidenburg *238*
Hemmaberg 240, *241*
Hérapel 233
Heruli 65, 269
Heumensoord *140*
Hideglelöskerestz 50, *183*, 184–5, 193, 259
Highdown 241
hilltop fortifications 189, 195, 215–6, 220, 226–244
  churches in 243–4
  dating 249
  in Spain 221–2, 240
Hoischügel 220, 240, *241*
Honorius 44, 119, 150, 221, 260
Horbourg 49, 101, 143, 158, *160*, 244
Horany *187*
Horncastle 132, 205, 212
Horwa 215
Housesteads 51
Hrušica 216–7, 218, *219*
Hüchelhoven *140*
Hungary 51
Huns 194
Huntcliff *213*

Ilchester 132
Ilerda (Lérida) 72
Iller (river) 72, 169–172
Inestrillas *46*, 46
Ingenuus 72
Inn (river) 175
Intercisa 51, 52, *186*, 188, 191
Irgenhausen 53, 164–6, *165*, 166
Iruena 240
Iruña (Veleia?) *46*, *127*, 128–9, 130, 131
Isny 46, 47, 53, 169, *170*, 171, 190, 249
Italica 20, 22
Italy 10, 11, 36, 72, 80, 135, 174, 189
  defence of 215–221
  hilltop sites 227
  town defences 117–121
Iuthungi 69, 78
Iznik see Nicaea

Jemelle 230
Jublains 53, *92*, 93, 94, 99, 116
Julian 32, 58, 61, 63, 74, 78, 79, 101, 114, 115, 116, 142, 143, 167, 193, 220, 221, 257, 259
Julian Alps 215–219, 227
Jülich *157*, 166
Junglinster 231
Jünkerath 155, *156*, 166, 218
Justinianic, law codes 59

Kaarlsbierg *228*
Kaasselbierg 231
Kaiseraugst 49, 53, 161, *162*, 166, 255
Karnburg 220, 240
Kärnten
  Extensions to Claustra Alpium Juliarum 220
  hilltop sites in 189, 195
Kasr Bser *28*, 61, 255
Kastel *43*, 138, 255
Katzenberg 241
Kellmünz 47, *170*, 172, 189, 190
Kempten 54, 72, 116, 121, 169, 171, 242
Kenchester 132
Kindsbach 233
Kisárpás 123
Kleiner Laufen 163

Klosterneuburg 179
Koblenz 151, 152, 167
Koi-Krylgan-Kala 30
Köln 16, *19*, 22, 35, 38, 40, 70, 101, 136, 142, 166, 167, 236, 249
Köln-Bavai road 138, 166, 249
Konstanz 163
Környe 123
Kranj (Karnium) 220
Kreimbach 233
Küchl 189, 242
Kümpfmuhl 68
Künzing (Quintanis) 173

La Calzada 240
Lactantius 57
Ladenburg 20
Laeti 135
  in Gaul 234–5
  in Spain 221
  settled in Roman Empire 246–7
Lake Constance (Bodensee) 169, 189, 190
Lambaesis 25
Lancaster 207, 208
Langres 82, 84, 234
Lanišče 218, *219*
Larçay 94, 99
Larga 138
Lausanne 116, 143, 242
Lavant 244
Leányfalu *187*
Lecco 215
Lech (river) 72
Lectoure 109
Le Mans 37, 38, *39*, *41*, 43, 45, 48, *89*, 89–91, 234
León *126*, 129, 130, 221
Lérida 125
Lescar 108
Libanius 65
Liberchies 138
Lienz (Aguntum) 20, *22*
Liesenich 136
Lillebonne 86, 88, 117
Limigantes 193
Limoges *103*, 104
Lincoln 20, *23*
Linz (Lentia) 175
Lisieux *87*, 88
Lixus (Aulucos) 223, 224
Ljubljana (Emona) 13, *14*, 15, 32, 119, 121, 216
Ločica 68
Lollianus (Laelianus) 72
Lompret 231
London 42, *43*, 132
Lorch (Lauriacum) 68, 178, 195
Lorenzberg 53, 175, *238*, 239, 249
Lugdunensis Prima 47, 82, 87, 88, 244
Lugdunensis II 86
Lugdunensis III 94, 113
Lugdunensis IV (Senonia) 99, 113
Lugo 45, *46*, *127*, 129, 130, 221
Lumbrales 240
Lympne 52, 53, 201, 204, *205*, 207
Lyons *14*, 82, 86
Lyons Medallion 42, *43*, 138, 255

Maastricht *147*, 167
Mâcon 82, 84, *85*, 244, 249
Macquenoise 231
Macrianus 79
Magnentius 109, 148, 193, 220

INDEX

Mainz 42, *43*, 69, 79, 101, 136, 138, *139*, 142, 150, 236
  sites round 152–3, 166, 167
Malalas 253
Manching 173, 190
Mandacher Egg 189, 236
Manfalut-Hieraconpolis 253
Marcellinus, Bishop of Milan 242, 243
Marcianus 74
Marcis 201, 209, 213
Marcomanni 1, 70, 221
Marcomannic wars 68, 174, 178, 182, 216
Marcus Aurelius 22, 68, 182
Martinj 218, *219*
Matrica *186*, 188, 192
Mauer an der Url (Locus Felicis?) 175
Mauretania Caesariensis 223–4
Mauretania Tingitana 72, 223–5
Mautern (Favianis) 46, *177*, 178, 247
Mauthen 215
Maxentius 44, 119
Maxima Caesariensis *42*
Maxima Sequanorum 101, 116, 143, 158, 164, 193, 215, 244, 253
  hilltop sites in 189, 236–40
Maximian 50, 63, 121, 130, 171, 172, 190, 199, 251
Maximinus 69, 70, 220
Maximus, Bishop of Turin, 80
Meaux *100*, 101
Melun *100*, 116
Memmingen (Cassiliacum?) 171–2
Mesopotamia 24, 61
Metz 143, *144*, 227
Milan 119
Milanovice 192
Mildenhall (Wilts) 132
Moers-Asberg *140*
Moesia 69, 190, 246
  fan-shaped towers in 190–1, 193
Mogorjelo *243*, 243
Montana (Michaelovgrad) 249
Mont-Sommerain 228, *229*, 230, 231
Moors 68, 72, 131
Moosberg 46, *47*, 48, 53, 175, 189, 238, 239–40, 249
Morlaix 94
Mosel 152
Muciaster 215
Mumpf 54, 163, *164*, 172
Municipium Bataviorum 145

Nabatean Judaea 27
Nadleski Hrib 216
Naher *42*
Nanniennus 78
Nantes 45, *47*, *89*, 92–3, 94, 113, 209, 251
Narbonne 104, 116
Narbonensis I 104
Narbonensis II 104
Neckarau *141*, 259
Negev 27
Neuberg 173, 190
Neumagen 48, 155, *156*, 155–8, 166, 167
Neuss 15, *18*, 148
Nevers 47, *103*
Newton Kyme 207
Nicaea (Iznik) 57, 63
Neiderlahnstein 155, 193, 259
Nijmegen 145, 242
Nimes 13, *14*, 15, *17*, 45, 104
Noricum 38, 70, 74, 121, 169, 221, 240, 249
  frontier in 175–159, *176*
  hilltop sites in 189
  Severinus in 195, 247
Notitia Dignitatum 41, 58, 59, 62, 101, 128, 130, 135, 171, 174, 178, 180, 188, 189, 192
  arms factories 79
  Germania I 150
  Norican frontier 175
  praefecti laetorum 234–5
  Raetian frontier 172–3
  Saxon shore 201, 209–10, 211, 213
  Spanish troops 221–222
  Tigitania 223–5
Notitia Galliarum 82, 84, 86, 88, 93, 94, 95, 98, 104, 108, 111, 130, 142, 143, 164, 244
Novempopulana 108–112, 114, 130, 244
Noyon 98, *98*, 114, 143, 234
Nyon 101

Oberlieserburg 192
Oberranna 172, 175, 194
Obersteufenbach *228*
Oberwinterthur see Winterthur
Odilienburg 235
Odruh 27
Oescus 255
Oloron 108
Olten *160*, 164, 167
Orange 14
Orléans *47*, 100, 101, 234
Orosius 69
Oudenburg 48, *49*, 52, *147*, 197, 201, 208

Pächten *156*, 166, 235
Pamplona 124
Panegyrists 199
Pannonia 38, 69, 70, 72, 74, 172, 193, 194, 246, 249, 251
  frontier in 179–188
  town defences in 121–123
Pannonia II 121, 128, 172
  frontier in 188–9
Paris 101, *102*, *103*, 116
Parthia 24
Passau (Batavis) 25, 173, 174, *178*, 191, 247, 253
Passau Innstadt (Boiodorum) 175
Pataulia 22
Paulinus 45
Pécs see Sopianae
Pella 45
Périgueux 45, 104, 106, *107*, 108, *109*, 117
Persian Empire 10
Peutinger Table 58
Pevensey 33, 36, *42*, *47*, 49, 204–6, *205*, 206, 212, 257
Pfalzel 243
Pfyn 164, *165*, 166
Philippopolis 22, 73–4, 78, 249
Piercebridge 207
Pilismarot (Ad Herculem) 46, 53, *183*, 185
Pilismarot-Malompatak 54, *187*
plague 69
Plato 32
Pöchlarn (Arelape) 178
Poetovio see Ptuj
Poitiers 106, *107*, 234
Pölch 233, 234
Polhov Gradec 216, 220
Portchester 36, *47*, 49, 53, *203*, 206–7
Portiflüh 189, 236

Portus Aepatiaci 209
Probus 63, 79, 88, 114, *115*, 130, 166, 171, 189, 190, 210, 211, 249–53, 259, 261
  invasion of AD 275–6, 74–5
Procopius 253
Postumus 70, 72, 73, 131, 249
Ptuj (Poetovio) 121

Quadi 193, 194
Quadrata 180
quadriburgium 94, 166, 168, 253–5, 259
Qualburg 146, 166, 242

Raetia (Bavaria) 53, 68, 72, 74, 78, 116, 121, 164, 169, 193, 194, 215, 249
  end of Roman rule 194–5
  frontier in 169–175
  hilltop sites 189, 236–240
Rapidense 255
Ravenna 15, *17*, 23, 247
Ravenscar 212
Reculver 68, 169, *198*, 201
Regensburg 23, 24, *25*, 35, 68, *170*, 173, 174, 253
Reims 97, 98, 114, 234
Remagen 150, *151*, 167
Renggen 189, 236
Rennes 38, *92*, 91–2, 251
Rheinberg *140*, 146
Rheinbrohl 155, 259
Rheinheim 163
Rheinzabern 154
Rhine 15, 40, 46, 58, 61, 69, 70, 72, 78, 169, 197, 198, 211, 245–6, 253, 257
  frontier along 136–168
Rijeka 216, 218
Risingham 25
Richborough 36, 38, 43, *48*, 49, 52, 53, 201, 202–4, *203*, 206, 207
'Robur' 61, 160–1, 257
Rochester 132
Rodez 104
Rome 15, 30, 32, 37, 40, 43, 44, *45*, 48, 60, 65, 69, *118*, 117–9, 130, 249
Rossatz 175
Rossum 145
Rouen 87, 209
Rouveroy 230
Rubricaire *92*, 94, 99
Rugii 247
Runder Berg 235

Saarbrücken 155, *156*, 166, 255
Ságvár 123
St. Bertrand de Comminges 109–10, *110*
St. Brieuc 94
St. Laurent 73, 138, *141*
St. Lizier *110*, 111
St. Quentin 101
Saintes 106, *107*, 113
Saldae (Africa) 80
Salona 15, *49*, 221
Saloninus 70, 99
Sapor 79
Sarmatians 69, 185, 187, 193, 194
Savaria (Szombáthely) 64, 123, 194, 253
Saverne 155, 157, 167, 257
Saxons 68, 69, 76, 94, 199, 213, 251
Saxon Shore forts 33, 36, 43, 201–7, 211–214, 230, 251, 257, 259
Scarborough *213*
Scarponne 156
Schaan 48, 50, 53, *165*, 166, 168, 189, 194,

215, *238*, 240, 259
  hilltop fort 189
Schierenhof 25
Schlögen 175–9, *178*, 253
Schlössle 215
Schwechat 179
Scotland 32
Scriptores Historiae Augustae 57, 58, 72, 73, 74
Sées 86
Seltz 154
Senlis 37, 95–7, *96*, 114
Senon 73, 138, *141*, 142
Sens 78, 99, *100*, 101, 104, 113
Septimius Severus 88
Serdica (Sofia) 22, 249
Severinus 195, 247
Severus Alexander 69, 70, 125
Sextus Aurelius Victor 57, 70, 76
Sidonius Apollinaris 57, 242
Sirmium see Sremska Mitrovica
Sisseln 54, 163, *164*, 172
Sisteron 242
Slaveni *26*
Soissons 95, *96*, 97, 98, 114
Solothurn 164, *165*, 167
Solva (Esztergom) 184
Sopianae (Pécs) 123, 253
Sopron (Scarbantia) 35, 121, *122*, 123, 194, 253
sources, contemporary for wall building 55–61
South Shields 51
Spain 10, 20, 35, 44, 45, 68, 69, 70, 72, 76, 117, *124*, 135
  frontier in 221–3
  Mauretania Tingitana 223–5
  town defences in 124–131
Spello 15
Speyer 142, 154
Spielberg 175
Split 38, 48, 49, 50, 54, *243*, 243
Sponeck 158
Sremska Mitrovica (Sirmium) 121, 193, 253
Stellfeder 215
Stillfried 192
Strasbourg 33, 35, 42, 136, *139*, *141*, 142–3, 154, 230
Straubing 70, 173
Stürmenkopf 189, 236, *238*
Sucidava 255
Suebi 135, 193
Suellacabras 240
Suiar 223
Susa 48, 121
Sveta Marjeta 216
Synesius 80
Syria 24, 27
  frontier 224, 255
Szentendre Dera-Patak *187*
Szentendre Hunka Domb *187*
Szombathely see Savaria

Tabernae 223
Tacitus 57, 78, 79
Tamuda 72, 80, *222*, 223, 224
Tarbes 108
Tarraco (Tarragona) 72, 130, 224

Taviers 138, *147*
Tetricus 91
Theilenhofen 24, *25*
Themistius 61, 246
Theodoric 119, 242
Theodosian law code 59, 64
Theodosius 133, 175, 221
Thérouanne 95
Thessaloniki 78
Thrace 22, 63, 69, 73–4, 246, 249
Tiberius 15
Tigris 27
timber, use of in construction 43–4
Tipasa 15, *18*
Tirol hilltop sites in 189, 195, 240–1, 243–4
Tokod 46, *47*, 53, 182–4, *183*, 193
Tomi 253
Tongres 20, 35, 72, 142, *144*, 234
Toul 47, 143, *144*
Toulouse 35, 37, 104
Tournai *98*
Tournus *85*, 86, 249
Tours 76, 78, 88–9, *89*, *90*, 94, 113, 117
towers
  external 38–40
  fan-shaped 38, 190–2, 253
  roofing of 40–3
Traismauer *177*, 178
Trajan 9, 20, 180, 182
Transaquincum 186, *187*
Transmarisca 253, 255
Trebius Justus, tomb of 36
Trier 19, *23*, 23–4, 35, 37, 51, 60, 65, 72, 101, 117, 136, 142, 143, 227
Trieste (Tergeste) 218
Tripolitania 61–2
Troesmis 257
Troyes 78, *100*, 101
Tscheltnigkögel 220
Tulln (Comagenis) 179, 247
Turin 15, 16, *17*, 23, 36, 119

Ulcisia Castra (Castra Constantia) 51, 185, *186*, 191
Ulm (Febiana?) 172, 173
Untersaal *164*, 172, 175
Ursulus 81
Utrecht 25, 145
Utrecht Psalter 40
Uxama Argaela see Burgo de Osma

Valcum (Fenekpuszta) *122*, 123
Valence 15
Valens 61, 246, 257
Valeria 172, 193
  frontier in Valeria and Pannonia 179–188
Valerian 65, 70
Valentinian 61, 64, 79, 123, 138, 142, 146, 153, 155, 158, 161, 163, 164, 167, 168, 171, 172, 178, 182, 184, 185, 192–3, 194, 246, 257–9
Valentinian III 119, 234
Valkenburg 138
Vannes *89*, 93, 94, 209
Vegetius 31, 32, 40, 58
Veleia see Iruña
Venantius Fortunatus 244

Venta Icenorum see Caistor St. Edmund
Ventimiglia 15
Verdun 143, *144*, 242
Vermand 95, 116, 242
Veroce *187*, 193
Verona 60, 119
Verona list 223
Verulamium 22, *23*
Vetus Salina 188
Victorinus 136
Vienna (Vindobona) 179, 195, 253
Vienna Genesis 40
Vienne 13, *14*, 104, 116
Viennesis 104, 113
Villenhaus 138, *140*
Virunum 220
Visegrád *183*, 185, 191, 255
Vitruvius 31, 32, 33
Völklingen 233, 234, 242
Vranje 220, 240
Vrknika 216, 217–8, *219*

Waldfishbach *228*, 233
Wallsee (Ad Iuvense?) 178
Wallsend 51
Walton Castle 201, 202
watchtowers
  in Pannonia/Valeria 193–4
  on British coast 212
  in Raetia and Noricum 172, 175, 194
  on Rhine 163, 168
  Valentinianic 257–9
Wellesweiler 231
Wels 19
Whylen 161, 255
Wien (Vindobona) 68 see also Vienna
Wiesbaden 152–3
Wilhering 175
Williers-Chameleux 231, 233
Wilten 49, 53, *170*, 175, 189, 194, 215
Winchester 134
Windisch (Vindonissa) 47, 49, 74, 101, 143
Winterthur 50, 163, 163–4, *165*, 166, 253, 259
Wittnauer Horn *48*, 50, 236–9, *238*, 249
Worms *139*, *141*, 142, 143, 154
Wroxeter 132

Xanten 19, 20, 142
  late Roman fort 146, *157*

Ybbs 175, 194
Yecla de Yeltes 240
York 38, 48, 132, 133
Yverdon 46, *47*, 101, 143, 164, *165*, 166

Zara 15
Zaragossa 126, 129, 130
Zeiselmauer *177*, 179
Zirl 175, 189, 215
Zonaras 57
Zosimus 57
  on Constantine and Diocletian 255
Zullestein *141*, 155, 259
Zürich 50, 164, *165*, 259
Zurzach 53, 161–3, *162*, 166
Zwentendorf 46, 51, *177*, 178–9